# NATURAL TECHNOLOGY

## The Theory of Everything

Gray Matter Publications

## S. A. Cooper

Natural Technology: The Theory of Everything

Copyright © 2025 by **S. A. Cooper**

For inquiries, please visit:

GrayMatterPublications.com

ISBN: 979-8-9921788-1-4

The United States of America

We take the side of science in spite of the patent absurdity of some of its constructs, in spite of its failure to fulfill many of its extravagant promises of health and life, in spite of the tolerance of the scientific community for unsubstantiated just-so stories, because we have a prior commitment, a commitment to materialism. It is not that the methods and institutions of science somehow compel us to accept a material explanation of the phenomenal world, but, on the contrary, that we are forced by our a priori adherence to material causes to create an apparatus of investigation and a set of concepts that produce material explanations, no matter how counterintuitive, no matter how mystifying to the uninitiated. Moreover, that materialism is absolute, for we cannot allow a Divine Foot in the door.[1]

—Richard Lewontin

*The king asked Daniel (also called Belteshazzar), "Are you able to tell me what I saw in my dream and interpret it?" Daniel replied, "No wise man, enchanter, magician or diviner can explain to the king the mystery he has asked about, but there is a God in heaven who reveals mysteries.*
Daniel 2:26-28

# Preface

In the fall of 2006, several years after graduating from high school, I entered college as a nontraditional student. The timing couldn't have been more significant in my life. My English class was tasked with drafting a research paper. The strength of our arguments? It wasn't mentioned. The instructor only emphasized MLA formatting, structure, and avoiding plagiarism at all costs.

Relying on Wikipedia for anything was an absolute no-no. He regularly slammed Wikipedia as a formidable threat to all of academia, as did my Spanish teacher with online translators. Back then, Wikipedia and online translators were considered diabolical. Ironically, these 'threats' are proving to be indispensable tools, despite the early skepticism from teachers like mine.

Unlike today, I'm not sure if cloud storage existed back then, and there was definitely no automatic cloud save like Google Drive in which saving occurs as you write. In hindsight, I could have saved my paper on a thumb drive or emailed it to myself. However, it never occurred to me that I might want to reference it in the future.

My paper's topic? After a recent skimming of Michael J. Behe's *Darwin's Black Box*, I decided to explore his concept of the irreducible complexity of the cell. As I dug deeper into his book, Behe's argument became more striking. The clarity and strength of his presentation blew my mind. I had never encountered anything like it before. The way he explained the fascinating concept in simple terms, using analogies, allowed me to intuitively follow—not all of it, but enough to grasp his argument. His claims were so extraordinary that I found myself wondering why I hadn't heard anything like it before. I searched but found no compelling evidence to refute it.

Growing up, however, I found science irrelevant, especially biology. To be honest, everything about school seemed pointless to me—except for the girls, who made going every day bearable. I hadn't yet grasped how special life was or the privilege of living in a world teeming with such diverse life. Not that most people ever do.

Early on, I picked up on the idea that really smart people viewed life and its origin as mere happenstance. On some level, I tacitly accepted this. Yet at the same time, having been raised in church until about the age fourteen—when my mother gave up on getting me up on Sunday mornings—I still understood that the Bible taught God created life. Beneath the surface, however, I assumed that neither my mother nor the people at church knew anything about science.

As far as I can remember, my teachers—whom I regarded as authorities on origins—explained that all life could be traced back to lightning striking a pond. Yet, whether life originated from God or from that "primordial soup," followed by evolution, it really didn't matter to me. At church, it was God; at school, it was the primordial soup. For me, it was just a matter of code-switching.

But everything changed after choosing to research Behe's concept of irreducible complexity in bacterial flagella. The existential question, particularly where life came from, began to matter. In fact, life in general began to matter. From reading Behe's intuitive arguments and explanations about molecular machines, I realized that science wasn't as boring as I had thought. My intrigue was monumental, and I put more effort into my research topic than any before.

It felt as if I had stumbled upon a deep secret. "How had I not realized how extraordinary life was compared to inanimate matter," I thought? How did cells come about—a world of functionality within a "simple" cell? I felt a little dumb for my oversight.

But when the time came to return our papers, as always, the instructor flipped them over as he placed them on our desks. When I turned over mine, my stomach dropped—there was a red "D" circled on it. No scribbled feedback, no explanations—just the letter grade "D."

"D? What did I do wrong?" I thought. Shaken, I requested a meeting with the instructor. I knew I hadn't used Wikipedia for supplementary sources, but I wasn't sure if I had mistakenly plagiarized.

In his office, the meeting was tense, as he made it clear he had no issue justifying his reasoning. "Overall, your paper's structure and formatting are fine," he explained somberly. "The real issue is the foundation of your argument," as if citing Behe's argument made it mine. "Your paper lacked sufficient evidence," he continued.

This led to a lengthy defense on my part. Though this was just a 5-to-7-page, double-spaced English assignment requiring one primary source and at least two supporting sources, it felt like I was defending a doctoral thesis on microbiology in front of an adjunct English instructor. No matter what objection I made, the response was, "But there is no evidence to support this."

I tried to cite the rubric, "But you never stated that the arguments of the authors we cite have to be valid."

He sighed, "Neither did I say you could argue for something fictional. It's not about an argument for an argument's sake; it's about engaging with credible, well-supported academic evidence. Evolution is a fact. Your paper is not based on any evidence."

Looking back, I think the meeting—for him—was less about explaining why the paper I had worked so hard on deserved a D but more about ensuring, with quiet certainty, that I understood Behe was wrong. Today, I'd call that paper a book review rather than an argument. I mainly focused on the task at hand: properly citing sources in MLA format. But this instructor had moved the goalposts, shifting the focus from accurately crediting others' ideas to defending the scientific claims of a seasoned biochemist.

My interest in the topic was purely academic, driven by fascination rather than deep conviction. Despite his insistence on discrediting Behe's claims, my primary concern was my GPA. I had returned to school with big goals, including grad school, and maintaining a strong GPA was essential to my plan. Had I known the topic was that contentious, I would've avoided it. A decisive position on the origin of life wasn't relevant to me or my goals.

But there's something about first experiences: they can stick with you. What lingered with me after that meeting wasn't so much the academic dispute itself but the instructor's reaction. I will never forget it. He was visibly disturbed by Behe's argument, as if it was the first time he had heard anything like it, just like me. I was astonished too, but my teacher and I processed it in entirely different ways.

Though I didn't fully grasp the implications of Behe's argument at the time, I now better understand how it had shaken something in my teacher's fundamental view of reality—threatening something deep within. I had been critiqued several times before, but I had never seen him, or any instructor, so rattled while offering no constructive feedback. For years, that teacher's words have echoed in my mind: 'There's no evidence… but it's not based on any evidence.' The way he looked at me—calm, assured, yet with nothing to substantiate his claim—just gaslighting. Clearly, he had no reasoning to offer.

Despite this, I still earned an A in the class, and the experience was enlightening. The greatest lesson I took from that class—and the experiences that followed—was that bias and predispositions aren't confined to any group or profession. Even otherwise astute educators can have their judgment clouded by personal convictions. This was an awakening for me, recognizing that everyone, in their way, clings to certain beliefs— religiously—when interpreting reality. As for me, believing that all life originated from lightning striking a pond no longer seemed like something a smart person would accept, let alone teach.

Even after the fallout from that 2006 paper, I still subconsciously gave some weight to my teacher's dismissal of Behe's legitimacy. There were many issues with mainstream origin stories that I had yet to discover. I largely shelved my interest in the topic of intelligent design. It wasn't until 2013— sparked by a lecture from Stephen C. Meyer and his compelling work, *Signature in the Cell: DNA and the Evidence for Intelligent Design*—that my curiosity was reignited.[2]

If Meyer's clear explanation of the biological functionality of sequential, syntactic, and structural information wasn't enough, the accompanying videos of cellular processes in *Signature in the Cell* made the case even more compelling, convincing me entirely. Similar to Behe, his concepts—central to understanding the informational basis of DNA—demonstrate the foresight required for life, and they left me with an appreciation for the compelling case he makes.

I couldn't help but notice, however, that the general public seemed largely unaware of these critical discussions. Whether due to a lack of exposure or the active suppression of such ideas, the depth and significance of these topics remained absent from mainstream understanding, despite their profound implications.

By 2016, I continued to read up on these topics. I gravitated toward YouTube videos and books by outspoken critics of intelligent design—figures like Lawrence Krauss, Neil deGrasse Tyson, Christopher Hitchens, Brian Greene, and Richard Dawkins—where I encountered extensive hand-waving. It was during this time that I gained clarity and began to really understand the gravity of origin explanations. I realized the extent to which atheistic scientists fought tooth and nail in the war of worldviews, striving to maintain control of the narrative in explaining reality.

I was both surprised and disappointed by how these thinkers resorted to projection and a multitude of inappropriate and flawed arguments. Even more surprising was how their attacks on Christian beliefs were lauded as exemplary scientific refutations. Their frequent dismissals of intelligent design—often devoid of substantial counterarguments and reliant on ad hominem attacks—revealed a troubling pattern.

This realization was another moment that contributed to a disheartening part of growing up: recognizing that the ideals I held about adults always doing what's right were shattered. Pastors, teachers, scientists, politicians, and even family members who taught me to uphold integrity all, in some way, failed to live up to the façade they presented, especially when a loss of money, power or ideological control were at stake. But what shocked me most was realizing that scientists—the people I once saw as the epitome of rational, unbiased truth-seekers—were sometimes even worse than the cult-like dogmatism they claimed to oppose, fiercely guarding their own orthodoxy with a religious-like zeal.

With these scientists, this wasn't just scientific disagreement or about presenting evidence: I began to see a broader cultural bias, a kind of gaslighting that was rampant and unchecked within the scientific community. Moreover, these scientists often argued in a childlike manner rather than as the mature, impartial thinkers I mistakenly thought most atheists were. Hitchens, though not a scientist, often represented the public face of atheism and scientific discourse but came across as overly brash and, at times, possibly intoxicated, such as in his debate with William Lane Craig. This further removed the impartial-minded pedestal I lifted scientists and scholars up to.[3] [4]

In search of more reasoned answers, I began reading peer-reviewed literature on cell biology: particularly, origin of life studies. That's when things got weirder. When it comes to explanations of life's origin, I found the origin of life papers deserving the critique of my English teacher: "There is no scientific evidence to support this claim."

Paper after paper churned out the same recycled, "it was this, not that," "it might have been that all along," impractical ideas: 'deep sea vents,' 'no, warm springs,' 'dry-wet cycles,' 'asteroids,' 'ice crystals,' 'RNA first,' 'clay minerals,' 'hydrothermal vents,' 'lightning strikes,' 'underwater volcanoes,' 'comet impacts,' 'pyrite formation,' 'submarine alkaline hydrothermal systems,' 'UV radiation,' and more. Each hypothesis is presented as significantly plausible, yet none provide substantial evidence. They all felt like the academic equivalent of clickbait, leading nowhere.

The more I investigated, the more I formulated my own hypotheses, expanding upon the assertions of Behe and Meyer. I began to uncover additional connections, for example, deeper links between the engineered universe and human ingenuity. This exploration led me to consider how the principles of design and technology observed in nature have informed and inspired our technological advancements.

Additionally, as I delved deeper, I began to uncover that gaslighting has long been a cultural tradition among Darwinists, dating back to Darwin's own circle, where dissenting views were routinely dismissed or misrepresented under the guise of scientific progress. Together, these reflections guided me down a less-traveled path. It brought me to a clarity that revealed deeper connections others may have overlooked.

From my initial encounter with Behe's argument and the rejection by my English teacher, to my expanding understanding over the past 18 years, this journey has been filled with challenges and enlightenment. Now, I want to share these discoveries with you.

In revisiting my experience with my English teacher from 2006, I must confess something that brings me no small satisfaction: some of the ideas in this book, concepts I'm excited to share, have been supported by none other than Wikipedia. Muahahaha! Good ideas can come from unexpected places, even from the tools scorned by the so-called gatekeepers of knowledge.

# Introduction

Through these pages, I confront conventional views of technology and nature, uncovering connections that challenge mainstream assumptions. Though this book draws inspiration from pioneering thinkers like Michael J. Behe and Stephen C. Meyer, it ventures beyond the traditional foundational focus of intelligent design theory. Proponents of intelligent design are widely recognized for their insights into the technological sophistication of the cell, describing molecular structures like DNA and proteins as genetic codes and molecular machines. Building on these insights, however, the perspective advanced here takes a step or two further.

This book proposes a deeper connection between human ingenuity and the engineered universe we inhabit, describing our technology as an extension from the natural technological framework of the universe. This convergence exemplifies an intrinsic technology transfer, where insights from nature are uncovered through our inherent connection to it and our instinctive drive to produce through this union. Like a seductive mating mechanism, nature entices us with its mysteries, releasing an allure, like pheromones, that draw us in through scientific intrigue and the desire for knowledge and progress.[5]

Through this process, we uncover nature's secrets and translate them into innovation.[6] This instinct to seek knowledge, to create, and to innovate feels innate, yet it is driven by forces outside our conscious awareness. Human technology is an echo of the far superior technology inherent in nature, which predates our existence.

This leads to a clarification: it's a mischaracterization to say natural phenomena resemble our technology; it's more accurate to say that our technology is inspired by the natural world. The former would be like saying a mother looks like her daughter, rather than recognizing that the daughter originates from the mother. This view provides a lens to fully conceptualize the universe, including its technological configurations, nurturing qualities, and beauty.

Further, proponents of intelligent design often state that "some" or "certain features of the universe and of living things are best explained by an intelligent cause, not an undirected process like natural selection."[7] This book extends this idea, arguing that not just certain, but every material component and process in the universe—along with its natural laws—constitutes the purest form of technology. Yes, even the design of a snowflake is included. This is the essence of *Natural Technology: The Theory of Everything*.

This presents a novel perspective on the universe's fundamental laws of physics. Rather than viewing them solely as rules governing cosmic behavior, this book proposes conceptualizing these laws, including gravitational, electromagnetic, strong, and weak nuclear forces, as core *specifications* of the universe's technological system, orchestrating a precise, ordered pattern throughout existence.[8]

Additionally, it argues that a universe like ours could not exist through a purely deterministic framework, as probabilities are embedded into its foundation. A deterministic universe would lack the richness and adaptation and recalibration required to support the diversity of life, conscious experience, and innovation we observe. Though this presents a hard problem for some, there is no equation to resolve it.

The universe's processes, from quantum mechanics to biological systems, are parts of an overarching natural technological blueprint. Embracing this perspective unlocks possibilities that transcend conventional deterministic models, revealing the universe as a system grounded in adaptability, recalibration, complexity, and intrinsic order.

This provides a pathway to monumental scientific progress, enabling the prediction of innovation while offering principles to guide it. These possibilities point to a future where technology mirrors the recalibrations of nature, harnesses quantum mechanics for systemic breakthroughs, and redefines humanity's role as both participants in and stewards of this interconnected system.

Those who question the status quo, whether deliberately or not, are poised to drive meaningful progress in science. Materialism has often limited scientific exploration by dismissing the theory of intelligent design and inflicting a materialistic-only interpretation of evidence on science, blocking educators and academia from engaging with the reality being revealed through modern knowledge.

## Materialism

I will largely use the term *Materialism* to critique the doctrine of the worldview that asserts matter and physical processes as the sole constituents of any reality. My primary focus is on radical Materialism—an extreme and explicitly militant adherence to this doctrine. However, this critique extends to all who, whether knowingly or not, operate within the same framework.

This distinction is subtle but vital: radical materialists wholeheartedly commit to their doctrine, fiercely dismissing even the strongest evidence that could challenge it. Meanwhile, more casual or agnostic materialists may claim a less rigid stance, but by implicitly adhering to the same foundational premise, they remain bound to the same dogma. In this sense, the "moderates" provide cover for this radical doctrine by refusing to confront compelling refutational evidence and its deeper implications.

I will also use the designation *materialist* interchangeably with *Darwinist* in some instances, though I recognize that not all Darwinists explicitly identify as materialists. However, as the true implications of Darwinism are understood, Neo-Darwinism has shifted well past its original intent of explaining the diversity of biological systems. Its significance now lies in a broader narrative of cosmic history—one that encompasses the universe's inception, the origin of life, and its ultimate descent into thermodynamic heat death.

My critique is primarily directed at the radicals because they claim to be skeptics yet crusade against scrutiny. This is not simply a debate over philosophy—it is a challenge to the fatally flawed foundation of this worldview that frames our reality as its cause and its created creator. This circular reasoning distorts understanding of existence and restricts one's ability to explore truths beyond its self-imposed boundaries. When I refer to Materialism throughout this book, I ask readers to keep these subtleties in mind.

The broader aim of this book is to contribute to the advancement of scientific knowledge, inviting scientific intrigue across these pages, while encouraging readers to explore perspectives outside the mainstream consensus. This marks a distinction between this theory and conventional cosmology: while I consider some ideas about the universe beginning with a singularity, I ultimately reject this notion. The concept of a universe traceable to a singularity is not rooted in evidence but arises out of necessity under materialist assumptions that allow no alternatives. It is often overlooked that the idea of cosmic inflation was invented to reconcile contradictions with materialist interpretations of existing data.

This book challenges the idea that these professionals, who attempt to explain reality, are the sole arbiters of such inquiries or that we must adhere to their post hoc constructs, such as inflation. On deeply philosophical topics, like contemplating the origin of reality, they are no better equipped than the general public, who frequently defer their critical thinking to these so-called experts—just as I once did, assuming them to be impartial authorities.

Based on my premise, there is no reason to assume that the universe's expansion implies it can be "rewound" back to a singularity. This assumption, accepted without sufficient scrutiny, has become a conventional rule, like many other unjustified notions. In fact, the idea of rewinding to a singularity is fundamentally flawed when considered in light of the inflation theory itself.

Inflation posits the abrupt expansion of a baby universe, possibly several light-years in expanse, and does not support rewinding to a singularity. The popular analogy of a movie rewinding to its beginning overlooks the fact that inflationary theory begins well into the "movie," not at its true inception. This theory glosses over the actual beginning with the euphemistic misnomer of "inflation." This concept is like trying to rewind a spliced tape that starts at the climax, with key scenes of the beginning missing or altered.

## Thomas Henry Huxley and Materialism:

It is well known that Thomas Henry Huxley, a close contemporary of Charles Darwin, was strongly opposed to religion. He is credited with coining the term *agnostic* to describe his view that certain truths—such as the existence of God or ultimate reality—are inherently unknowable. Derived from the Greek *a-* (meaning "without") and *gnosis* (meaning "knowledge"), *agnostic* encapsulated Huxley's belief in the limits of human understanding.[9] Less frequently acknowledged or even known, however, is his sharp critique of Materialism, which he considered philosophically untenable. In his essay *Science and Morals* (1886), Huxley elaborates on his rejection of both Materialism and Spiritualism:

> I have more than once taken pains to say in the most unadorned of plain language, I repudiate, as philosophical error, the doctrine of Materialism as I understand it, just as I repudiate the doctrine of Spiritualism ... and my reason for thus doing is, in both cases, the same; namely, that, whatever their differences, Materialists and Spiritualists agree in making very positive assertions about matters of which I am certain I know nothing, and about which I believe they are, in truth, just as ignorant.
>
> And further, that, even when their assertions are confined to topics which lie within the range of my faculties, they often appear to me to be in the wrong. And there is yet another reason for objecting to be identified with either of these sects; and that is that each is extremely fond of attributing to the other, by way of reproach, conclusions which are the property of neither, though they infallibly flow from the logical development of the first principles of both. ...
>
> I understand the main tenet of Materialism to be that there is nothing in the universe but matter and force; and that all the phenomena of nature are explicable by deduction from the properties assignable to these two primitive factors. That great champion of Materialism whom Mr. Lilly

appears to consider to be an authority in physical science, Dr. Büchner, embodies this article of faith on his title-page. Kraft und Stoff—force and matter—are paraded as the Alpha and Omega of existence. This I apprehend is the fundamental article of the faith materialistic; and whosoever does not hold it is condemned by the more zealous of the persuasion (as I have some reason to know) to the Inferno appointed for fools or hypocrites…

…[I]t seems to me pretty plain that there is a third thing in the universe, to wit, consciousness, which, in the hardness of my heart or head, I cannot see to be matter, or force, or any conceivable modification of either, however intimately the manifestations of the phenomena of consciousness may be connected with the phenomena known as matter and force. In the second place, the arguments used by Descartes and Berkeley to show that our certain knowledge does not extend beyond our states of consciousness, appear to me to be as irrefragable now as they did when I first became acquainted with them some half-century ago. All the materialistic writers I know of who have tried to bite that file have simply broken their teeth.[10]

# Table of Contents

# TABLE OF CONTENTS

## Relationships

A defining idea of this book is relationships—the interactions between systems throughout the universe. The universe operates as a finely tuned system in which every component—whether natural or human-made, including our actions—influences the balance or imbalance of the whole. These relationships demonstrate how systems interconnect and function, analogous to a vast technological facility in which every part has an integral purpose.

Natural Technology challenges materialistic theories by emphasizing this relational purpose as the foundation of the cosmos. It argues that everything is best understood through its interactions and role within a larger system. Exploring these relationships offers deeper insight into the universe's design and our place within it, revealing the interconnectedness between human creativity and the fabric of existence.

The universe's content is like the pages of a well-crafted novel, where every detail—from setting to narrative—is interwoven with relationships that extend even to the Author.

CHAPTER 1

# Irreducible Complexity

*If you search the scientific literature on evolution, and if you focus your search on the question of how molecular machines—the basis of life—developed, you find an eerie and complete silence.* —Michael Behe, *Darwin's Black Box*, 2nd ed. (2006)

To understand the roots of this book, we should revisit irreducible complexity, a concept that, over the last 30 years, has sparked intense debate and curiosity. Coined by Dr. Michael J. Behe, who received his Ph.D. in Biochemistry from the University of Pennsylvania in 1978, the concept has become central to discussions of biological complexity and what best explains the design seen in nature.[11] Behe's work calls into question the adequacy of Darwinian explanations for speciation and the plausibility of abiogenesis, directly challenging the idea that gradual speciation can be explained by Darwinian or any other natural mechanism.

Consequently, for those who are dogmatically opposed to the concept of inherent design and purpose in nature, Behe's theory is often reduced to a straw man version, misrepresented as a simplistic "God of the gaps" argument. The critique typically goes something like this: "[Irreducible complexity (or your ID argument of choice)] is a pseudoscientific belief based on the idea that life on Earth is so complex that it cannot be explained by evolutionary science and therefore must have been designed by a supernatural entity" (see note, for link to an example).[12] However, after 30 years of Behe's argument standing largely unscathed, this misrepresentation persists, with critics preferring to evade his claims rather than directly address the crux of his arguments.

In reality, Behe does not argue, as these critics often misrepresent, that life is 'just too complex to understand, so God must have done it.' Instead, he presents irreducible complexity in a laser-focused way that points out the insufficiency of Darwinian mechanisms in accounting for the origin of certain biological systems, irrespective of how these systems ultimately came to be. There are many biological examples of irreducible complexity that challenge Darwinian speciation, but Behe has largely stuck to the one he first presented over twenty years ago—one that remains unanswered. By his premise, if a single example cannot be refuted, then none can.

Behe defines an irreducibly complex system as "a single system composed of several well-matched, interacting parts that contribute to the basic function, wherein the removal of any one of the parts causes the system to effectively cease functioning."[13] If his premise holds true, then the idea considered foundational to modern biology has been fundamentally wrong for almost two centuries.

Imagine if models of cars were theorized to have developed through evolution. In a similar way, Darwin, in his time, could only observe the superficial structure of organisms, the visible 'body plans'—like the outward features of vehicles, such as wheels, doors, or headlights. What he couldn't see were the elaborate inner workings: the microscopic engines, fuel systems, and circuitry driving these biological machines. Without access to modern microscopy, Darwin was unaware of the sophisticated systems beneath the hoods. Imagine someone explaining the origin of a car solely by its shape and resemblance to other vehicles—without even knowing that a motor exists.

Modern advancements in microscopy allowed Behe to examine the sophisticated inner workings of the cell, the simplest biological unit. Through his observations, he developed the concept of irreducible complexity, arguing that even this 'simple' structure is composed of multiple interdependent systems that could not have arisen gradually. To illustrate, Behe used the E. coli cell's motor as an example.

For anyone familiar with car troubles, it's easy to see how the essential components of a motor—such as the cylinder block, pistons, crankshaft, connecting rods, camshaft, valves, spark plugs, fuel injectors, and timing belt—must all be present and functioning together for the engine to run. Each part is interdependent, meaning the failure or absence of even one component renders the entire system inoperable.

No engineer would design any one of these parts in isolation without the concept of a motor in mind. While belts exist in different contexts, no engineer would specifically develop a timing belt without a preexisting system requiring it, nor could just any belt be adapted to function as a motor's timing belt.

Similarly, biological systems exhibit this kind of interdependence, where individual components make sense only within the fully formed system they serve. The flagellar motor of the E. coli cell demonstrates this irreducible complexity—its parts could not perform a role within a cell unless fully integrated, and they would not contribute to cellular function in isolation. This makes it implausible for such a structure to arise without foresight or intentional design. For critics to overlook this arrangement and its

3

implications is to fundamentally misunderstand the nature of the system as a whole.

This would be like reading individual words of a novel, learning vocabulary along the way, without grasping that the words are part of a story they collectively convey. Yet even today, with all our advancements, Natural Technology predicts that we are still only scratching the surface of the natural, technological sophistication of life and the cosmos—sophistication that will one day make our current understanding seem as limited as Darwin's observations.

However, intelligent design hypotheses, such as Behe's, have sparked significant controversy, as their implications point toward the idea of a Designer—a concept often considered taboo in modern science. For example, in 2005, the Kansas Board of Education proposed changes to the state's science standards that included critiques of evolution, sparking widespread controversy.[14] Although the Board did not explicitly mandate the teaching of intelligent design, many—including 38 Nobel laureates— interpreted the decision as a threat to undermine evolutionary theory and subtly introduce intelligent design, leading to a fervent outcry evocative of a town square rally against heretical ideas.[15]

Amidst the controversy, Bobby Henderson, a recent physics graduate, wrote a satirical letter to the Kansas Board demanding equal time for an "equally scientific" theory in which a "Flying Spaghetti Monster" created the world.[16] Henderson's satire, as described in *Wired magazine*, parodies the idea of an intelligent creator, stating that "The Flying Spaghetti Monster," a deity of "The Church of the Flying Spaghetti Monster… is devoted to spreading the Pastafarian gospel of the true origins of life."[17]

Many have since embraced this trend, dismissing Behe's and Meyer's theories by equating them to "magic" or a "Flying Spaghetti Monster," mocking their faith rather than directly addressing the scientific reasoning they present. However, these dismissals stand out when contrasted with the many speculative explanations their critics promote as frontier science—for example, the multiverse hypothesis.[18]

Though initially introduced to address a different set of challenges to materialist explanations, the multiverse theory has become a versatile, all-purpose tool. What a divine agent cannot accomplish, the multiverse can. Under this framework, even a Flying Spaghetti Monster becomes tenable. A major implication of the multiverse asserts: *Though this (fine-tuning or materialist-gap-filler of choice) doesn't seem naturally plausible in our universe, it must be possible in one of the infinite possibilities—somewhere out there.*

The multiverse theory, which theoretical physicists such as Brian Greene embrace—despite Greene himself describing it as a "strange" line of reasoning—allows for any type of universe to exist supernaturally, meaning, even with laws that would be unnatural or impossible in ours.[19] In the infinite expanse of the multiverse, all possibilities are realized. Somewhere, there is a world where every word of this book is celebrated as unassailable truth, and critics of its arguments have retreated into quiet contemplation, acknowledging the validity of its challenges to Materialism.

As physicist Brian Greene has noted, resisting ideas solely because they seem strange does not disprove them. And because quantum physics is strange, he suggests, whatever we can conceive is possible. Brian Greene explains to Joe Rogan:

> There have to be Realms out there that duplicate ours as well. Many can be different, but there have to be versions of this reality that are also

instantiated, occurring out there in other Realms. So, you come to these crazy-sounding, sci-fi-sounding ideas that you and I are having this conversation—out there—in other distant Realms—an infinite number of times. And moreover, small differences can also arise in these other Realms, where maybe our positions are interchanged at the table—or, you know—maybe your name is Joe Green and I'm Brian Rogan, or there's strange realities that can be taking place. And this is not an overworked theorist's imagination; this is the careful, dispassionate analysis of the mathematical equations.[20]

In a nutshell, under the multiverse hypothesis—or, in a related concept, the Many-Worlds Interpretation, terms physicists like Greene and Sean M. Carroll often misleadingly use interchangeably—there exists a universe where Christopher Hitchens, Lawrence Krauss, Neil deGrasse Tyson, and Sam Harris are monks in a secluded monastery, painstakingly transcribing religious texts and debating the mysteries of faith by candlelight.[21] In yet another universe, Dawkins is a shaman deep in a rain forest, waving smoking leaves over a backpacker who has traveled a great distance seeking healing.

Conversely, Behe's concept of irreducible complexity is clearly derived from empirical observations, while his critics rely on speculative narratives often found in comic books—yet they accuse Behe of lacking scientific grounding. In doing so, they effectively dodge his argument, warding it off with rhetorical wizardry.

So, when Behe's critics dismiss his questions—such as how systems like the flagellar motor could arise through gradual Darwinian processes—as 'unscientific' while relying on an infinite, undetectable set of 'realms,' it's hard to imagine any thoughtful person not raising an eyebrow at such a double standard. At some point, the burden of proof has to shift—or at least be applied consistently.

# The Bacterial Flagellar

**Figure 1.1** This 3D model offers a conceptual sketch of the bacterial flagellum motor. The bacterial flagellum, a molecular motor, is composed of multiple interdependent protein structures that must function together as a whole. **Image source:** Wikimedia Commons.

The flagellum mechanism, scientifically known as the bacterial flagellar motor, is a whip-like structure that helps certain cells, such as bacteria, move through liquid environments.[22] Although bacterial flagella and sperm tails are structurally and functionally different, the comparison provides a familiar way to understand how cells move through fluid. Just as a sperm tail propels the cell forward by undulating in a wave-like motion, the bacterial flagellum enables movement—but instead of undulating, it rotates like a propeller.

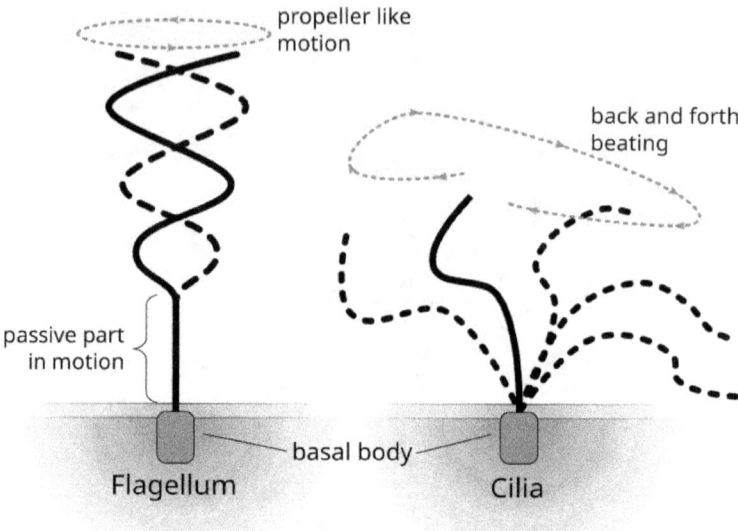

**Figure I.2** Beating pattern of eukaryotic "flagellum" and "cilium." **Image source:** Urutseg, Wikimedia Commons.

Figure 1.2 illustrates that bacterial flagella function by rotating like a propeller, whereas eukaryotic flagella and cilia generate movement through a whip-like or back-and-forth beating motion. This diagram highlights the fundamental mechanical differences between the two, reflecting their distinct structural compositions and cellular roles.

## Human-Made Motors

Like human-made engines and motors, the flagellar motor relies on a precisely organized set of components to convert energy into motion, allowing bacteria to navigate their environments efficiently. By examining engineered motors—systems that we know require foresight, planning, and precision—we can draw important parallels to biological machines like the bacterial flagellum. With this in mind, let's take a look at some standard motors in engines and key concepts.

Fig. 340.  ROTOR OF MOTOR OF A.E.G. ZOSSEN
MOTOR CAR.

Fig. 341.  ROTOR OF MOTOR OF A.E.G. ZOSSEN CAR IN PLACE ON AXLE.

**Figure 1.3** This image shows the rotor of the A.E.G. Zossen Motor Car—an early electric motor component, shown separately and mounted on an axle. **Image source:** "Electric railway engineering" by H. F. Parshall and H. M. Hobart. Published 1907. Public domain.

Figure 1.3 shows the rotor assembly of an early motor used in the A.E.G. Zossen motor car. The visible components include the rotor, which spins inside the motor to generate mechanical motion, and the shaft, which transmits this motion to the car's axle. The two circular components at each end of the shaft help support and stabilize the rotor while also potentially housing bearings, electrical connections, or mechanical linkages that facilitate smooth operation within the motor system. These components work together to convert electrical energy into the mechanical energy required to drive the car.

**Figure 1.4** A turbine engine, specifically a turbofan engine. Turbofan engines are a type of jet engine used primarily in aircraft. **Image source:** Ariadacapo.

Figure 1.4 shows the core of a sectioned Rolls-Royce Turboméca Adour turbofan engine. This cutaway view of a jet engine reveals key internal components, including the fan (blades on the right), compressor stages, a combustion chamber, and turbine. Air is drawn in through the fan, compressed by multiple axial compressor stages, mixed with fuel and ignited in the combustion chamber, and then expelled through the turbine, generating thrust. The precise arrangement and interaction of these components allow the engine to efficiently convert fuel into thrust, powering aircraft at high speeds.

**Figure 1.5** A view of the internal structure of the 1963 Chrysler Turbine Engine. **Image source:** Tm, Wikimedia Commons/ Flickr via Flickr2Commons.

Figure 1.5 shows the internal components of a 1963 Chrysler turbine engine, designed for use in the Chrysler Turbine Car. The image displays the turbine rotors and stators arranged along the central shaft. The rotors convert high-energy gas flow into mechanical energy, while the stators direct airflow through the engine to optimize efficiency. The system relies on precise alignment between these components to maintain smooth operation.

Unlike traditional piston engines, which generate power through controlled explosions in cylinders, the turbine engine operates by continuously expanding heated gases through the turbine stages.[23] The resulting rotational force is transmitted through a reduction gearbox, which lowers the high RPMs of the turbine to a usable speed for driving the vehicle's wheels.[24]

This turbine engine represents a unique adaptation of aerospace technology in automotive design, showcasing Chrysler's experimental propulsion systems during the 1960s. Although this particular engine was ultimately not adopted for widespread use in cars over traditional piston engines due to efficiency concerns, its design illustrates an essential principle: the interdependence of components.[25] Each part plays a necessary role, and the system as a whole cannot function if any critical element is removed.

## The Flagellar Motor

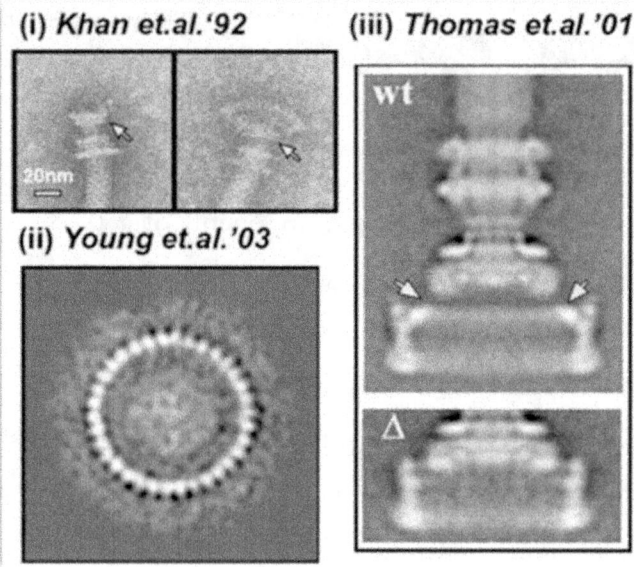

**Figure 1.6** Cropped from the original image, (i) shows the isolation of the extended cytoplasmic structure linked to the MS-ring (yellow arrow), identified as part of the switch complex (Khan et al., 1992). (ii) Overproduction and assembly of components allowed determination of C-ring stoichiometry (Young et al., 2003). (iii) Single-particle analysis resolved FliG substructure differences between wild-type and mutant reconstructions (Thomas et al., 2001). **Image source:** Epipelagic, Wikimedia Commons.

Figure 1.6, based on electron microscopy data, shows the bacterial flagellar motor's architecture, with comparisons between wild-type and modified versions from multiple studies. The top-down and side views illustrate the concentric rings formed by the motor's structural proteins, which are essential for the motor's function in converting chemical energy into mechanical force, enabling bacteria to swim and navigate their environments. The yellow arrows point to the switch complex, a structure that controls the direction of rotation.[26] When the flagellum rotates in one direction, the bacterium moves forward; when it reverses, the bacterium changes direction. This allows the bacterium to navigate its environment in response to signals.[27] In our human-made vehicles, the transmission or differential determines the direction of rotation, which in turn dictates how the car moves—whether forward, backward, or turning.

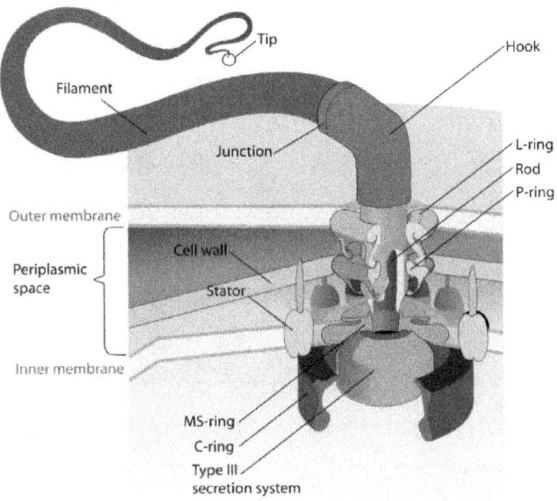

**Figure 1.7** The depicted type of flagellum is found in bacteria such as E. coli and Salmonella, and rotates like a propeller when the bacterium swims. The bacterial movement can be divided into 2 kinds: run, resulting from a counterclockwise rotation of the flagellum, and tumbling, from a clockwise rotation of the flagellum. **Image source:** Wikimedia Commons, Public Domain.

1. **MS Ring and C Ring (Rotor):** The MS ring, made up of the transmembrane protein FliF, is an essential part of the rotor, integral to the functional flagellar system.[28] The C ring, attached to the cytoplasmic face of the MS, ring is essential for generating torque

and regulating the motor's switching behavior.[29] The combined MS-C ring configuration functions as the motor's rotor, central to its operation.

2. **Stator:** The stator in the bacterial flagellar motor is a complex of protein subunits, MotA and MotB, that forms a ring around the rotor.[30] Embedded within the cell's inner membrane, the stator actively participates in the motor's function. Each MotA and MotB pair in the stator unit forms an ion channel.[31] These channels are necessary for the motor's operation as they allow ions (protons or sodium ions, depending on the bacterial species) to flow through. This ion flow is directly converted into mechanical torque that drives the rotor's rotation.[32]

3. **Drive Shaft (Hook):** The hook functions as a universal joint within the bacterial flagellar motor, linking the flagellar filament to the rotor.[33] This component is designed to flexibly transmit rotational force from the rotor to the flagellum, enabling multidirectional movement.

4. **Bushing (LP Ring):** In the bacterial flagellum, the LP ring functions as a molecular bushing, crucial for enabling the high-speed, low-friction rotation of the flagellar motor.[34] This ring showcases intricate interactions and charged surfaces that align precisely to minimize friction, allowing the flagellar rod to spin efficiently at speeds of up to 1,700 revolutions per second—far surpassing the performance of common electric motors, which typically operate at speeds up to 60 revolutions per second.[35]

There is no scientific evidence that any of these parts can form gradually or exist independently outside of their integration into a similarly irreducibly complex system. The implications are clear: there is no way to arrive at this system step by step. The system itself is one step. Just like a car engine cannot run without its cylinder block, crankshaft, or spark plugs, this motor depends on all its components working together.

After examining the design and functionality of human-made motors and the bacterial flagellar motor, it seems obvious that if the flagellum operates at magnitudes higher in efficiency, both systems must be works of engineering. The precision and purpose in their construction compel us to consider their origins. While its origin may be debatable, the claim that proponents of design in nature are simply promoting religion out of ignorance or confusion about the complexity found in nature is only a rhetorical defense mechanism. That assertion isn't an inadvertent strawman; it's a blatant one.

Moreover, the common premise that some assert—that such an assembly is beneficial and will therefore self-assemble—does not explain its feasibility without agency, foresight, or deliberate planning. Consider this scenario: If Person A has $100 and Person B has $10,000, and someone asks how Person B obtained so much more money, it would be irrelevant to simply respond that $10,000 is more beneficial than $100. Acknowledging that $10,000 is more advantageous does nothing to explain how Person B acquired it.

Accordingly, if a bank robbery involving $10,000 had recently occurred and an investigator found you near the scene with $10,000 in your possession, his natural question would be, 'Where did you get this money?' Now imagine responding to the detective with:

> 'I have this exact amount because it's more beneficial than being broke. But because you detectives have an ulterior motive, you let your assumptions cloud your judgment instead of considering this—and other possibilities.'

Such a response avoids the real question entirely. Rather than clarifying the source of the money, it shifts the focus to accusing the investigators of bias, leaving the issue unresolved. It would be far more straightforward to simply clarify the source of the money and move on.

Similarly, when discussing the origins of systems made up of purposely arranged parts that function together for a higher purpose, simply pointing out their benefits or usefulness does not speak to how they came into existence, nor does it reveal the processes that led to their development and integration into a functional system. The motives or implications of an investigation have no bearing on its veracity. Likewise, after 150 years of asserting that all life on Earth diversified from a common microbial ancestor, asking how this process accounts for biological motors should not be met with hostility, as if the question were unwarranted.

A biological motor appears to be an ancient technology that existed eons before humans conceptualized the first mechanical motor. Impressively, it continues to operate with nearly 100% efficiency. The flagellum incorporates components strikingly similar to those we've only devised within the last century. Yet, by building one precept upon another, we ultimately succeeded in engineering the motor—a feat that would not have been possible without human ingenuity, foresight, and careful planning.

# CHAPTER 2

# Ancient Wisdom

*One thing that will quickly become clear is that the prevalent 'big picture' of history—shared by modern-day followers of Hobbes and Rousseau alike—has almost nothing to do with the facts. But to begin making sense of the new information that's now before our eyes, it is not enough to compile and sift vast quantities of data. A conceptual shift is also required.*[36] —David Graeber and David Wengrow

Popular culture often reinforces a stereotype of ancient people as dull, caveman-like figures—relying on rudimentary tools, speaking crude languages, and living in an uncouth manner. However, this view is fundamentally flawed. This misconception persists in part because the dominant evolutionary perspective frames human intelligence as a linear progression tied to cultural and technological development, dividing it into staged increments of cognitive ascent. The traditional evolutionary model is often infused into historical accounts by scholars and theorists, suggesting a progression of human cognitive abilities that moves societies from simple hunter-gatherer groups to complex industrial civilizations.

The assumption that technological advancement directly correlates with cognitive development has reinforced the erroneous belief that ancient people were intellectually inferior. If humans today, however, were removed from their cultural frameworks—without access to accumulated knowledge, tools, or societal structures—their cognitive abilities would remain unchanged. Without these familiar reference points their capacity to express and apply their intelligence in recognizable ways would be diminished.

What if art had never existed? In that case, cave paintings would be seen as extraordinary. For example, Middle Stone Age engravings discovered in Africa, often estimated to be around 70,000 years old, demonstrate early symbolic thought and artistic expression, which are key traits of our cognition.[37] What if the concepts of reading, writing, or books had never existed? Even Isaac Newton—brilliant as he was—relied on cultural foundations.

Without reading the works of Euclid and Kepler, without the mentorship of Isaac Barrow, or without his education at Trinity College, Cambridge, Newton likely would not have developed calculus, optics, or the laws of motion.[38] If such cultural reference points had never existed—with no written mathematical tradition, established universities, or system of scientific inquiry—Newton's genius would have remained latent. In many ways, unless a person has a learning disability, the idea that one person is inherently more intelligent than another is an illusion. Everything anyone understands is largely shaped by cultural reference points and experiences.

Consider our greatest feat: space travel. This marvelous achievement was not the result of increased human cognition but rather a series of innovations building upon one another. The ability to send astronauts to the Moon depended on computers, but these were developed from earlier calculators. In fact, the term 'computer' originally referred to a person who performed calculations, not a machine. Over time, this task was mechanized, then fully automated, allowing digital computers to evolve from simple number-crunching tools into essential systems that made space exploration possible.

Likewise, the Wright brothers' first flight was made possible by prior advancements in engineering. It depended upon the combustion engine, a technology first developed for automobiles. Before automobiles, a car was simply a horse-drawn carriage, and the transition to motorized vehicles was made possible by centuries of mechanical progress.

This becomes increasingly evident as technological advancement accelerates at an exponential rate. Even within a single lifetime, older generations often struggle to keep pace with new developments. This is not because grandpa is cognitively inferior, but because the foundational knowledge required to understand technologies is shifting ever faster. Yet, despite basking in modernity and relying on these technologies daily, the average person does not understand how they work. Technological progress is a collective and cumulative process, not a reflection of an increase in innate intelligence.

Assuming that the absence of early cultural and technological artifacts indicates lower cognitive capacity reflects a Darwinian confirmation bias. On one hand, Darwinians insist that humans are just another species within the animal kingdom, framing our intelligence as something that must be learned rather than innate. However, they describe animals as possessing *instinctive* behaviors, portraying them as machines running on pre-programmed software while framing human cognition as an acquired accumulation rather than a natural function.

But if intelligence must be learned, when did species start learning—and why did they stop? Are we really just another animal? Birds instinctively know how to build nests. Spiders weave webs without instruction. Yet we are not born knowing how to build a house or even how to survive without guidance. Why is design language used to describe animal behavior, yet dismissed when discussing human intelligence? What does instinct really mean? And what happened to our specialized skills—those that required, from birth, no learning.

If intelligence must be built up incrementally, how do crows and ravens possess the natural intelligence to manufacture tools from leaves and twigs, shaping them into hooks and spears to extract insects?[39] How did they learn to plan for the future, selecting and storing tools they won't need until later?[40] Some of these birds have been observed using three tools in sequence to solve a problem, demonstrating an understanding of cause-and-effect relationships. If human intelligence required a gradual, linear ascent, how did these birds develop such advanced cognitive abilities outside of that framework?

In reality, as with those animals, archaeological and anthropological findings indicate that humans have always possessed the same cognitive abilities seen today. For humans, it is culture that serves as the medium for cognitive abilities; it provides the framework for the *natural learning* process through which *natural intelligence* is expressed and applied.

People mistake increasing *familiarity* with certain knowledge for being inherently intellectually elite, but it is through culture that cognitive abilities are translated into perceived levels of intelligence, creating a false sense of pride.

In their book, *The Dawn of Everything*, David Graeber and David Wengrow challenge this misconception with compelling evidence.[41] [42] This work reveals that ancient societies were far more cognitively sophisticated and diverse than commonly assumed and that their technological limitations did not hinder their cognitive capabilities. They present a radical departure from the traditional model, which typically views human societies as progressing through predetermined stages of social and political development. In essence, they argue that the common depiction of history does not reflect the actual diversity and complexity of early human societies. They reveal how we've been conditioned to believe that as soon as agriculture and technology developed, hierarchies and social inequality were the necessary outcomes. They offer evidence to the contrary: many early societies practiced agriculture for centuries without adopting centralized power structures or inequality.[43]

A key implication of their work is the overestimation of external pressures (e.g., environmental or technological) as drivers of social and political evolution, at the expense of human agency. Graeber and Wengrow argue that ancient societies were not simply subject to the environment or resource constraints. Instead, they had the imagination and flexibility to shape their social structures according to cultural and ideological choices. The book presents the view that history isn't an inevitable march toward social or political progress, even though technological advancements have consistently shaped human societies.

Though "a few privileged experts" know the evidence does not support the traditional narrative, this knowledge remains largely hidden from the public, much like the myth of Santa Claus is preserved for children.[44] It is deemed too disruptive to challenge the simplified and comfortable perception of history.

As Graeber and Wengrow put it,

> The pieces now exist to create an entirely different world history – but so far, they remain hidden to all but a few privileged experts...By the late

nineteenth century, it was becoming clear that the original sequence as developed by Turgot and others – hunting, pastoralism, agriculture, then finally industrial civilization – didn't really work. Yet at the same time, the publication of Darwin's theories meant that evolutionism became entrenched as the only possible scientific approach to history – or at least the only one likely to be given credence in universities.[45]

Graeber and Wengrow argue that early human societies were not simply primitive bands but rather exhibited a wide range of social structures and political arrangements. "One problem with evolutionism," they write, "is that it takes ways of life that developed in symbiotic relation with each other and reorganizes them into separate stages of human history. This framework falsely equates societal complexity with hierarchical structures, ignoring the existence of complex, egalitarian societies"[46] This critique points out how Darwinian narratives have been woven into historical accounts, leading to the misrepresentation of the complexity of ancient societies by *forcing* them into an artificial linear progression.

Moreover, the achievements of ancient civilizations in engineering, social organization, science, and mathematics contradict the traditional evolutionary models that depict human progress as a linear advancement from primitive to complex societies. Ancient peoples demonstrated advanced cognitive abilities and intellectual sophistication. Though Graeber and Wengrow challenge traditional assumptions about societal arrangements, the rapid progression from stone tools and early writing systems to modern technological advancements is difficult to reconcile with linear models based solely on cognitive happenstance or environmental pressure. If biological evolution took billions of years to produce the human brain, the technological leap of the past few thousand years suggests influences beyond what traditional evolutionary narratives typically address.

Despite the notion that modern humans are more intelligent due to technological advancements, there is an argument to be made that cognitive abilities may have, in fact, diminished comparatively today due to advancement in technology and our ever-growing reliance on it. The thing that we see as a signature of our cognitive advancement may in fact have dumbed us down. Ancient civilizations required a high level of cognitive abilities for daily survival, social organization, and technological innovation. They were required to develop their innovations, cultures, and skills more independently, without the interconnected knowledge networks of today.

If we were suddenly transported back in time and dropped into cultures from thousands of years ago, we would quickly realize that advancements in technology and an increasingly complex society do not necessarily equate to intelligence. Without the skills and knowledge needed to navigate those ancient worlds—and without our modern resources—it would quickly become clear that intelligence is relative to environment and time. In the past, tasks that we now perform with the aid of technology had to be done manually, often requiring complex knowledge and problem-solving skills. In ancient times, hunting, fishing, and practicing mental arithmetic and memory were not just hobbies but essential skills for survival.

Graeber and Wengrow discuss the use of these mental skills such as early societies relying heavily on memory and oral tradition to transmit knowledge across generations. The authors note that there may be misinterpretations of the "simple" art of the people such as the Chavín, an extinct, pre-Columbian civilization in the northern Andean highlands of Peru. They argue that the Chavín's artistic depictions might "not be meant to illustrate or represent but instead serve as visual cues for extraordinary feats of memory."[47]

Graeber and Wengrow also note:

> Up until recent times, a great many indigenous societies were still using systems of broadly similar kinds to transmit esoteric knowledge of ritual formulae, genealogies or records of shamanic journeys to the world of chthonic spirits and animal familiars. In Eurasia, similar techniques were developed in the ancient "arts of memory," where those trying to memorize stories, speeches, lists or similar material would each have a familiar 'memory palace'.[48]

In ancient times, it was not unusual for entire book-length histories and stories to be stored in the practiced memory of many. However, over time, reliance on memorization began to fade with the advent of technologies such as writing and print, but it wasn't until the rise of personal computers that the decline became pronounced. The final phase began with the spread of cell phones in the 2000s, further diminishing the need to retain even basic information. A seven-digit number was the last remnant of everyday memorization before technology supplanted human memory, eliciting increasing dependency and eroding autonomy.

What we create does not remain a passive tool; it takes on a life of its own, shaping our habits and cognition in ways we must remain aware of and actively navigate. Our creations' influence can be deceptive—while we believe we are ascending, our lack of awareness often leads us to relinquish fundamental cognitive abilities, while technology's insidious side compels us to adopt detrimental habits.

In contrast, prehistoric peoples walked frequently. They ate with family members while engaging in deep discussions. They harvested and prepared their own food and often navigated vast territories without modern mapping technologies, requiring exceptional spatial awareness. Today, many people lead largely sedentary lives, often eating alone or distracted by screens.

Processed and pre-packaged foods have replaced the need for direct involvement in food gathering and preparation. GPS and other navigation technologies have largely replaced the need for spatial awareness, leading to a decline in our ability to navigate without assistance. Many who frequently use these technologies rely on them even for short trips to familiar places, and some lack the confidence to find their way back home without them.

Interestingly, while much has been speculated about our (supposedly) humble cognitive beginnings, little attention is given to the future potential of human intelligence. While evolution's journey toward intelligence is celebrated up to the present, it is rarely projected into the future. Among scientists and the general public alike, there seems to be an implicit disconnect with the plausibility of what such a view would entail. This cognitive gap is particularly odd for a theory centered on continuous change. If our cognitive abilities evolved from stardust, to Earth, to bacteria, to primates, and finally to where they are now, it stands to reason that the future should hold nothing less than brains capable of quantum computing and telepathy becoming commonplace.

# CHAPTER 3

# Technology

*The technium also wants what every living system wants: to perpetuate itself, to keep itself going. And as it grows, those inherent wants are gaining in complexity and force.*[49] *— Kevin Kelly*

Seeing technology as only practical tools has evolved as thinkers contemplated its broader implications. The Greek word *technê* is rooted in the meanings of art, skill, or craft, whereas *logia* refers to study.[50] Together, these terms represent human ingenuity and mastery of natural forces. In his dialogues, Plato elaborates on the relationship between knowledge and craft.[51] He believed technê involves both practical skills and theoretical understanding. Over time, the concept of technê expanded to include *epistêmê*, which refers to reflective, theoretical knowledge.[52] This duality can be seen in biological processes, which also demonstrate a sophisticated coordination of components guided by underlying principles.

Moreover, Plato introduces the concept of the *demiurge*, or cosmic craftsman, in his dialogues.[53] This figure is a creator who molds the universe with deliberate skill, mirroring a master artisan's craftsmanship. Plato's demiurge doesn't just randomly assemble the cosmos; instead, it uses an esoteric understanding of eternal forms, which are ideal and timeless patterns, to impose order on chaos.[54] In this philosophical framework, Plato argues that the universe itself operates with a kind of inherent "technê." This implies that the cosmos is structured with an order and precision of the finest human artistry.

As civilizations advanced, so did our conceptualizations of technology. During the Enlightenment, the scope of technology broadened significantly. Figures like Francis Bacon, seeking to fulfill man's role in the book of Genesis, advocated for using science and technology to control and dominate nature. This theme, stripped of its religious undertones, persists in modern discussions about technology's role in human progress and environmental impact.[55] The Industrial Revolution marked a significant moment in the history of technology, shifting the focus from individual tools to complex machines and systems that transformed social and economic structures. A theme explored by philosophers such as Karl Marx and later Martin Heidegger, this period positions technology in a dual role as both creator and disruptor. Marx discussed technology as a fundamental factor in societal change, whereas Heidegger considered the essence of modern technology and its metaphysical role in shaping human understanding of the world.[56]

Heidegger offered a new way to think about technology in his work *The Question Concerning Technology*.[57] He believed that technology does more than just help us achieve goals; it also helps us understand and organize the world around us.[58] According to Heidegger, technology shows us truths about how everything connects and works together, organizing and ordering the world for human use.[59]

According to Heidegger, however, this "mode of revealing" has a downside: it frames our view of the world in a way that reduces everything to mere resources, things to be used. Heidegger calls this "Gestell" or "enframing."[60] Different eras and technologies bring forth different modes of revealing; ancient crafts or art, for instance, disclose truth differently than modern technology. In the modern age, enframing reveals the world as "standing-reserve"—nature and resources are perceived primarily in terms of their utility.

As we better harness technology to navigate and rearrange the world around us, the fabric of nature recedes into the background, its inherent value eclipsed by its utility. This shift, driven by a relentless quest for efficiency, subtly diverts our attention from the complex and intimate relationships that underpin the natural world. Increasingly, we view our environment through the narrow lens of what it can provide or how it can be reshaped to suit our purposes. This reframing weakens our capacity to connect with nature's more fundamental realities and diminishes our sensitivity to its subtle wisdom. The gradual detachment from this life force that nourishes and enriches us may, in essence, be making us less alive, diminishing the vibrancy and richness of our existence. Heidegger warned of this ensnarement that technology brings.

More recently in this discourse, thinkers such as Don Ihde and Bruno Latour have expanded our understanding of the philosophy of technology.[61] Don Ihde's perspective, known as postphenomenology, examines how technology reshapes human experiences and perceptions. Analyzing the relationships between people and various technologies, Ihde reveals that technology is not just a tool but an active mediator in our daily interactions.[62]

On the other hand, Latour's "actor-network" theory proposes a broader networked perspective, where humans and technological artifacts are interconnected, each influencing and shaping the other.[63] Latour's theory pays special attention to the "agency" of non-human elements, suggesting that objects and technologies play almost as significant a role in societal dynamics as humans do.

Building on these ideas, Jacques Ellul's work in *The Technological Society* delves into the concept of the autonomy of technology.[64] Ellul introduces the term "technique" to describe the operation of machines and a broader philosophical principle.[65] According to Ellul, "technique" represents a pervasive drive toward efficiency that governs all aspects of modern life.[66] This drive, inherent in the fabric of technology, dictates the ways technologies are developed, utilized, and integrated into society. For Ellul, technique is a guiding force that shapes our institutions, cultural norms, and even personal behaviors. He pointed out how this autonomy evolves independently of human values and is often at odds with ethical or life-enhancing priorities.[67]

Kevin Kelly, in *What Technology Wants*, presents a dynamic view of technology as an almost living system, which he calls the "technium."[68] Kelly suggests that technology inherently seeks to proliferate, diversify, and enhance its efficiency, much like biological life.[69] This perspective is similar to Dawkins' concept of the selfish gene, but instead of genes, Kelly attributes a "selfish" drive to technology itself.

Kelly sees technology as living and evolving similarly to Darwinian evolution. I extend that concept to include a duality. Like Kelly, I view technology as living, but I see it as a purposeful unveiling, aligning with Heidegger's idea that technology reveals truths about the world. This unveiling is not random but unfolds through human desire and engagement, linking our ingenuity to the universe's deeper structure. Technology is not just alive—it is actively drawn out by human will, revealing itself in alignment with our pursuit of discovery.

## Nat Tech Universe

When woven together, the ideas of these thinkers offer a thought-provoking perspective that frames technology as both a revealer of cosmic truths and, equally, a potential ensnarer of human perception and freedom. This frames the creation and the use of our technology as a deciphering of nature that reveals existential truths. It also means that while we consider what technology does for us and the existential revelations it brings, we must acknowledge the insidious pressures it places upon us.

This prepares us for exploring Natural Technology, which we will primarily refer to as NatTech. But first, it is important to define what is meant by 'natural' and 'technology.' 'Natural' refers to all phenomena and processes in the universe that occur independently of human intervention or artificial influences. This includes everything from the movement of celestial bodies to the functions within living organisms. 'Technology,' on the other hand, encompasses the tools, systems, and methods created to solve problems, facilitate processes, and enhance life. By combining these definitions, Natural Technology is the sophisticated designs and systems found in our cosmos that serve existential purposes.

This leads to an inevitable question: if everything, including the physical makeup of humans, can be considered technology, what sets our human-made technologies apart from Natural Technology? The answer lies in free will. While every component of the universe can be seen as a form of technology, governed by natural processes, humans possess the unique ability to actively repurpose and reconfigure these processes through boundless abstract thinking. We manipulate and redirect nature to serve specific purposes, transcending the inherent limitations of its natural course.

31

This capacity to manipulate and redirect nature reflects within us an essence similar to The Designer of the technology from which our bodies and cosmos are crafted.

Though rarely acknowledged, this distinction between conscious creations and natural processes essentially dissolves under Darwinian materialism. This doctrine frames everything within this system—including human consciousness and even its technological creations—as just products of natural evolutionary processes, erasing the boundary between technologies born from conscious intent and those naturally occurring in nature. It encompasses the evolution of everything. It fundamentally is a theory of everything.

The narrative allows for no exceptions: inanimate matter is said to have evolved from a singularity into life, encompassing everything—animals; the zoos and veterinary technology used to care for them; the universe, through human beings, debating its own origins; and even self-organizing wars between itself. All of our technology, from medical devices that heal to weapons that kill, are simply products of evolution. Even this book is considered part of the evolutionary continuum. My role in writing this book is an illusion. My thoughts, my choices, and even the act of arranging these words are not products of free will but only reactions within an ongoing evolutionary sequence. Furthermore, under Materialism, the distinction between *natural* and *artificial intelligence* is an entirely meaningless construct. Any distinction between artificial and natural intelligence is artificial itself. In this context, evolution can be seen as an ongoing sequence of conflicts, each struggle a battle in a war in which evolution defeats natural laws and logic.

Notably, the standard description of the universe and its origin, such as originating from a singularity, in a way undoubtedly suggests technology. Consider kirigami, an art form where paper is both folded and cut to create intricate patterns that remain hidden until the paper is unfolded.

**Figure 3.1** Chinese kirigami pop-up artwork featuring a detailed pavilion: precise cuts and folds that reveal an intricate design when unfolded. **Image source**: fdecomite, Flickr.com

When you expand it, as illustrated in Figure 3.1, the paper unfolds like a pop-up book, revealing an elaborate design. This is analogous to how the universe, as hypothesized by the traditional model, unfolds, revealing its cosmic design.[70]

Many people, however, focus on the beauty of the paper structure as it unfolds, believing it is becoming more complex than when it was originally cut and folded. In reality, it becomes less structured as it experiences the natural forces of wear and tear. Similarly, if the traditional model holds true, what we see in the universe is not new complexity emerging, but the unfolding of beauty from a state that was far more tightly packed and highly ordered from the beginning.

Moreover, the paper structure does not simply hold its form by chance—it is held together by the precise cuts and folds that allow it to expand while maintaining its stability. Without these built-in design features, the structure would collapse into disorder rather than revealing an elegant design.

Similarly, while the universe experiences entropy according to the second law of thermodynamics, its fundamental order persists much longer than a purely entropic system should allow. Galaxies hold their shape, the laws of physics remain constant, and life actively resists decay. Materialism takes these laws as given but does not explain their origin, stability, or why they persist unchanged. This persistence suggests that, like a kirigami structure, the universe was not a chaotic system that happened to unfold into order, but one that was set in motion with the necessary components to sustain it, at least for a time. Through this analogy, it becomes clear that complexity does not increase over time, but instead regresses—revealing an initial, structured complexity that slowly unravels in one direction: toward an overall loss of order.

But if the laws of physics truly govern everything, if methodological naturalism is the rule, why should exceptions exist? Despite this fundamental principle, Darwinian Materialism uniquely contradicts it. Unlike every other system observed in nature, Materialism assumes that order, complexity, and even consciousness can manifest from entropy-driven processes—without any external insight or guiding force. This is not just an assumption; it is a necessary claim of Materialism, not of science. In doing so, Materialism imposes itself upon science.

By doctrine, Materialism must assert, allow, and explain self-organizing niches of complexity (such as stars, galaxies, life, and intelligence) as exceptions to entropy, rather than accounting for them within established natural law. However, if entropy dictates that all systems tend toward disorder, the persistence of order—on cosmic, biological, and intellectual levels—suggests something deeper at work, something Materialism fails to explain: phenomena that, under the laws we know, should not even exist, let alone persist.

CHAPTER 4

# Fire to AI

*Any sufficiently advanced technology is indistinguishable from magic.*

—Arthur C. Clarke

Standing on Earth with limited tools and knowledge, Isaac Newton unraveled the laws of motion and gravity. He applied these laws to explain the orbits of planets, the behavior of light, and the principles of optics. This alone challenges the idea that human ingenuity is simply a product of random chance. Newton's discoveries point to a mystery—a natural provision that shapes our understanding and fuels our capacity for innovation. One way to conceptualize this inherent provision is by drawing an analogy with machine learning.

Machine learning systems improve by recognizing patterns in data, adjusting their outputs through repeated exposure and optimization. However, they require pre-existing algorithms to function at all. Similarly, human cognition follows a structured learning process, where discoveries build upon an embedded foundation of knowledge. But unlike machine learning, our innovation is driven by an innate essence, a natural ambition that compels us to create. As we accumulate a technological lineage, artificial intelligence takes shape as a reflection—an image of the natural technology embedded within us.

This structured learning process is evident in the trajectory of human technological advancement—from the harnessing of fire to the development of AI. Nowhere is this pattern more apparent than in biotechnology. From ancient fermentation practices to modern genetic engineering, humans have consistently demonstrated an ability to understand, manipulate, and optimize biological systems in ways that suggest more than just chance.

Entwined with some of our earliest technological breakthroughs, our roots in biotechnology stretch far back into human history. From bread-making to brewing alcohol, humans have long exploited biological processes to enhance everyday life. By using yeast to ferment carbohydrates, ancient civilizations pioneered techniques that form the basis of modern biotechnology, which continues to advance and drive innovation today.[71]

It's remarkable how quickly fermentation practices arose independently across culturally and geographically disconnected societies. Within a few millennia, various civilizations, each driven by local resources, developed fermentation techniques independently and without direct contact. By 3500 BCE, the Sumerians in Mesopotamia were already brewing beer, and the Egyptians began producing beer and wine by 3000 BCE.[72] Meanwhile, Chinese evidence suggests that rice-based alcoholic beverages were being made as early as 7000 BCE, and indigenous communities in the Americas had developed maize-based chicha by 3000 BCE.[73] These American societies also created maize-based flatbreads as far back as 10,000 BC.[74]

These near-simultaneous but isolated developments, including pottery, metallurgy, agriculture, and writing, represent a *big bang* of innovation that cannot simply be reduced to reactions to external pressures or a process of trial and error. This ability to engage with and transform natural processes was driven by more than survival needs. The refinement of fermentation and other processes suggests early humans had an understanding that cannot be explained by serendipity alone. The fact that these innovations occurred independently yet within a narrow timeframe, despite geographical isolation and a lack of cultural exchange, points to a natural human aptitude for innovation.

Another consideration, however, is that the migration of peoples and their knowledge from a shared origin occurred at a much faster rate than previously understood, enabling these early technologies to spread widely. As these groups migrated, they adapted their ancient knowledge to their local environments and available resources. This adaptation would allow similar innovations to emerge across geographically distant civilizations within a relatively compressed timeframe, possibly suggesting that the shared origin of knowledge was not only widespread but also incredibly versatile in how it was applied to different ecosystems and resources.

As civilizations progressed, so did our techniques and understanding of biotechnological processes. What once began as the mysterious cultivation of yeast and bacteria became a science as scholars unraveled the agents behind fermentation and disease. This expanding knowledge paved the way for transformative developments in biotechnology.

By the 19th and 20th centuries, the groundwork for modern biotechnology had been firmly laid by pioneers like Louis Pasteur and Gregor Mendel. Pasteur's advancements in microbiology and Mendel's revolutionary work in genetics reshaped our ability to manipulate biological systems at a molecular level.[75] These two scientific visionaries, both born in 1822, contributed independently to two fields that would forever alter the course of human understanding. Their work laid the foundation for more advanced discoveries in both biological sciences and computational technologies, ultimately contributing to the discovery of DNA. These breakthroughs directly paved the way for modern advancements in genetic engineering and biopharmaceutical innovations, fields that have transformed the world we live in today.

## Language

Yet, all these innovations rest on the clearest manifestation of human cognitive ability: *language*. The human capacity for abstract speech and creative thought is evidence that we live in a fundamentally relational universe—one in which meaning and communication are intrinsic, rather than incidental, to existence. Long before scientific experiments were documented in notebooks or other written records, language served as the primary vehicle for transmitting knowledge, culture, and innovation.

Noam Chomsky, known as the father of modern linguistics, proposed that language is an innate universal property, hardwired into the human brain.[76] He introduced the concept of a Language Acquisition Device (LAD), a natural biological mechanism that enables humans to instinctively learn language, regardless of cultural or environmental conditions.[77]

This innate mechanism explains why children across all societies effortlessly acquire complex linguistic systems without formal instruction. Moreover, research suggests that there is a critical period during early childhood when the LAD is most effective, after which language learning becomes significantly more difficult.[78] This implies that language acquisition is not a sequential, external process but rather a systemic one that relies on a biologically timed mechanism, which activates during a "critical period" in early development.[79]

Chomsky's theory of Universal Grammar (UG) posits that all human languages share an underlying structure, meaning that the differences between languages are superficial.[80] [81] This preprogrammed cognitive architecture allows for the acquisition of our diverse languages, but limits the possibilities of syntax, grammar, and phonology that can be realized across all human languages. This means, although individual languages may vary in terms of vocabulary or specific rules, UG provides a universal set of principles that underlie all possible human languages.[82]

Under the standard evolutionary framework, language was traditionally seen as the product of gradual adaptation, shaped by environmental pressures. However, this view fails to explain why the precise genetic, neurological, and anatomical structures required for speech appear fully integrated in humans with no clear evolutionary precursors. The human brain is uniquely wired for complex language, the vocal tract is finely tuned for articulation, and the genetic components linked to speech, such as FOXP2, do not show a gradual developmental pathway.[83] [84]

Moreover, speech production depends on intricate coordination between the respiratory system, larynx, and vocal tract—along with precise motor control for articulation.[85] *Serial ordering* in utterance planning ensures smooth and efficient speech, while real-time feedback mechanisms continuously adjust articulation.[86] However, the precise coordination of these processes is explicitly recognized as the "serial order problem" in modern science.[87] The serial order system in speech functions like a real-time processing system, similar to a predictive AI model or a robotic control system, as it simultaneously anticipates future sequences and references past patterns to maintain coherence.

How such a highly structured sequencing system could have arisen incrementally remains unresolved in evolutionary models, as no known animal exhibits comparable hierarchical control over vocalization. If language evolved incrementally, we should expect to find partial linguistic abilities in other species or transitional anatomical features, but none exist. Instead, human speech appears as an all-or-nothing system, requiring seamless integration of multiple complex components to function. Some might argue that parrots and other mimicking birds provide evidence of partial linguistic abilities in non-human species. However, this comparison fails to address the fundamental distinction between mimicry and language.

The *gesture-first theory* of language suggests that early humans communicated through gestures before developing speech. It was not, however, originally an evolutionary idea. It was first proposed by Étienne Bonnot de Condillac, an 18th-century French philosopher and Catholic priest, as part of his broader study of how human thought develops from sensory experience.[88] Influenced by John Locke's empiricism, Condillac believed early humans first relied on gestures before associating sounds with ideas, gradually leading to spoken language.[89]

Though Condillac's theory was not evolutionary, it was later adopted into Darwinian linguistics as a way to explain how language could have developed gradually. However, evolutionary models require incremental changes, but spoken language appears too complex and too universal to have emerged suddenly through small, unguided steps. This issue was not fully understood in Condillac's time, nor was it a major concern for evolutionists until more recent advances in linguistics and cognitive science. Scholars like Michael Corballis and David McNeill have since argued that gestures laid the foundation for language, pointing to primate behavior and early hominin hand use as stepping stones toward speech.[90] [91]

Beyond the neurological challenges already discussed, the gesture-first model overlooks a key issue: if language evolved from gestures, why has no known human society remained gesture-based? Given the immense diversity of spoken languages, we would expect at least some cultures to have developed fully sign-based systems as their primary mode of communication—but none have. Under a blind evolutionary process, different environmental and social pressures should have produced multiple adaptive solutions. Yet across all known societies, spoken language dominates, and no group has ever relied solely on gestures for communication.

Evolutionists often counter that spoken language was simply more efficient, outcompeting gesture-based communication. But this assumes rather than proves that speech was superior. If spoken language were just one random adaptation among many, why did it always win? Evolution does not favor universal solutions—it favors diverse adaptations suited to different conditions. Walking may be better than slithering in some cases, yet snakes persist in many varieties. If biological evolution produces radical differences in form and function, why is every known human society bound to the same spoken-language paradigm?

This contradiction exposes a larger inconsistency in evolutionary reasoning. Diversity is often used to explain and justify biological evolution—why different species develop different traits, why some lineages survive while others disappear. But when it comes to language, this diversity conveniently vanishes. If communication systems evolved like biological traits, at least one culture should have fully developed and sustained a sign-based linguistic system. The fact that none did suggests that language is not a random adaptation but an inherent cognitive faculty unique to humans.

Chomsky's theory of Universal Grammar and the Language Acquisition Device presents a formidable challenge to evolutionary explanations of language. His model posits an innate, preprogrammed linguistic structure that defies gradualist, adaptive models. While Chomsky himself does not frame his work in terms of intelligent design, the highly structured and universal nature of human language acquisition raises questions that materialist explanations struggle to answer. As with Chomsky's concept of the LAD, scholarly literature is filled with descriptions that reflect NatTech. The LAD functions as an embedded biological technology—an innate framework that enables language acquisition, mirroring computational architectures designed for structured data processing.

This preprogrammed linguistic framework, which Chomsky identified while overlooking the full significance of his own research, inadvertently positions him as a pioneer in the modern understanding of intelligent design, one marked by a technological and computational perspective. In fact, his preprogrammed LAD and UG led many of his contemporaries to reject his ideas well before intelligent design became a formalized theory, even though he was simply presenting the conclusions of his extensive research on language.[92] According to Chomsky's theory, language development would require saltation rather than gradual changes, much like what is observed in the fossil record. His theory also explains that this linguistic framework is universal, consistently appearing among isolated groups, further challenging the assumption of gradual evolutionary adaptation.

Chomsky's ideas reveal a parallel in software. Software mirrors the technological framework embedded within the brain, which primes individuals from birth to encode, store, and transmit linguistic information. This is similar to computer programming languages: no matter the language—whether Python, Java, or C++—they all operate within universal frameworks of logic and syntax defined by the architecture of computers.[93]

In both human and programming languages, context-free grammars establish fundamental structures. A context-free grammar is a type of formal grammar in linguistics and computer science that generates languages using rules that do not depend on the context of surrounding symbols.[94] For instance, in English, a simple sentence like "The dog runs" follows a basic Subject-Verb (S-V) structure. Expanding it—such as "The big dog runs quickly"—maintains the underlying grammatical hierarchy, where adjectives modify nouns and adverbs modify verbs.

Similarly, in programming, context-free grammars and other formal systems define the permissible structures of code.[95] These grammars play a critical role in determining the syntactic correctness of programming languages. A simple arithmetic expression, like (2 + 3) * 5, follows strict operator precedence, ensuring the addition happens before multiplication. These structured rules are akin to linguistic syntax, where sentence formation is constrained by grammatical rules regardless of the words used.

## Fire

Beyond our ability to communicate through a vocal code, our mastery of fire was perhaps the most foundational technology humans learned to harness.[96] Observing and interacting with nature laid the groundwork for more complex inventions, leading to early tool-making as humans shaped stones, sticks, and bones into functional objects. These tools weren't simply products of necessity; they were responses to environmental challenges, showing early forms of problem-solving and adaptability.

As our natural intelligence advanced, the rise of metalworking became a pivotal milestone. It enabled us to transform natural resources into more functional forms, marking a shift toward sophisticated methods of shaping and repurposing elements. Metalworking not only laid the foundation for countless innovations but was also critical for the world we live in today.

**Figure 4.1** Ancient Metalworking Depicted in the Tomb of Rekhmire: This detailed artwork from the Tomb of Rekhmire illustrates ancient metalworking techniques, showcasing craftsmen at work in Egypt. **Image source:** Pharos, Wikimedia Commons.

Interestingly, there was no guarantee that metallurgy would arise naturally. Humanity could have missed or delayed this discovery indefinitely. Yet, driven by an intuitive grasp of natural processes, metalworking became one of the earliest expressions of our ability to manipulate the material world with profound precision.

**Figure 4.2** An ancient Egyptian painting depicting the domestication of cattle, showcasing early agricultural practices. **Image source:** Public Domain.

Agriculture marked another pivotal leap in our machine-like learning journey. We didn't just adapt to our environment; we began actively shaping it through the domestication of plants and animals, showcasing a major step forward in our natural learning process. As this mastery over nature grew, a new need arose. Ancient civilizations developed writing systems to permanently capture the complexity of spoken language. These early scripts of specified information were representations of the natural linguistic technology embedded in human cognition.

**Figure 4.2** Cuneiform Inscription at Erebuni Fortress, Armenia: This ancient cuneiform inscription, dating back to the Urartian period, is located at the Erebuni Fortress in Yerevan, Armenia. **Image source:** Bot (Magnus Manske), Wikimedia Commons.

These writing systems transcended the limitations of memory and time. Writing became the *first data storage system*, preserving and transmitting knowledge across generations.[97] Similar to how human DNA provides a historical record, our earliest written records, which date back roughly 5,000 years, mark the beginning of recorded history, encoding the thoughts and knowledge of early civilizations. Everything before the advent of writing is termed prehistoric.[98] As an early form of data processing and management, writing served as a precursor to the information age.

Johannes Gutenberg's introduction of the movable type printing press around 1440 was another monumental advancement. His invention drastically expanded the availability and spread of written materials, accelerating the preservation of knowledge on a global scale through precise duplication. Centuries later, scientists uncovered a similar principle in biology—DNA, the cell's own printing system, which encodes and replicates genetic instructions with remarkable accuracy. Just as printing allowed human ideas to be preserved and transmitted, DNA ensures the continuity of life's blueprint across generations. This parallel between technological and biological information storage reflects the fundamental principle woven into both human innovation and the natural world.

Around the same period, the formalization of the scientific method in the 17th century revolutionized how humans acquire and apply knowledge. This systematic approach—observation, hypothesis, experimentation, and the formulation of laws—dramatically enhanced our capacity for learning and technological progress. It was a pivotal moment in our natural learning process, formalizing the principles that had guided human advancement for centuries.

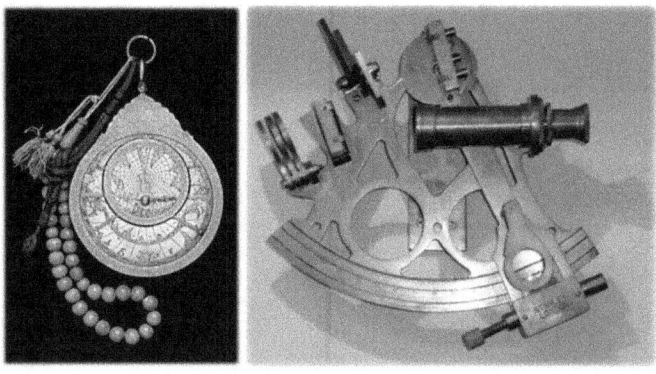

**Figure 3.3, (Left),** Brass Astrolabe, c. 14th Century, housed in the Museum of the History of Science, Oxford. **Image Source:** Martin Poulter, Wikimedia Commons.

**Figure 4.4, (Right),** Sextant, a Navigational Instrument: This brass sextant, an essential tool for celestial navigation, allows mariners to measure the angle between celestial objects and the horizon to determine their position at sea. **Image source:** Fotokannan, Wikimedia Commons.

Improved maps and precise instruments, like the astrolabe and sextant, enabled explorers to navigate the globe with accuracy, facilitating commerce and cultural exchange on an unprecedented scale.[99] These tools were expressions of humanity's capacity to impose will upon the natural world. Our innate curiosity seems to be nudged by a hidden force, compelling us to push boundaries.

The Industrial Revolution marked another significant leap forward. Humanity's understanding of natural forces expanded. We began replicating and harnessing these forces, driving the creation of machines that could perform tasks once limited to human or animal effort. Then came the Information Age, a period that revolutionized how we handle and process information.[100]

The Digital Age transformed communication and data management, but the real shift came with the development of AI and neural networks. From the computer to artificial intelligence, these technologies emulate the processing capabilities of the human brain, revealing our ability to simulate the most sophisticated natural processes.

Our intelligence has remained constant: it is our growing understanding of the environment and our ability to manipulate it that has advanced, allowing us to develop increasingly sophisticated technologies. This learning pattern stems from our *natural intelligence* (NatTel), which is more than data processing or problem-solving alone. It involves emotions, lived experiences, cultural influences, and subconscious elements that give rise to genuine intuition, making human thinking adaptive, culturally embedded, and context-aware. Artificial intelligence, however, cannot exist independently of the data it processes and remains fundamentally constrained by it.

The notion that AI could eventually mirror human-like consciousness often leads to comparisons with Darwinian processes, as if AI's development reflects how consciousness might have arisen randomly. However, this analogy is misleading. AI does not arise through random processes but requires intentional, sophisticated programming by human minds. If AI achieves human-like consciousness, it will be a product of deliberate design by intelligent beings. Likewise, if life were created in a lab, it too would be the product of an intelligent mind, resulting from purposeful scientific intervention not spontaneous evolution. Both would be reflections of the ancient technology already embedded in the universe. One of the ambitions behind efforts to create life or consciousness is an unspoken need to prove that life can emerge without design—by designing it, thereby unwittingly validating the principle it seeks to deny.

As we stand at the threshold of the Artificial Intelligence and quantum computing era, these technologies—met with both awe and caution—symbolize a modern-day Tower of Babel, a revolution poised to reach unprecedented heights and reshape our world in unimaginable ways.[101] [102] Humanity's innate capacity for learning, adaptation, and problem-solving has brought us to the brink of limitless knowledge, with the sky as our only boundary.

The progression from the intuitive discovery of fermentation to the sophisticated engineering of genomes and the monumental achievement of moon travel presents profound philosophical questions and reveals intriguing connections about the origins of innovation. How could such intricate systems—capable of producing precise, transformative outcomes while developing across disparate civilizations—arise from a reality devoid of agency or design? This pattern of development, seen repeatedly throughout history, suggests that our capacity for learning and innovation is more than a product of random chance; it reflects the inherent provision within the fabric of existence.

CHAPTER 5

# Natural Technology

*[T]he commonalities between biology and digital technology—code is code, after all— have inspired a new generation to reach across specialties and create a range of new cross-bred disciplines: bioinformatics, computational genomics, synthetic biology, systems biology. All these fields view biology as a technology that can be manipulated and industrialized. As Rob Carlson, founder of Biodesic and a pioneer in this arena, puts it, 'The technology we use to manipulate biological systems is now experiencing the same rapid improvement that has produced today's computers, cars, and airplanes.' Thomas Goetz*

In 2012, a captivating segment from *60 Minutes* showcased a groundbreaking leap in prosthetic technology.[103] The story revolved around an individual who, after losing his hand in an industrial accident, became involved in a groundbreaking prosthetics project. It resulted in a prosthetic hand that stands out for its ability to be controlled and felt through thought, marking a significant advancement in prosthetic technology. The human ingenuity revealed in this segment reflects our desire and need to mimic technology found in nature. This development was a milestone in the millennia-long evolution of prosthetics. Tracing back to the wooden and leather prosthetics of ancient Egypt and the iron hands of Roman warriors.

The *60 Minutes* segment emphasized this development as an engineering triumph and a milestone in restoring crucial aspects of the human experience. Despite the remarkable advancements in prosthetics, however, the natural human hand, with its elaborate assembly of bones, muscles, nerves, and tissues, operates with an elegance that continues to set a benchmark for our technological aspirations. Although prosthetic hands have made remarkable progress, they still have a considerable way to go to match the integrated functionality and complexity of the natural hand.

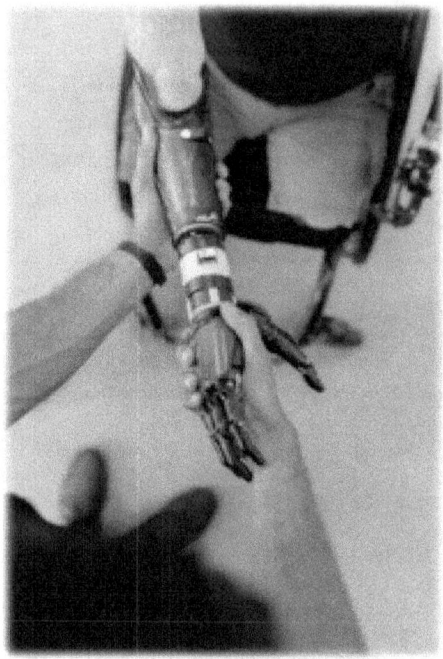

**Figure 5.1** Engineer Fitting Prosthetic Arm. **Image source**: ThisIsEngineering from Pexels

Take a moment to look at your own hand. Open it wide, then slowly curl your fingers into a fist. Now, open your hand again, spreading your fingers wide, and observe the seamless coordination involved in this simple action. This movement, guided not by intense concentration but by passive thought, is a marvel of NatTech. Consider the interplay of muscles, tendons, and nerves, all working together in a symphony of biological processes.

## Wrist and Hand
Deeper Palmar up Dissection at Right Hand

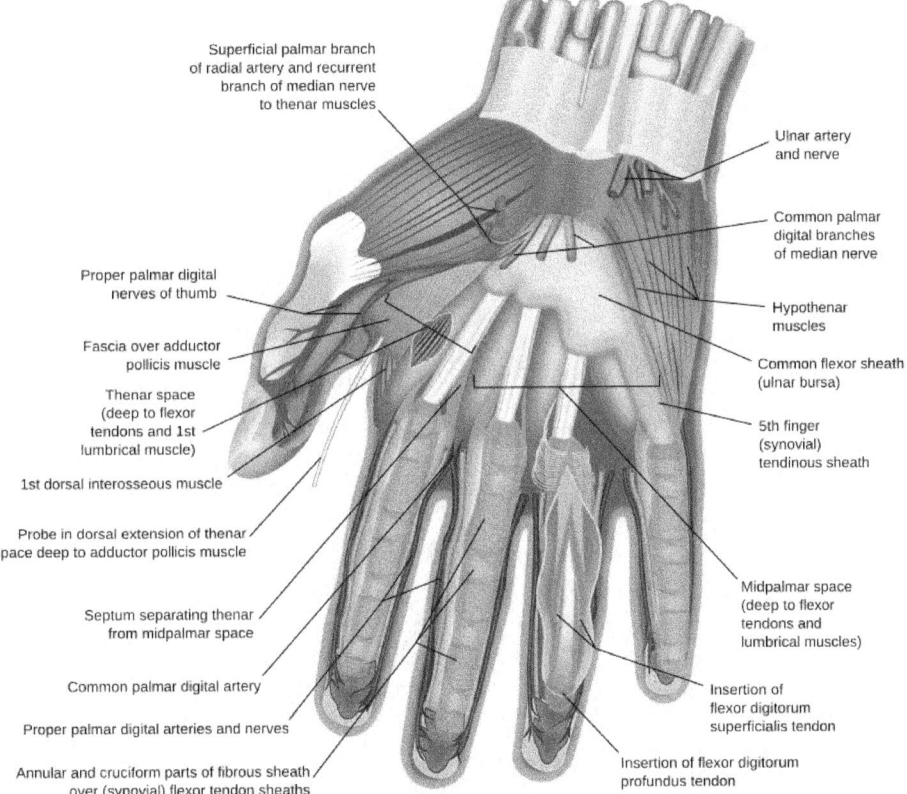

**Figure 5.2** This illustration provides a detailed view of the deeper palmar structures of the wrist and hand. It depicts the arrangement of muscles, tendons, and nerves of the hand's anatomy. **Image source:** Wilfredor.

This sophisticated mechanism operates with an ease and efficiency we often take for granted. Each movement of your hand demonstrates the subtle yet sophisticated capabilities of the biological systems that are an integral part of us. Extend this sense of wonder beyond your hand. All around us are marvels of NatTech.

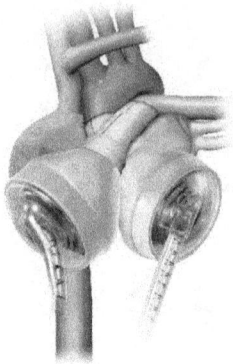

**Figure 5.3** The SynCardia Total Artificial Heart (TAH). **Image source:** Luciasyncardia, Wikimedia Commons.

Consider the human heart, a tireless pump capable of lasting over a century, continuously circulating life-sustaining fluids throughout our bodies. The device in Figure 5.3 is an artificial heart—a remarkable feat of human engineering. Yet, as a temporary stand-in for those awaiting a transplant, it remains a significant downgrade from the natural heart we are born with.

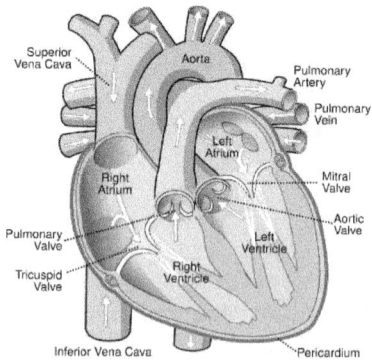

**Figure 5.4** The anatomy of the human heart: this image depicts the internal structures including chambers and valves. **Image source:** Wapcaplet.

Figure 5.4 illustrates the sectional anatomy of the human heart. The standard package comes complete with chambers and valves that work tirelessly to keep us alive. This elaborate system showcases the remarkable efficiency and precision of the heart's natural design.

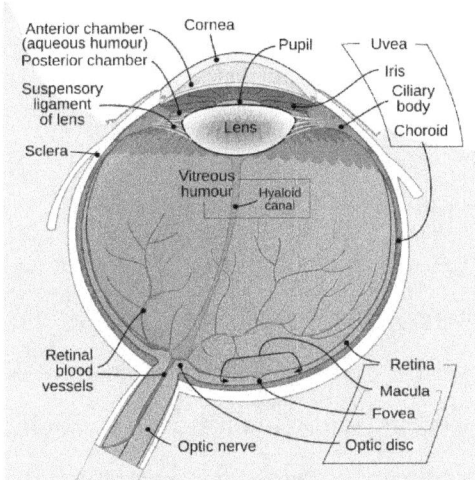

**Figure 5.5** Sectional anatomy of the human eye, illustrating the internal structures such as the cornea, lens, retina, and optic nerve. **Image source:** Rhcastilhos and Jmarchn.

Now, think about our eyes, which process images with a precision and efficiency that surpass even our most advanced optical systems.

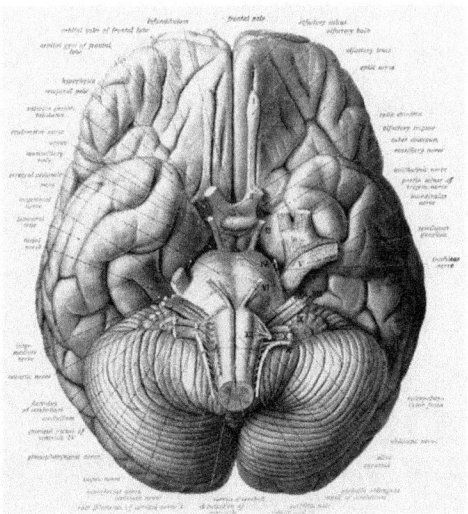

**Figure 5.6** Illustration of the human brain from an inferior view, detailing major nerves and structures involved in sensory and motor functions. **Image source**: Wikimedia Commons, Public Domain.

Then there's the human brain, a marvel of organic technology; its capabilities in parallel processing, learning, and adaptability outshine our artificial computers.[104] Despite the brain's extraordinary computing power and its ability to repair itself, many scientists consider it taboo to call it a 'biological computer,' even though the brain possibly makes a hundred quadrillion computations every second—and yes, that's a quadrillion with a Q.[105] For these scientists, resistance to the "biological computer" analogy is philosophical. Some researchers argue that calling the brain a computer implies an architect or programmer, which raises uncomfortable questions about origins. They recognize the brain's superior architecture by trying to replicate it, yet they dismiss the notion that it could be the product of intentional engineering, even while their best efforts fail to match its capabilities.

The principle that "familiarity breeds contempt" may also explain why the marvels of technology found in the natural world often go unnoticed.[106] Our constant exposure to these wonders has led to a desensitization, causing us to overlook the sophisticated systems. Though we might casually pass by a flower or observe a bird in flight with fleeting attention, we frequently disregard the exquisite craftsmanship of engineering that grants the ability to soar and the capability to convert sunlight. These instances offer just a glimpse into the expansive universe of natural technologies all around us, each representing a masterpiece of design and functionality.

## Natural Information

Another key component of technology is information. Information is generally seen as human-made data or facts.[107] Expanding on this theme, however, it's important to acknowledge that natural information (Ninfo) forms the basis by which the universe functions and provides meaning to everything we identify. Anything that is identifiable is, by definition, information because identification requires detection and processing, which only happens if there is something to process. If something had no information at all, it would be indistinguishable from nothingness. Existence itself is information. In the beginning, matter was structured by words. This positions our reality as fundamentally language-based, where everything we perceive and interact with follows structured, recognizable patterns—much like a coded system that can be read, processed, and interpreted.

Even so-called randomness or noise carries informational content because it can be measured, distinguished, and processed by our brains. Since our senses interact with the world by detecting and interpreting patterns, everything we perceive or recognize must contain information in some form—whether structured (like language or DNA) or unstructured (like white noise).

Think about when you look outside and perceive it to be sunny. This observation represents a set of data systematically processed by our sensory systems.[108] It's not just a fact; it's a piece of Ninfo informing your understanding of the day ahead. We process this natural information, often subconsciously, and use it to plan our activities, like deciding to wear lighter clothes or carry sunglasses. This adaptive response to environmental information showcases the sophistication of our Ninfo processing capabilities. Similarly, consider the feeling of a chilly evening creeping in, prompting you to adjust the heat in your home.[109] This too is information, naturally occurring and detected and processed by our bodies.

These one-way communications show that our environment is not indifferent, but instead indicate a purpose relevant to our lives.[110] [111] They provide information essential for conscious beings to understand and interact with their environment. Take, for instance, a dog that senses its owner's return even before any visible sign. This is information processed by the dog, a response to subtle cues in its environment. This sensory ability isn't just a trait of higher animals; even plants detect and respond to information. They react to changes in light levels, water, and nutrients, engaging in complex interactions with their environment.

Even basic chemical reactions occur due to the inherent properties of substances, revealing the embedded information in their behavior.[112] Similarly, meteorological phenomena like wind and rain follow predictable atmospheric patterns, which reflect the universe's reliance on an informational blueprint.[113] Thermal reactions, such as freezing and burning, are governed by the laws of thermodynamics, demonstrating the structured and systematic nature of the universe.

# CHAPTER 6

# Scientific Miracles

*The universe could so easily have remained lifeless and simple — just physics and chemistry, just the scattered dust of the cosmic explosion that gave birth to time and space. The fact that it did not — the fact that life evolved out of nearly nothing, some 10 billion years after the universe evolved out of literally nothing — is a fact so staggering that I would be mad to attempt words to do it justice. And even that is not the end of the matter. Not only did evolution happen: it eventually led to beings capable of comprehending the process, and even of comprehending the process by which they comprehend it.* [114]—Richard Dawkins

In the early 2000s, a television program embarked on a quest to find natural explanations for biblical miracles, examining phenomena like Moses parting the Red Sea and Jesus walking on water.[115] To explain the parting of the Red Sea, the host proposed finely tuned environmental and physical factors, such as extraordinary water levels and uncommon natural events.[116] Although the presenter attempted to be scientifically grounded, he inadvertently underscored the implausibility of this miracle within natural laws, thereby reinforcing its extraordinary, divine nature.[117]

## Noah's Ark

Likewise, in 2010, Ken Ham, a well-known Biblical creationist, embarked on a distinctive yet related endeavor.[118] His initiative to construct a full-scale Ark aimed to demonstrate the physical possibility of the Noah's Ark narrative.

From Ham's perspective, this effort is an understandable extension of his belief that Genesis, as a literal account, is foundational to the Bible's coherence, notably since Jesus Himself recognizes these narratives as literal in the Gospels. However, while Ham's intention is to lend tangible credibility to the scriptural account, this endeavor might somewhat overlook the inherently miraculous elements of the story.[119] In my personal opinion, Ham's focus on the feasibility of a physical construct paradoxically reduces the essence of faith, reframing what is traditionally viewed as a divine miracle into an engineering feat of human capability.[120]

These efforts reveal the broader struggle to reconcile faith with a materialistic perspective, whether through a scientific approach, like the TV program I once saw, or Ken Ham's attempt to model a miracle; although they aim to demystify and rationalize miracles by offering natural explanations, they may inadvertently detract from the extraordinary nature of these events, which seem to rely on the power of the transcendent Biblical God. I'm not sure if attempting to rationalize supernatural events through natural explanations is the best way to convey the intended perspective of these accounts.

Imagine explaining how the widow in 1 Kings never ran out of oil or how Jesus turned water into wine in the Book of John, offering natural explanations such as the pressing of olives or grapes. Such explanations reduce acts of divine will to mundane, explicable events, as if attempting to validate them through natural means, thereby diminishing the need for faith. One may rebut that the Bible states Noah built the Ark by hand, but by the same logic, the Bible also claims that Samson killed a thousand men with the jawbone of a donkey—yet few would attempt to explain how that could have happened through ordinary circumstances.

## An Ark in a Quark—And Much More

However, the appeal to rationalize miraculous events is not limited to biblical narratives but extends into the scientific realm, particularly in efforts to explain the universe's origins and how life came to be.[121] These scientific ventures often encounter a series of extraordinarily improbable events.[122] Consider the Big Bang theory, the prevailing scientific explanation for the universe's inception. This narrative goes even further than the idea of fitting two of every species into an Ark, suggesting that the entire universe, with all its species, vessels, and structures, was once contained in a space smaller than a quark.[123]

This theory hinges on specific conditions and occurrences that follow a fantastically fine-tuned trajectory. Additionally, elements like cosmic inflation and the multiverse theory are necessary to make the theory work.[124] [125] These improbable conditions, invoked to explain the extraordinary phenomena observed throughout nature, are often overshadowed by meretricious equations and models.[126] The suggestion that something as complex as modern cities like New York City could evolve from a state of nothingness to such order only intensifies the intrigue surrounding the miraculous nature of these claims.

No matter your worldview, whether theist, agnostic, pantheist, materialist, or otherwise, introducing any cosmogony or explanation of how the universe came to be involves a creation story and, therefore, a form of creationism. Whether or not an agent is named (e.g., 'nature' or 'physical laws'), describing a process of origin does not escape this fact. To claim otherwise is to misrepresent what these explanations actually entail.

Whether we acknowledge it or not, when we posit an explanation about the origins of life or the universe, we inevitably enter the realm of faith that underpins all creation stories, as they seek to explain what cannot be empirically observed, tested, or known. As the philosopher Karl Popper argued, science is fundamentally about falsifiability and empirical testing, not about providing absolute certainties or definitive truths about origins.[127] Incorporating a little more humility in the face of these mysteries is perfectly reasonable. Such moments remind us of our small place in the vastness of existence, urging us to stay grounded and accept the reality we perceive.

# CHAPTER 7

# Cosmic Novel

*[W]e have measured the curvature of the universe and found it to be zero. ... In one sense it is both remarkable and exciting to find ourselves in a universe dominated by nothing. The structures we can see, like stars and galaxies, were all created by quantum fluctuations from nothing. ... But no one ever said that the universe is guided by what we, in our petty myopic corners of space and time, might have originally thought was sensible. It certainly seems sensible to imagine that a priori, matter cannot spontaneously arise from empty space, so that something, in this sense, cannot arise from nothing. But when we allow for the dynamics of gravity and quantum mechanics, we find that this commonsense notion is no longer true. This is the beauty of science, and it should not be threatening. Science simply forces us to revise what is sensible to accommodate the universe, rather than vice versa. ... Does this prove that our universe arose from nothing? Of course not. But it does take us one rather large step closer to the plausibility of such a scenario.*[128] —Lawrence Krauss

In the beginning, there was nothing. The beginning was nothing. Despite some speculations about nothing, when we say 'nothing,' we mean an absence so absolute that it eludes our very capacity to conceptualize it.[129] This notion of 'nothing' should not be confused with the concept of a vacuum within space-time; it refers to the state before the existence of space and time themselves. In this context of the universe's inception, there is no 'here,' or 'there,' or any 'where' that exists, nor a 'when,' when anything could pop in or out.[130]

And yet, from this sheer absence, something—somehow—*emerged*. What many scientists describe as an infinitesimal point, said to be brimming with infinite potential, is theorized to have given rise to all that we see today.

But what does it mean for time to begin?

For causality itself to be born?

The mind struggles to grasp these questions because everything we know—every cause, every effect—exists within time. How can we comprehend a reality where even the potential for existence had yet to exist? It is here, in this paradox of naught, that the very essence of 'to be *or* not to be' lay dormant—until a moment manifested, and nothing became something.[131]

An eruption of cosmic creativity transformed a canvas-less canvas into an array of galaxies and planets, ultimately leading to the phenomenon of life and consciousness. This consciousness, in turn, gave rise to public education, judicial systems, Broadway shows, and daytime "reality" television with its iconic chants of "Jer-ry, Jer-ry, Jer-ry," and even artificial consciousness. This transition from *nothingness* to *being* encapsulates one of existence's greatest mysteries.[132]

It was not a gradual beginning but an explosive manifestation of cosmic forces and natural laws. Upon its birth, the universe took no tentative steps nor had a moment of infancy to find its bearings. Instead, it burst forth, dashing from the starting line with precision and unmatched vigor. Every particle, every force came into play with instant prowess and coordination. The cosmos embarked on a teleportation into development, grounded in cosmic perfection from the get-go.[133] Under the standard model, the universe was developed so rapidly that it would be like a mother just giving birth and then turning over, asking to see her baby, only for the doctor to say, "I'm sorry, but your baby isn't a baby anymore. He's an adult at home with his own kids."

Consider this: an atom is typically around $10^{-8}$ centimeters in diameter.[134] Within an atom, protons and neutrons make up the nucleus, and these are composed of even smaller particles called quarks. According to the standard cosmological model, the radius of a quark is constrained to be less than $0.43 \times 10^{-16}$ centimeters.[135] This means a quark is about 100,000 times smaller than an atom.

Consider this: You, everyone you know or have ever seen, and everything you've ever encountered—whether in person, on TV, in books, or magazines—even these words you're reading right now—all evolved from an unimaginable space of next to nothingness. According to the standard model, all the sophisticated matter in the universe, everything we consider something—including every institution of higher education—expanded from this location of unthinkable density and evolved into everything, like a pop-up gift card playing music. Compared to that, arranging any number of living animals on a boat is quite doable.

# CHAPTER 8

# Cosmic Child

*Big Bang cosmology described a universe with a beginning and a history, so it turned cosmology into a historical science, an account of change and evolution. According to this view, the universe began as an infinitesimally small entity, which expanded rapidly and continues to expand today. In form, at least, this account is similar to the traditional creation myths known as emergence myths. In such accounts, the universe develops, like an egg or an embryo, through distinct stages from a remote and perhaps undefinable point of origin, and under the control of internal laws of development.*[136] —David Christian

In entertaining the narrative of the standard cosmological model, a striking parallel unfolds when likened to the process of embryonic development. The Big Bang, viewed as the universe's point of origin, mirrors a seed or embryo, brimming with latent potential. In this primordial state, all astronomical and physical phenomena remained unrealized, much like the potential of a seminal cell before it reaches the egg. Following the Big Bang, the universe entered its own post-fertilization phase, transitioning from latent potential to active formation.[137]

Similar to how a zygote undergoes rapid division and differentiation to form an embryo, where foundational structures and, eventually, organs begin to develop, the universe's foundational components rapidly materialized and organized during exponential growth and formation never seen again.[138] [139] This stage in the cosmos could be seen as the rollout of a "technological framework" inherent in the singularity.

This subsequent *expansion*—a term I'm using here quite loosely and with reservation—of the universe, which mirrors the dramatic growth of an embryo in its early stages, often referred to as inflation in cosmology, represented a physical expansion and a rapid actualization of pre-encoded natural technological systems.[140] The sheer magnitude and rapidity of this growth, when viewed this way, make the creation of our universe more aptly described as a *Big Conception* rather than a 'Bang.'[141]

Consider the grandeur of the cosmos as the nurturing environment inside a mother, filled with intentionality. In the mother's womb, a baby is cradled in a self-sustaining environment, seemingly autonomous yet resulting from careful planning and nurturing. Through her diet and habits, the mother consciously cultivates an environment conducive to the baby's development. This is not a random occurrence but a purpose-driven act of prenatal care.

The voices and sounds experienced by the baby in the womb are like cosmic whispers, omnipresent and mysterious. The steady rhythm of the mother's heartbeat, the muffled rush of blood through her veins, and the vibrations of her voice as she hums or speaks form a soundscape that envelops the child. These sensations hint at a realer reality yet to be discovered, similar to the unseen forces such as 'dark matter' and 'dark energy' that shape the universe.[142] (We'll delve deeper into *dark matter* and *dark energy* in Chapter 12.) Though not directly observed, their presence is inferred, shaping the universe much like the parents' unseen efforts—nutrition and environmental conditions—quietly shape the baby's development. As the baby grows and tests the limits of the womb, it mirrors the universe's expansion driven by dark energy. The child and the cosmos are on trajectories of growth that test the boundaries of their respective confines.

However, the baby's perception of the womb's expansion is limited, similar to how we perceive the universe's growth as an intrinsic phenomenon.

The revelation awaiting the baby upon birth, where it encounters a world beyond the womb, parallels our quest for a greater cosmic understanding. Once hidden, the intentionality behind the baby's nurturing environment becomes apparent, similar to how we may one day encounter the mysteries behind the universe's finely tuned conditions. This narrative is not to suggest the existence of cosmic parents but to inspire us to push the boundaries of our understanding of what constitutes 'natural' within our universe, and to contemplate what might lie beyond.

Unlike the baby in the womb, oblivious to the deliberate intentions shaping its environment, we, as observers of the universe, have the cognitive ability to detect and analyze purposeful patterns.[143] In this context, investigating concepts like dark matter and dark energy becomes a philosophical journey to understand the underlying mechanisms at play in the cosmos.[144] Similarly, the quest to explain the origin of DNA and the bacterial flagellum delves into foundational questions of life's design.

The experience of a newborn transitioning into a world far greater than its earlier confines reflects how transcending materialistic views of the universe may reveal glimpses of an unseen reality. In all its mystery and majesty, the universe still holds secrets waiting to be unveiled, like the unknown wonders beyond the womb.

## CHAPTER 9

# Cosmic Dawning

*The champions of Intelligent Design make two mistakes when they claim that the SETI enterprise is logically similar to their own: First, they assume that we are looking for messages... In fact, we're on the lookout for very simple signals. That's mostly a technical misunderstanding. But their second assumption... that complexity would imply intelligence, is also wrong. We seek artificiality, which is an organized and optimized signal coming from an astronomical environment from which neither it nor anything like it is either expected or observed: Very modest complexity, found out of context. This is clearly nothing like looking at DNA's chemical makeup and deducing the work of a supernatural biochemist.[145] — Seth Shostak*

Let's examine the speed of light, denoted as 'c.'[146] This constant isn't just a figure in physics; it functions as the universe's cosmic governor, setting the pace for its operations.[147] It anchors the famous equation $E = mc^2$, showing the foundational relationship between mass and energy.[148] The speed of light is a fundamental law of the universe—a cosmic speed limit that defines the structure of spacetime and governs the behavior of energy and matter.

The staggering speed of light in a vacuum is 299,792,458 meters per second—a value, to some, might seem incredibly random. If the universe were the result of random chance, it would be logical for this universal parameter to seem arbitrary. Imagine a speed limit in your neighborhood set at exactly 29.9792458 miles per hour, and speeding tickets were issued the moment you approach the limit—at 29.9792457. Let's say you live in New York, and your friend lives in London. After talking to your friend, you discover that, whether measured in miles per hour in New York or kilometers per hour in London, by coming within a millionth of a unit triggers an immediate pull-over by the police and penalty.

Now consider that the speed limit of the universe is known to incredible precision—down to parts per billion. Naturally, you'd question: why is the speed limit set that way?

Though some view this fine-tuning as arbitrary, others explain that the speed of light only appears that way because of the measurement system we use, such as meters per second. As physicist Christopher S. Baird argues, for instance, the seemingly odd numerical value of the speed of light arises from our human-defined units of measurement.[149] According to Baird, switching to natural units, where $c = 1$, the arbitrariness disappears entirely.[150] This explanation, though compelling, only addresses the numerical appearance of arbitrariness, not the deeper question of why the speed of light exists as a precise and unwavering constant in the first place.

I agree that the speed of light appears arbitrary, but only in a materialistic framework. The issue is that this view conflates the measurement system of space and time with the fundamental constants that govern their interaction—two distinct concepts. From a NatTech perspective, there is a clear distinction between structural dimensions and operational constants.

To illustrate, consider a TV. It has physical dimensions—like width and height in inches or centimeters—that are intuitive to the consumer. But it also has technical specifications, such as refresh rate (in GHz) or resolution (e.g., 4320p), which may seem arbitrary to some consumers—just numbers without obvious meaning.[151] [152] Yet these operational values are essential to the TV's function, not its size. To suggest that these numbers are interchangeable just because they can be converted between units misses a key point: they serve fundamentally different roles within the producer-consumer system. Similarly, reducing $c$ to 1 in natural units might simplify its numerical representation, but it fails to capture the deeper significance of $c$ as the boundary *specification* that governs spacetime itself.

Operational constants like $c$, $G$ *(Gravitational constant)*, and $\hbar$ *(Reduced Planck's constant)* define distinct operational settings of nature, and conflating them with the physical properties of the universe for the sake of convenience obscures their unique yet technological roles. These constants underpin reality, embodying indispensable yet unforeseeable characteristics that cannot be explained away by numerical conversions.

Baird also argues that speeds beyond the speed of light are not physically real, claiming that the phrase *faster* than light "makes no sense" and likening it to ideas such as "*darker* than absolute darkness."[153] While intriguing, the concept of "darker than absolute darkness" is not entirely comparable, as the speed of light involves the intrinsic structure of spacetime rather than the absence of a quantity. In other words, "absolute darkness" refers to the absence of light, whereas the speed of light is not about the absence of something but rather a specified limit inherent in the structure of spacetime.

Baird's analogy is akin to saying that it makes no sense for a computer's clock speed to function as a limit because it is not physically real. A computer's clock speed, however, is a specification—it defines how many cycles per second the processor can execute (measured in Hz or GHz). It's a fundamental parameter of the system's operation, much like how $c$ functions in physical theories, defining spacetime's operational framework.

As astrophysicist Martin J. Rees explains, "The speed of light thus has fundamental significance. It fixes the 'conversion factor:' it tells us how much each kilogram of matter is 'worth' in terms of energy."[154] This underscores that c is more than a unit-dependent value: it is a core specification of the universe's structure. Even if we set its value to 1, we'd still face very large or small numbers in other practical contexts, making conversions necessary and often cumbersome.

Often, when scientists contemplate physical constants as mysteriously arbitrary, they are tacitly grappling with the profound philosophical implications of why these constants exist with such precise values. Thus, these cosmic settings become problems to explain away. Some questions raised become gnawingly troubling for those grounded in the belief that the universe was created without purpose or foresight. Their discomfort stems from questions like: Why are these constants so finely tuned to allow the universe to function as it does? Why do these values seem 'chosen' in a way that supports life and order?

Roger Penrose's discussion on the speed of light in his book *The Road to Reality* notes that c is not arbitrary or a product of human-defined conventions like rapidity.[155] Instead, it is a physical constant intrinsically woven into the fundamental laws of electromagnetism and the fabric of spacetime. By linking c to the forces that bind matter and govern the universe's structure, Penrose implicitly points to a deeper significance behind its value. As Penrose notes, quoting Maxwell, "The theory of electromagnetism is also the theory of light."[156]

For everyday applications, educational purposes, and most engineering uses, such as civil and mechanical engineering, we often simplify the speed of light by rounding it from 299,792,458 meters per second—to 299,792 kilometers per second. This simplification works because the slight difference does not significantly impact outcomes or understanding in these contexts.

But in high-precision scientific areas, such as GPS satellite technology, particle physics, or cosmological models, every fraction of a meter per second counts. The precision of the speed of light opens an intriguing question: Why isn't it slightly slower or faster, like 299,792,448 or 299,792,468 meters per second? Even a slight change could disrupt the delicate balance of forces within atoms, alter particle behavior, affect the universe's expansion, and throw off instruments and technologies that rely on this exact value.[157]

## Network Initiation

According to Einstein's theory of relativity, the speed of light sets the maximum rate for information, energy, and matter to travel across the universe.[158] The speed of light and other constants aren't *just numbers*; they are the universe's way of ensuring everything runs smoothly, even our communication. These constants are fundamental, from how we interact with satellites to how we observe distant galaxies. Imagine tweaking that speed just a bit; it would be like changing the clock speed in a computer. Suddenly, everything's out of sync, and the whole system crashes.

Here, we see another clear link between the NatTech of light and our technology. For example, consider bandwidth in our everyday communication networks.[159] Whether it's streaming a video or observing distant galaxies, human technology and the cosmos rely on the principle of bandwidth to regulate communication and transmission.

But there's more to this cosmic network than just speed. Much like our internet—designed with multiple data pathways to ensure resilience—the universe's pathways of light ensure that cosmic information is reliably conveyed, even amidst obstacles.[160] The redundancy in the universe's design speaks to an underlying sophistication, a system not left to chance but carefully structured for stability and resilience. *In this light*, the universe becomes a finely tuned cosmic system, where every parameter is calibrated for harmony and purpose.

Additionally, we can envision the appearance of light not as a gradual process, but as a moment of activation—like flipping a switch in a complex device or the sudden expression of a gene in biological development.[161] Just as a dormant device springs to life when powered on, the early universe may have existed in latent potential until the right conditions aligned for the triggering of light to manifest and propagate.

From this perspective, the universe's development is not driven by slow evolution but by critical activations. Applying this to the horizon problem (discussed further in the next chapter), we can hypothesize that the uniform energy observed across disconnected regions resulted from the instantaneous activation of light, offering a more compelling explanation for the uniform energy distribution than the traditional model.

# CHAPTER 10

# Semantic Games

*Look, it's important to remind ourselves over and over again if necessary that our intuitions, our predilections for how we assess reality, they have been shaped by hundreds of thousands of years of evolutionary history in which the focus was on successfully navigating the everyday world, and that formative goal is oblique to the far more recent goal of understanding the true nature of reality. So perhaps we should expect that, when confronted with the true nature of reality, our intuition will not be prepared to easily accept it. Now look, this doesn't by any means establish that the Many Worlds approach is right, but it does make clear that our inclination to resist such a strange idea is by no means evidence that it is wrong.[143]* —Brian Greene

## Problems, Problem, and more Problem

The "horizon problem" arises from the necessity, within materialist cosmology, to explain how all observable regions of the universe—from the nearest to the farthest—exhibit nearly identical temperatures and physical properties.[162] According to the Big Bang model, regions separated by billions of light-years—beyond each other's observable horizon—should not have had enough time to interact due to the finite speed of light.

This creates a significant issue for the standard model, which assumes a universe that evolved gradually over billions of years. The striking uniformity of these distant regions, rather than supporting an unguided evolutionary process, suggests properties more consistent with a *special creation*. Thus, the horizon "problem" is a *euphemism*, since in other theoretical contexts, such a discrepancy would be labeled more strongly as a contradiction.

Like the horizon problem, the "flatness problem" exposes another inconsistency in the Big Bang model. Whereas the horizon problem challenges how distant regions of the universe share uniform properties, the flatness problem concerns the universe's observed geometry, which appears to be spatially flat. This requires the universe's density in its early stages to have been incredibly close to what cosmologists call the critical density—an exact value needed to balance expansion and gravity. Such precision seems highly improbable, raising further questions.[163] The critical density is the precise amount of matter and energy needed to balance the universe's expansion. Any slight deviation from this critical density would have resulted in a vastly different curvature than what we observe today.[164] Such fine-tuning is highly unlikely to have occurred by chance. Thus, the contradictions introduced by the horizon and the flatness problem in the Big Bang model prompted materialists to develop a theoretical solution to *fix the evidence*, which they saw as *a problem*.

Imagine someone coming up with a theory that claims to solve the "round-Earth problem," providing equations and physics that exist outside of the reality we know. What if they also introduce explanations for the "solid-Earth problem," proposing that our planet is hollow with vast civilizations within, or they tackle the "linear-time problem," suggesting that time flows in cycles rather than a straight line. To address the "physical-Moon problem," they argue the Moon isn't a solid object but a projection or hologram hiding secret truths about the cosmos.

Consider how we approach well-established facts: How much sense would it make to develop models to "fix" so-called problems with the idea that the Earth is round, solid, that historical eras are fully accounted for, or that the Moon is a physical object? Wouldn't it make more sense to accept the extensive evidence supporting these realities?

This is exactly the question we should ask when materialists refuse to accept the evidence that the universe came into existence all at once, matured, and finely tuned. So, when you hear scientists refer to something as a "problem," it's typically a signal that they are struggling to shoehorn certain evidence into their existing framework. From just a few questions, you might quickly find "supernormal" phenomena that exist only in equations, with no basis in the physical reality we know.

## We Have Evidence

First proposed by physicist Alan Guth in his 1980 paper, "The Inflationary Universe: A Possible Solution to the Horizon and Flatness Problems," cosmic inflation suggests that a *brief* but intense burst of exponential growth was necessary to explain the observed uniformity and flatness of the universe.[165] Guth's "solution" was to propose that the early universe expanded faster than the speed of light, driven by a hypothetical scalar field he called the "inflaton."[166] [167] This was his scientific answer to the horizon and flatness *problems*.

A 2002 article quotes Guth as stating:

> It's not a coincidence that the Bible starts with Genesis...I like to strongly push the scientific answer. We have evidence. We no longer have to rely on stories we were told when we were young.[168]

His comments reflect irony and hypocrisy within materialistic pursuits. Even as they advocate for a departure from the traditional narratives of Genesis and argue for a reliance on empirical evidence, Guth and proponents of inflation theory themselves have crafted a new kind of 'story.'

This story, designed to uphold materialistic cosmogony, aims to replace theistic explanations with scientifically framed narratives, ironically using fantastically speculative elements like the inflaton field, which lacks empirical support aside from being a necessity.[169] Any supposed "evidence" for inflation—such as the uniformity of the CMB—is entirely circular. Inflation was not proposed because of independent empirical observation but because the CMB's uniformity contradicted Big Bang expectations. Calling this "solution" evidence for itself is the definition of retrofitted reasoning. Guth's commitment to a "scientific answer" seems to allow for anything—so long as it doesn't lend support to the Genesis account.[170]

The available evidence for drawing conclusions about the genesis of everything we observe is always a combatant. Given the numerous conflicts between materialist theory and this evidence, there appears to be a war between radical materialism and science itself. The general narrative of how inflationary theory gained traction centers on its ability to "fix multiple problems" in the Big Bang model that had no other plausible solutions.[171] Guth's theoretical constructs ostensibly saved materialistic explanations from their conflict with physics and natural laws, relieving the discomfort of lacking a substantial explanation for why the universe appears to have come into existence all at once.

Even Paul Steinhardt, one of the original architects of inflationary theory, became one of its foremost critics over time due to several concerns.[172] He points out the theory's lack of predictive power and its excessive flexibility, making it nearly impossible to falsify, thus weakening its scientific validity.[173] Steinhardt reinforced this critique with Guth's own statement:

> In an eternally inflating universe, anything that can happen will happen... an infinite number of times.[174]

By his own admission, Guth acknowledged that "inflationary universes need not be natural," effectively conceding that the theory is, by definition, supernatural (from the Latin *super*, meaning 'above' or 'beyond').[175] [176] This flexibility eventually gave rise to the inflationary multiverse concept—a version proposing countless universes generated by eternal inflation—which Steinhardt criticized for introducing excessive randomness and undermining any coherent explanation for the universe's properties.[177]

Furthermore, Steinhardt pointed out that despite numerous refinements, inflation theory struggled with fundamental issues, such as the initial conditions required for inflation to start and the "measure problem," which involves defining probabilities within an infinite multiverse.[178] Frustrated by these unresolved challenges, Steinhardt has pushed for a shift towards more fundamental theories that could offer clearer, more testable predictions, ultimately leading him to investigate alternative explanations for the early universe's conditions.[179]

To better understand a key issue in the standard cosmological model, imagine waking up in front of what you are told was a bonfire at the center of a football field. As you regain awareness, you notice what appears to be embers radiating a pulsing heat, making you sweat. It's surprisingly warm form such a cool day. You question people around you, and they confidently explain that there isn't a heating system.

They assure you they've checked with the officials, who have calculations accounting for all the heat in the stadium as a result of this once-fierce bonfire.

However, as you walk toward the goal line, you continue to feel the same intense warmth, unaffected by your increasing distance. Even as you ascend to the bleachers, the pulsing heat remains constant, despite the cool weather.

Curiously, you climb all the way to the nosebleed section—the highest bleachers—where you expect the temperature to drop, given the bonfire's distance and the time that has passed. Yet, it's just as warm there. You ask the people sitting in the highest seats how is it possible to still feel the same pulsing heat from the extinguished bonfire, even so far away. They confidently assure you that the warmth undeniably comes from the bonfire, which they've been told was lit and burned out only 13 minutes ago.

This situation defies the laws of thermodynamics, which state that heat should dissipate as it spreads over time and distance. There hasn't been enough time for the heat from the bonfire to reach every corner of the stadium and maintain such a uniform temperature. As they offer theories of unusual phenomena to support their belief, you begin to suspect that the warmth may not have come from the bonfire at all, but rather from a more complex, unseen heating system warming every part of the stadium simultaneously and equally.

Despite this, everyone around you insists that their theory proves the uniform warmth originates from the bonfire and that it's the only possible explanation. They confidently present this as fact, pointing to the apparent evidence while dismissing any alternative possibilities. But you begin to wonder if you're still half asleep—or if there was even a bonfire in the first place.

## Once Upon a Brief Time

Let's quietly shift our focus from the uniformity *problem* and examine a more discreet yet far more intriguing conundrum: the actual time frame commonly referred to as the 'inflation epoch.' This shift is critical, as it reveals the stark difference between the conventional terms used and the reality of these events. Words like "inflation," "brief time," "this period," and "epoch" are frequently employed to describe the early moments of our universe. However, these terms misrepresent the true nature of what they attempt to describe, acting more as euphemisms. The term "inflation epoch" is particularly misleading. According to Guth, the universe underwent what he describes as "rapid" expansion, but based on his own calculations, the term 'instantaneous' better captures the speed and scale of this event. To put it plainly, the inflationary period is said to have lasted approximately from $10^{-36}$ seconds to $10^{-32}$ seconds after the Big Bang.

> In a moment so fleetingly, immeasurably small, scientists theorize that the Big Bang was followed by an 'Inflationary Period.' In a billionth of a trillionth of a trillionth of a second, the Universe grew by a factor of 10^26, comparable to a single bacterium expanding to the size of the Milky Way. —*The Center for Astrophysics, Harvard & Smithsonian*[180]

At such scales, even the notion of 'after' becomes peculiar. But what is the real issue here? Behind the euphemisms, proponents essentially suggest that the universe appeared fully formed and matured in a single, sudden event. To explain away the evidence that pointed to the instantaneous appearance of our universe, the *super*natural event, inflationary theory was proposed. Consider what that means: if the universe expanded to a size comparable to the Milky Way in that instant, and the Milky Way spans roughly 100,000 light-years, then inflation theory effectively places the universe at tens of thousands of light-years across—instantly.

Written out, the 'duration' of the inflation epoch spans from

0.000000000000000000000000000000000001 seconds
to 0.00000000000000000000000000000001 seconds.

In everyday terms, the time it took for the universe to come into existence—spanning 10,000 light-years—was so brief that it defies human ability to conceptualize or assign meaningful context. In fact, words like 'short,' 'instantaneous,' or even the term 'timeframe' fail to capture the scale and nature of this event. And this is typical of materialist semantic games: they downplay the sudden, inexplicable nature of the event by labeling it "inflation." To counter the idea of it originating from a specific point, many argue it was not an explosion in one place but an expansion that occurred everywhere at once, a circular attempt to deny a fully-formed creation.

In standard English, however, the word "inflation" brings to mind gradual expansion, like a balloon filling with air, which subtly masks the concept's radical nature. Describing the time scale proposed by inflationary theory ($10^{-36}$ seconds to $10^{-32}$ seconds) using any terms of standard time scales downplays the sheer immediacy and magnitude of what occurred according to this model.

Let's evaluate how many times the inflationary timeframe fits into a blink of an eye:

$$\frac{10^{-36} \text{ seconds}}{10^{-32} \text{ seconds}} = .02 \times 10^{34} = 2 \times 10^{33}$$

The inflationary "period" fits into the blink of an eye 2,000,000,000,000,000,000,000,000,000,000,000 times.

So, when comparing the inflationary period to something as simple as blinking, this equation shows just how far this scale exceeds our usual understanding of time. To be frank, how logical is it to claim a scientific explanation while positing that the universe *inflated* on such an extreme scale that even the word 'instantaneous' feels too slow from a human perspective?

In a way, inflationary theory reads like a tale of a cosmic superhero—a member of the materialist Justice League. Imagine the hero, Inflation, dramatically swooping in to save the universe in the blink of an eye:

> In a flash. No, much faster than that. In the tiniest slice of time you can imagine—Inflation swoops in. With not even a billionth of a second's hesitation, it stretches the universe from the size of a quark to 10,000 light-years across.
>
> After a grueling fraction—of a fraction—of a fraction—of a second—or even less—just as Inflation realizes the hurtling freight train of a cosmos has reached its critical point, it shifts gears.
>
> In a cosmic feat of control, Inflation leaps ahead to stop the freight of hyper-expansion, its feet becoming the brakes. There's a deafening sound of steel grinding as the universe screeches and squeals to a halt. From inflation's heroic expansion, which broke the cosmic speed limit, to once again defying natural laws by coming to a rolling stop—all within the tiniest fraction of a second.
>
> As the immense train of expansion has slowed to a crawl, Inflation nudges the cosmos to speed up once more, while perfectly balancing spacetime—smooth as a billiard table—ensuring that everything falls into place, just right for stars, galaxies, and life to self-organize itself billions of years later.
>
> And then, just as quickly as it arrived, Inflation was gone. No fanfare, no trace—just the perfectly expanded cosmos, quietly set in motion.
>
> As we look out into the cosmos, studying the stars, and puzzled over the flatness of the universe, a collective thought dawns on us—a question so simple, yet profound:
>
> "Who was that?"
>
> From the back of the room, a quiet voice answers. Alan Guth, with a wry smile, a tumbler of Coke Zero, and a twinkle in his eye, swirls his tumbler and says, "That... that... was Inflation."

Despite its beyond-natural aspects, proposing the inflationary hypothesis is fair game. But here's the puzzle: Why do these speculations, which go beyond the laws of nature, get accepted while alternative perspectives, like those acknowledging apparent design, are dismissed out of hand, even though they are more clear, coherent, and observation-based? Why does a paradigm that prides itself on rejecting metaphysical ideas consistently resort to them to solve "problems" that are, in fact, just observations of reality? Why, in the end, does it have a problem with reality itself? Is this not hypocrisy?

# CHAPTER 11

# Cosmic Specs

*These six numbers constitute a 'recipe' for a universe. Moreover, the outcome is sensitive to their values: if any one of them were to be 'untuned', there would be no stars and no life. Is this tuning just a brute fact, a coincidence? Or is it the providence of a benign Creator? I take the view that it is neither. An infinity of other universes may well exist where the numbers are different.[171]* — Martin J. Rees

It's hard to understand how someone who recognizes the delicate balance of millions of interdependent factors—from the cell to the cosmos—could still endorse theories rooted in chaos, all the while mocking those who point to that order as if it were pseudoscience. When considering the complexity of even a single eukaryotic cell, there are easily millions of coordinated processes, reactions, and regulatory mechanisms at play. And when you expand that to the cosmic scale of physical constants, gravitational dynamics, and particle interactions, "millions" quickly becomes a conservative estimate.

Much of the public, often presented with theoretical façades, do not fully grasp the discrepancies and perplexing paradoxes associated with the concept of our universe founded on randomness.[181] Many accepted theories are not built on evidence, but on the desire to circumvent certain implications about the nature of reality. In claiming to be guided by reason, these thinkers advance convoluted ideas that strain logic—presenting themselves as wise while proposing intellectual nonsense. Their models consistently fail to confront the reality that the universe is held together by a precise framework of integral constants—each one fine-tuned, interdependent, and essential to its operation.

In addition to the constancy of the speed of light, the universe contains up to 26 fundamental constants that are foundational to its structure.[182] Each presenting additional problems with the Big Bang and inflationary models, but we don't have time to discuss them all here. These constants influence everything from subatomic particles to the expanse of galaxies—much like the components of a finely tuned device, but with an integration and precision beyond anything we could ever achieve technologically.

## Cosmic Calibrators

Dark matter and dark energy can be likened to the process of calibration in technology. Calibration adjusts and verifies the accuracy of a system to ensure it operates within specified tolerance levels, much like how dark matter and dark energy maintains balance in the cosmos. Dark energy adjusts the expansion rate of the universe, similar to how dynamic calibration ensures accurate responses to changing inputs in sensors. Likewise, dark matter anchors gravitational interactions, stabilizing galaxies and large-scale structures, just as master calibration calibrates systems against reference standards to ensure reliability.

A master calibration process uses reference standards—predefined, accurate measurements (like voltages, weights, or timing standards)—to make sure all related instruments or systems are correctly aligned and functioning consistently.[183] This alignment ensures accuracy across the system, even when components are complex, sensitive, or operating under different conditions.[184]

The persistent discrepancies revealed by the Hubble tension, however, extend this analogy beyond calibration to include recalibration.[185] This "tension" arises from a fundamental misinterpretation of the cosmos. In technology, recalibration becomes necessary when systems face complexities or conflicts that single-level thinking cannot resolve, requiring dynamic adjustments to account for relationships operating across multiple levels. Similarly, these unresolved *tensions* reveal the failure of current models to account for the universe's fundamental, technological aspects. Reductionist equations and explanations cannot describe their true nature.

Just as calibration is a behind-the-scenes technological process, invisible but vital for accurate and consistent operation, dark matter and dark energy remain unseen yet play a critical role in shaping the universe's behavior. As unseen forces that calibrate the cosmos, dark matter and dark energy ensure the order and precision necessary in a universe so deeply relational that its complex interdependencies are often mistaken for 'noise.' In such a system, recalibration isn't a response to failure, but a built-in feature—an ongoing adjustment that sustains cosmic balance.

In the world of technology, a device's specifications are critical to its performance and effectiveness. Take, for instance, a high-end smartphone. Every aspect, from its processor speed, measured in gigahertz, to its memory size, measured in gigabytes, is meticulously calibrated. A processor speed of 3.2 GHz may seem arbitrary at first glance, but it is essential for balancing the device's computational power with considerations like heat generation and energy efficiency.[186]

Figure 11.1 Intel Core i7-8700 processor with a clock speed of 3.2 GHz. This processor balances computational power with heat generation and energy efficiency. **Image source:** Kzuo, Wikimedia Commons.

Similarly, the universe exhibits this level of precision with its fundamental constants, each constant a finely-tuned specification in our cosmic spacecraft. The gravitational constant, for instance, is not just a number; it's like the exact pressure needed in a fuel line, essential for the engine to function without failure.[187] Consider also the fine structure constant, often denoted by the Greek letter alpha ($\alpha$).[188] [189] It characterizes the strength of the electromagnetic interaction between elementary charged particles, like electrons and protons.[190] This constant is like the perfect balance in a navigation system, ensuring the spacecraft stays on course.

The universe is not just a random collection of cosmic parts and pieces where specific phenomena happen to emerge. The universe is not just highly structured but "programmed" with a set of parameters that ensure its harmonious operation and the possibility of life.[191] Instead of viewing these constants as arbitrary figures in the equations of physics, we should see them as key indicators of a universe that operates with foresight, much like the most advanced technological systems we can conceive.

A tablet's screen resolution (pixels), pixel density (PPI), display type (LCD or OLED), and refresh rate all work together to ensure sharpness and clarity.[192] At the same time, battery life depends not just on capacity (mAh), but also on processor efficiency, display power use, and software optimization. None of these specs are random—they reflect a deliberate design calibrated for performance.[193] [194] Each specification is carefully crafted to optimize the device's efficiency, enhance the user experience, and ensure overall functionality.

In considering the universe as a kind of cosmic device, with its own set of specifications that determine its structure and behavior, another such *spec* is the cosmological constant, denoted as Lambda ($\Lambda$).[195] [196] This constant is associated with the concept of dark energy, which plays a critical role in the universe's expansion. Much like the finely tuned specifications of a pacemaker, $\Lambda$ is a value so precisely set that even a slight deviation would lead to a dramatically different universe, rendering vivi-technology (biological technology) impossible.[197]

According to the standard Big Bang model, if $\Lambda$ were too large, the universe would have expanded too rapidly after the Big Bang, preventing the formation of stars and galaxies.[198] [199] If it were too small, the universe might collapse back in on itself.[200] This framing not only assumes that clouds of gas collapse into stars—it imagines that complexity emerges from conflict, as if explosions can organize structure billions of times over. It treats space itself as a kind of battleground between invisible forces. These forces are described as oppositional pressures with no clear source or mechanism, yet the results they produce—stable galaxies, consistent expansion, and large-scale structure—suggest something far more ordered.

But from the perspective of Natural Technology, these interpretations miss the structure at work. Even if we accept the materialist reading—describing these phenomena as impersonal forces—it still inadvertently reveals a logic. What's interpreted as push or pull is better explained as a dynamic unfolding of a memory-structured system, expanding not through force but through distributed regulation. The expansion is not chaotic. The proportions are preserved. The balance is exact. This is not the behavior of randomly interacting energies; it is the behavior of an embedded system sustaining coherence at every scale.

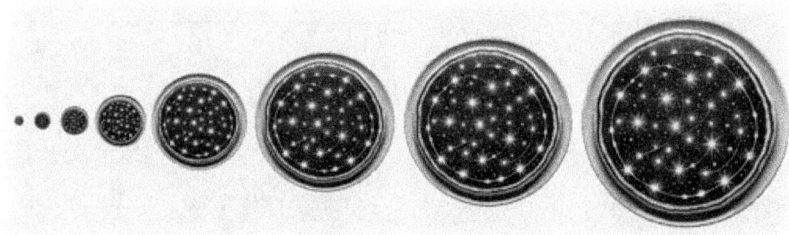

**Figure 11.2** Illustration of cosmic expansion, analogous to a cosmic pancake with stars embedded. **Image source:** S.A. Cooper

For the sake of clarity, let's allow the standard model to speak on its own terms. Figure 11.2 illustrates the universe as a cosmic pancake, with stars baked in like toppings. No one offers a clear explanation of where the batter came from, but in its earliest stage, the pancake is dense—stars and galaxies tightly packed, as the standard model describes.

As the pancake expands, those toppings move farther apart, representing the universe's ongoing expansion. Yet the proportions remain: galaxies don't fly apart internally. Like the toppings keeping their form while the pancake stretches, cosmic structures stay intact. This is where dark matter comes in—it acts like the binding that holds galaxies and clusters together, preserving structural relationships even as galaxies themselves move away from each other.[201]

Meanwhile, dark energy drives that expansion, much like heat causes an angel food cake to rise. Without this interplay, the universe could easily dissolve into chaotic disproportionality—but instead, it expands with balance embedded in its very design. Unlike heat causing a cake to rise, we don't know what this mysterious force is. Despite common theories about the expansion of the universe, we cannot even say for sure whether it is the universe itself expanding or if galaxies are simply moving away from each other.[202] As the image suggests, however, everything is spreading out in an organized manner, held in place—like toppings on a pancake.

## Perfect Balance

The role of dark energy, as reflected in the cosmological constant, points to the extraordinary precision required for the universe's balance. To appreciate this, it's essential to understand just how finely tuned the cosmological constant ($\Lambda$) is. It is expressed in terms of energy density, approximately $10^{-29}$ grams per cubic centimeter.[203] [204]

Though technically positive, this number is unimaginably close to zero; this is a fact that cannot be overstated, given the scales and forces at play in the universe.[205] [206] In fact, I hesitate to refer to it as positive at all—there's no term in English that accurately captures how vanishingly slight this value is in cosmic terms. Think of the old-school balance beam scale at the doctor's office, where you have to start by ensuring it's set to zero for accuracy. Imagine trying to measure a number this small:

0.00000000000000000000000000001.

Even the most advanced digital scales, capable of detecting nanograms, fall short by more than twenty orders of magnitude. The precision of this balance exceeds anything achievable with current instruments, yet it is critical to the formation and structure of the universe.

Imagine measuring the mass of the entire Earth with a scale so sensitive it could detect the weight of a single feather added to or removed from the planet. The cosmological constant is like the weight of a feather compared to the mass of the Earth; it is present and measurable, but extraordinarily slight in relation to the scale of the entire system.[207] Now, imagine adding a feather to the surface of the Earth—only for it to send our planet into total chaos. Likewise, despite its minuscule calculated size, any alteration in the cosmological constant could dramatically affect the universe.

We can go further. Comparing $\Lambda$ to a feather's weight relative to the Earth still vastly understates its precision and significance.[208] For a more fitting analogy to illustrate the cosmological constant's fine-tuning relative to the universe, imagine comparing not just a feather to the Earth but to something far more extreme: a single grain of sand to the Earth's entire mass. Even this comparison, while staggering, fails to fully capture the disparity between the predicted and observed values of the cosmological constant.[209]

The term *fine-tuning* is often used to describe this remarkable precision, but the term fine-tuning itself is a euphemism. In reality, the cosmological constant represents *perfect* balance, not fine-tuning. It is more perfect than anything we've ever described or even conceived of as "perfect"—short of a deity in the English language. And keep in mind, the universe has at least 26 fundamental constants, each of which must be precisely calibrated to sustain the cosmos as we observe it. This isn't just a technical detail for physicists—it's the key to everything we hold dear. The stars we admire at night, the lives we cherish, and the entire web of existence we inhabit all depend on this exactness. This level of balance, underpinning the structure of the cosmos, cannot be hand-waved away by superficial explanations or casual assumptions. The very nature of this technological precision transcends the physical.

In reality, though the term fine-tuning is often used to describe this remarkable precision, it's a misnomer: a category error. In technology, we don't look at a functioning processor and say, "Wow, it seems finely tuned to function." We recognize it as structured: built to perform that way. When a system operates with this level of integrated precision, we don't attribute it to chance. We call it design. Even the brightest minds that have ever lived can't fully conceptualize the necessary, integrated processes required for the universe and life to exist and thrive. To describe the cosmological constant in terms of fine-tuning is to imply a narrow statistical escape from chaos. But this isn't chance that survived—it's structured logic sustaining the system itself.

## Network Protocols

Another way to illustrate how the cosmological constant may seem arbitrary to those who overlook the technological aspects of the cosmos, we can draw an analogy to standard protocols in computer networking, such as the Transmission Control Protocol/Internet Protocol (TCP/IP).[210]

These protocols define how data packets are structured, how devices are addressed (like IP addresses), and how data is transmitted across the internet.[211] They are fundamental to the internet's functioning, allowing for the interconnection and communication of diverse and disparate devices and networks. Without the standards set by TCP/IP, the global internet, as we know it, would not operate effectively, if at all.[212]

The role of $\Lambda$ in the universe's expansion is comparable to how TCP/IP protocols enable the expansion and interconnectivity of the internet.[213] This expansion in the cosmos allows for the vast distances to be *bridged*, facilitating the movement and interaction of matter and energy over time.[214] Similar to how a flawed protocol could disrupt the functionality of the internet, without $\Lambda$—just as without TCP/IP—the system would be at risk: the universe could either collapse on itself or expand too rapidly.

## The Cosmological Constant "Problem"

It is essential to address why the cosmological constant is often labeled a "problem" within the scientific community and why the standard paradigm may mistakenly view it as arbitrary, finding it confounding.[215] The cosmological constant "problem" arises from the staggering discrepancy between the theoretically postulated magnitude of $\Lambda$ in quantum field theory and the constant inferred from observational measurements of cosmic expansion.[216]

$\Lambda$ represents an adjustment made to bridge the gap between theoretical expectations and observed cosmic behavior. Think of the cosmological constant as a hypothetical dial on the universe's control panel. Originally, Einstein introduced this 'dial' as a mathematical adjustment to force a static universe, consistent with the then-prevailing belief in an eternal, unchanging cosmos.[217] But when evidence showed the universe was expanding, $\Lambda$ was discarded—only to be reinterpreted decades later to explain the unknown force, now called dark energy, which appears to accelerate cosmic expansion.[218]

Here's where materialist assumptions run into trouble: quantum theoretical constructs predict a staggering value for this cosmic dial that does not match the extraordinarily tiny value we observe in reality. This immense discrepancy is what's known as the cosmological constant 'problem.' It is often regarded as "the worst theoretical prediction in the history of physics"—a blunt acknowledgment of just how far off quantum field theory is from the observed value.[219] This is a familiar scenario. Instead of reconsidering their foundational assumptions, the dominant scientific paradigm resorts to crunching the numbers and adding layers of theoretical complexity to force the data to fit their models.

This disparity has puzzled materialist cosmologists for decades.[220] Yet, much like the "horizon problem," the term "problem" is itself misleading—it reinforces dogmatic assumptions rather than prompting the necessary paradigm shifts. Models built on entrenched assumptions impose artificial restraints, limiting our ability to grasp the universe's technological features. The standard origins model has become the science itself, taking precedence over the evidence it was designed to explain. Failing to recognize this truth only delays our ability to unlock a deeper understanding of the universe's underlying structure.

Martin J. Rees, the prominent British cosmologist and astrophysicist mentioned earlier, is known for his contributions to our understanding of the universe.[221] In his book, *Just Six Numbers: The Deep Forces That Shape the Universe,* Rees discusses the profound significance of the universe's fundamental constants, emphasizing their extraordinary nature.[222] Rees suggests that these constants form a "recipe" for the universe, and the outcome is remarkably sensitive to their values.[223] He argues that even the slightest deviation in the values of these constants would lead to drastic changes in the universe's structure and evolution.[224]

In this light, the contrast between the public's perception of science as a steady march towards complete understanding and the reality of scientific inquiry is glaring. Advancements in certain technologies and reductive forms of knowledge have fostered the false perception that science is steadily progressing toward answering existential questions—reinforcing the assumption that materialist frameworks are inherently correct. Paradoxically, the more we learn about nature and the more our technological capabilities grow, the more they reveal the limitations of these assumptions, as the extraordinary technology within nature itself becomes apparent. Rethinking how science is discussed and engaged with is essential, because as this misconception deepens, it threatens to undermine the credibility and integrity of science as a whole.

While conventional models treat cosmic expansion as the effect of an unknown repulsive force labeled "dark energy," Natural Technology approaches it differently. Expansion, if real, is not a force-driven phenomenon but a structural expression of increasing coherence. As the universe stabilizes more complex relationships, its memory grows—space itself becoming a record of alignment.

Dark energy can be understood as a patterned adjustment within the system's underlying structure—a large-scale behavior that maintains balance without invoking new forces. This mirrors the exponential growth of human information—described in knowledge-doubling models, where understanding accelerates with each generation, demanding ever more structure to store and organize it.[225] The universe, likewise, extends from order—an expansion of stored relational logic. What expands is not just matter, but memory—none of which is ever lost.

## Intelligent design

**Intelligent design (ID)** is a pseudoscientific argument for the existence of God, presented by its proponents as "an evidence-based scientific theory about life's origins".[1][2][3][4][5] Proponents claim that "certain features of the universe and of living things are best explained by an intelligent cause, not an undirected process such as natural selection."[6] ID is a form of creationism that lacks empirical support and offers no testable or tenable hypotheses, and is therefore not science.[7][8][9] The leading proponents of ID are associated with the Discovery Institute, a Christian, politically conservative think tank based in the United States.[n 1]

Although the phrase *intelligent design* had featured previously in theological discussions of the argument from design,[10] its first publication in its present use as an alternative term for creationism was in *Of Pandas and People*,[11][12] a 1989 creationist textbook intended for high school biology classes. The term was substituted into drafts of the book, directly replacing references to *creation science* and *creationism*, after the 1987 Supreme Court's *Edwards v. Aguillard* decision barred the teaching of creation science in public schools on constitutional grounds.[13] From the mid-1990s, the intelligent design movement (IDM), supported by the Discovery Institute,[14] advocated inclusion of intelligent design in public school biology curricula.[7] This led to the 2005 *Kitzmiller v. Dover Area School District* trial, which found that intelligent design was not science, that it "cannot uncouple itself from its creationist, and thus religious, antecedents", and that the public school district's promotion of it therefore violated the Establishment Clause of the First Amendment to the United States Constitution.[15]

ID presents two main arguments against evolutionary explanations: irreducible complexity and specified complexity, asserting that certain biological and informational features of living things are too complex to be the result of natural selection. Detailed scientific examination has rebutted several examples for which evolutionary explanations are claimed to be impossible.

ID seeks to challenge the methodological naturalism inherent in modern science,[2][16] though proponents concede that they have yet to produce a scientific theory.[17] As a positive argument

**Figure 12.3** Wikipedia's misframing of Intelligent Design as pseudoscience reflects a broader pattern in which agency-based explanations are both dismissed and misframed—despite relying on the same inferential logic accepted in materialist contexts such as dark matter and cosmic structure.

## Irreducible complexity

**Irreducible complexity (IC)** is the argument that certain biological systems with multiple interacting parts would not function if one of the parts were removed, so supposedly could not have evolved by successive small modifications from earlier less complex systems through natural selection, which would need all intermediate precursor systems to have been fully functional.[1] This negative argument is then complemented by the claim that the only alternative explanation is a "purposeful arrangement of parts" inferring design by an intelligent agent.[2] Irreducible complexity has become central to the creationist concept of intelligent design (ID), but the concept of irreducible complexity has been rejected by the scientific community,[3] which regards intelligent design as pseudoscience.[4] Irreducible complexity and specified complexity, are the two main arguments used by intelligent-design proponents to support their version of the theological argument from design.[2][5]

Behe introduced the expression *irreducible complexity* along with a full account of his arguments in his 1996 book *Darwin's Black Box*, and he said it made evolution through natural selection of random mutations impossible, or extremely improbable.[2][1] This was based on the mistaken assumption that evolution relies on improvement of existing functions, ignoring how complex adaptations originate from changes in function, and disregarding published research.[2] Evolutionary biologists have published rebuttals showing how systems discussed by Behe can evolve.[6][7]

**Figure 12.4** Wikipedia's treatment of irreducible complexity continues the pattern of framing design-based reasoning as pseudoscience—despite relying on inferential logic that mirrors accepted reasoning in other scientific domains, such as the inference of unseen forces like dark matter.

WIKIPEDIA
The Free Encyclopedia

# Specified complexity

**Specified complexity** is a creationist argument introduced by William Dembski, used by advocates to promote the pseudoscience of intelligent design.[1] According to Dembski, the concept can formalize a property that singles out patterns that are both *specified* and *complex*, where in Dembski's terminology, a *specified* pattern is one that admits short descriptions, whereas a *complex* pattern is one that is unlikely to occur by chance. An example cited by Dembski is a poker hand, where for example the repeated appearance of a royal flush will raise suspicion of cheating.[2] Proponents of intelligent design use specified complexity as one of their two main arguments, along with irreducible complexity.

Dembski argues that it is impossible for specified complexity to exist in patterns displayed by configurations formed by unguided processes. Therefore, Dembski argues, the fact that specified complex patterns can be found in living things indicates some kind of guidance in their formation, which is indicative of intelligence. Dembski further argues that one can show by applying no-free-lunch theorems the inability of evolutionary algorithms to select or generate configurations of high specified complexity. Dembski states that specified complexity is a reliable marker of design by an intelligent agent—a central tenet to intelligent design, which Dembski argues for in opposition to modern evolutionary theory. Specified complexity is what Dembski terms an "explanatory filter": one can recognize design by detecting **complex specified information (CSI)**. Dembski argues that the unguided emergence of CSI solely according to known physical laws and chance is highly improbable.[3]

The concept of specified complexity is widely regarded as mathematically unsound and has not been the basis for further independent work in information theory, in the theory of complex systems, or in biology.[4][5][6] A study by Wesley Elsberry and Jeffrey Shallit states: "Dembski's work is riddled with inconsistencies, equivocation, flawed use of mathematics, poor scholarship, and misrepresentation of others' results."[7] Another objection concerns Dembski's calculation of probabilities. According to Martin Nowak, a Harvard professor of mathematics and evolutionary biology, "We cannot calculate the probability that an eye came about. We don't have the information to make the calculation."[8]

**Figure 12.5** Wikipedia labels the specified complexity argument as pseudoscience, even as it outlines a form of reasoning that closely mirrors many accepted inference methods in other fields—such as detecting intention through patterns or ruling out chance based on statistical improbability.

# CHAPTER 12

# Cosmic Network

Throughout our analysis of the universe as a manifestation of NatTech, we've drawn meaningful parallels to human-engineered systems—systems that arose in direct response to the structure of the reality they reflect. These analogies help us see cosmic components as more than just their surface functions, linking them to things like the internet, data transmission, and computing. In that spirit, consider again how the internet works: a vast, distributed network of servers and data centers operating in coordination. At first glance, many of these components may seem redundant; yet their presence is vital for the network's resilience and reliability.

Similarly, in the cosmic framework, the numerous stars and planets might support life indirectly, without an apparent relationship. Analogous to the role individual servers and data centers play in the broader functionality and resilience of the internet, the seemingly redundant number of stars and planets likely plays an essential role in maintaining the universe's overall stability and balance. In this analogy, each star and planet functions like a server or data center, contributing to the overall operation and stability of the cosmos. Their roles, though not always directly connected to life or visible phenomena, are essential to maintaining the universe's "operating system."[226]

Just like the unseen processes in technology that ensure the smooth functioning of complex systems, these celestial bodies, in their various capacities, might contribute to the workings of the cosmic machinery. In complex software systems, numerous background processes operate unnoticed but are essential for functionality. These processes manage memory, monitor systems, and allocate resources, much like the roles played by stars in the cosmic operating system.

The concept of the Internet of Things (IoT) furthers this analogy.[227] The IoT encompasses a vast network of physical objects that are embedded with sensors, software, and connectivity technologies, enabling them to communicate and exchange data with other devices and systems across the internet.[228] These objects range from everyday household appliances like refrigerators and washing machines to more complex industrial tools.[229] A key feature of IoT is its ability to connect diverse devices, which can then operate interactively and autonomously.[230]

Devices in an IoT system can automatically collect and transmit data, making decisions and performing actions based on predefined algorithms and programming, often without direct human intervention.[231] This level of integration and automation allows for more efficient, real-time responses within systems, enhancing the functionality and utility of ordinary and sophisticated devices.[232] This myriad of devices connect and operate, though inconspicuously, collectively forming a network that enhances efficiency and functionality.[233]

Each device may seem trivial or purposeless on its own but contributes to a larger, interconnected system. The universe, with its multitude of stars, resembles a cosmic IoT. Each star forms part of a greater cosmic network.

**Energy Transmission**

A deeper look at the principles behind electricity transmission reveals a complexity that mirrors the cosmos, challenging conventional views. People often think of electricity transmission as water flowing through pipes, with power flowing directly from plants to homes. However, this analogy oversimplifies the reality. Electricity is transmitted via a complex network of power grids, transformers, and substations, involving principles far more intricate than a straightforward flow.[234]

In an electrical grid, electrons don't simply travel one way from power plants to consumers. Instead, the process is dynamic: electricity is generated, stepped up in voltage for long-distance transmission, and stepped down at substations before reaching homes.[235] Electrons oscillate back and forth rather than flow in a single direction, and the power we use results from this oscillation.[236] Central to this is the concept of electric fields, where power generation and transmission rely on creating and manipulating these fields, not just moving electrons.

The electricity in your home might come from a plant hundreds of miles away, part of an interconnected grid that dynamically balances power generation and consumption over vast areas. Similarly, celestial bodies function within a larger system, influencing the universe's stability and structure. Just as electrical grids use controlled gaps in transformers to regulate power flow, the universe operates as a dynamic, self-regulating system, where gravitational and electromagnetic behavior reflects deeper orientation dynamics that sustain cosmic equilibrium.[237] Like substations in a power network, celestial bodies shape the structure and motion of galaxies and star systems, ensuring stability through their gravitational influence.[238]

In an electrical grid, multiple transmission routes and redundancies prevent failures, maintaining efficiency. Likewise, the gravitational forces of celestial bodies create pathways that regulate the movement of matter, keeping the universe in precise order. Transformers and substations manage energy distribution across gaps, much like stars, planets, and possibly dark matter work together to maintain the cohesion of galaxies and larger cosmic structures.[239]

When we think about stars and galaxies, we can see them as part of a massive cosmic 'circuit,' where visible matter and dark matter interact in an interconnected system that governs the universe's structure and movement. While the full nature of these interactions remains uncertain, the universe functions as a vast, interdependent network—one that mirrors the principles of technological infrastructure.

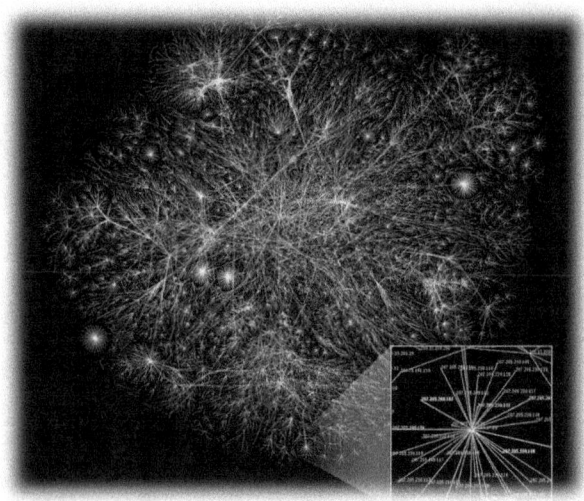

**Figure 13.1** A partial map of the Internet based on January 15, 2005, data from opte.org. Each line connects two nodes (IP addresses), with line length representing network delay. The structure reveals a web-like topology, resembling natural systems such as neural networks and the cosmic web. **Image source:** The Opte Project, Wikimedia Commons.

This mirroring becomes even clearer when we consider network topology, the structural design that defines how systems, technological or natural, are interconnected.[240] For instance, mesh topologies resemble fungal mycelium and neural connections, while hierarchical tree structures mirror river systems and vascular networks.[241] These aren't just metaphors—they're shared organizational logics. Among the various network structures found in nature and technology, star topology stands out as a particularly direct parallel to the organization of galaxies.[242]

In technological systems, star topology is a design in which all nodes connect to a central hub that directs communication and traffic. It's common in computer networks, electrical grids, and even transportation systems. Though the term was likely chosen for its visual resemblance to a star, not its cosmic accuracy, the similarity runs deeper than appearance. What seems like a practical engineering choice is actually a subconscious reflection of a structure already operating at the galactic level, where a supermassive black hole at the center governs the movement of billions of stars. However, unlike human-engineered star topologies, which are vulnerable to central failure and require substantial infrastructure, natural star systems—such as galaxies—exhibit the same structural organization without those inefficiencies.

Gravity, like a transmission system, maintains structure across vast distances. The Internet's network of central hubs, which routes data across a global web, mirrors this organization on a smaller scale, reflecting a design already embedded in the cosmos. This is not an arbitrary feature of the universe but an ancient framework that long predates human innovation.

## The CMB

When we contemplate the Cosmic Microwave Background (CMB) within these parallels, it fits with the perspective of the CMB as an artifact of the universe's initial 'activation' or 'switching on.'[243] In the chapter

'Cosmic Dawning,' the analogy of a bonfire at the center of a football field was employed to parallel this uniformity found in the CMB. The CMB is like a faint, uniform glow left over from the energy that radiated from the creation of the universe. It is the background energy that permeates the entire cosmic 'grid'—a remnant of the system's initial orientation parameter: its foundational 'activation' or 'switching on.'

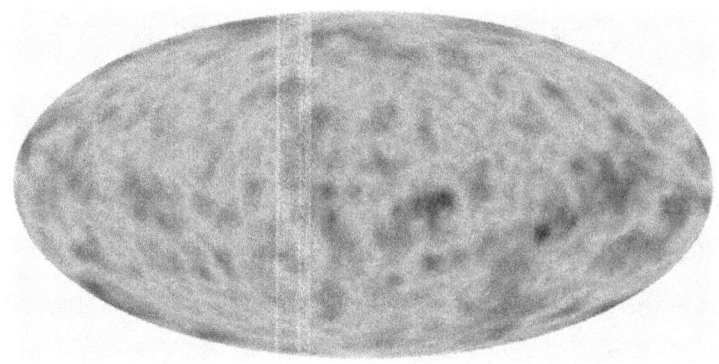

**Figure 13.2** Polarization of the Cosmic Microwave Background (CMB) detected by ESA's Planck satellite over the entire sky.  **Image source:** NASA/JPL-Caltech/ESA/Planck Collaboration, NASA ID: PIA18916.

This ancient radiation interacts with the 'circuitry' formed by stars and galaxies, influencing and being influenced by the distribution and movement of energy across the universe. The CMB can be understood in the same way that identifying baseline energy levels in an electrical grid enables more effective management and optimization of power flow. Electrical grids rely on constant baseline power (or "base load") to maintain stability and balance demand.[244]

These ideas provide a framework for reimagining the roles of celestial bodies and the CMB in the universe's grand scheme. Every celestial body functions like a character in a complex story, each playing a distinct role in the universe's narrative. Within this interconnected system, nothing is superfluous; and every star and planet contributes to the harmony and balance of the cosmos. With this, I challenge the idea of 'waste' in space, moving away from the perception of these bodies as purposeless.[245]

# CHAPTER 13

# Magnetic Fields

*Despite decades of intense interest and research, the origin of these cosmic magnetic fields remains one of the most profound mysteries in cosmology. In previous research, scientists came to understand how turbulence, the churning motion common to fluids of all types, could amplify preexisting magnetic fields through the so-called dynamo process. But this remarkable discovery just pushed the mystery one step deeper. If a turbulent dynamo could only amplify an existing field, where did the "seed" magnetic field come from in the first place?[246]*

Another fascinating aspect of cosmic mechanics is the role of magnetic fields.[247] While mainstream physics treat forces like dark energy and dark matter as foundational, this framework sees magnetic fields not as prime movers but as stabilizers within a coherent system of orientation and balance. These fields, essential to human-engineered technology and natural cosmic phenomena, epitomize the harmonious interplay of forces, such as attraction and repulsion.

In Maglev technology, magnetic repulsion counteracts gravity, enabling stable levitation.[248] This principle lifts and propels vehicles like trains without physical contact with the ground, thereby reducing friction and enabling higher speeds.[249]

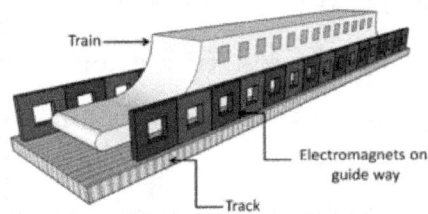

**Figure 14.1,** This illustration depicts the basic components of a magnetic levitation (maglev) train system. It features electromagnets mounted on a guideway that levitate and propel the train without physical contact with the track, reducing friction and allowing for higher speeds. **Image source:** Yasugawa, Wikimedia Commons.

**Figure 14.2,** The Maglev train at Longyang Road Station in Shanghai, photographed on November 15, 2014. **Image source:** Alex Needham, Wikimedia Commons.

Magnetic fields are often interpreted within mainstream astrophysics as regulators in star formation, influencing the collapse of gas clouds and the dynamics of accretion. However, this remains a materialist interpretation. We detect dense gas clouds and energetic emissions that are interpreted as sites of stellar birth, but we have not observed this process.

Magnetic fields are among the most pervasive yet mysterious forces in the universe. They weave through galaxies, shape the motion of plasma, and influence cosmic structures on every scale. Despite their undeniable effects, their fundamental nature and origins remain unknown. Modern astrophysics can describe how magnetic fields behave, but when it comes to their origins, there is only speculation. Some models propose that turbulence in primordial plasma spontaneously generated weak "seed" fields, later amplified by cosmic dynamos.[250] However, this explanation does not address fundamental questions: What caused the initial turbulence? Why did these fields persist and strengthen instead of fading away?

Unlike gravity, which is directly tied to mass, magnetism has no clear source on cosmic scales. There are no visible wires stretching between galaxies, no giant bar magnets shaping interstellar space. Yet magnetic fields are everywhere, governing the motion of gas, influencing galaxy rotation, and structuring the cosmos in ways we have yet to fully understand.

In their 2001 study, P. Heinzel and U. Anzer examined the role of magnetic fields in stabilizing quiescent solar prominences, large plasma structures suspended in the Sun's atmosphere. Published in *Astronomy & Astrophysics*, their research explores how these magnetic fields maintain prominences in magnetohydrostatic (MHS) equilibrium, balancing gravitational and pressure forces to prevent the plasma from collapsing.[251] Similar to how magnetic fields stabilize plasma structures in the Sun, as demonstrated by Heinzel and Anzer, they also serve as fundamental regulators of larger cosmic and technological systems. This balance of forces in solar prominences mirrors how engineered systems like Maglev use magnetic logic to maintain precise stability.

This idea of magnetic fields shaping the universe can also be seen in areas closer to our everyday experience, such as biological research. Magnetic beads, tiny particles used in labs to separate and purify biological materials like DNA and proteins, become magnetic only when exposed to an external field, preventing clumping and allowing precise control. This helps researchers isolate specific molecules more efficiently from complex mixtures.[252]

In certain advanced lab techniques, such as those examined by Venkataragavalu Sivagnanam, magnetic beads are used in microfluidic devices, which are tiny lab-on-a-chip systems that handle minuscule liquid volumes.[253] Similar to how magnetic fields in space help organize plasma and gas, Sivagnanam leverages the unique properties of these beads to create precise patterns within microfluidic systems, facilitating highly sensitive immunoassays for diagnosing diseases or understanding biological pathways.[254] By using magnetic fields to control the arrangement of these beads, researchers can quickly and accurately analyze biomolecules, leading to significant advancements in medical diagnostics and biological research.

In certain advanced lab techniques, like those involving magnetic beads, time-resolved luminescence detection is often used to enhance the accuracy of results. This technique involves measuring the luminescence (light emitted by a sample) over time after an initial excitation light is turned off.[255] The key advantage of this approach is that it allows researchers to distinguish between the desired luminescent signal and the surrounding background noise.[256] By focusing on the emission at specific time intervals, the sensitivity and specificity of the detection process are greatly improved. In this sense, the natural technology of the cosmos has been translated into a powerful tool for medical diagnostics and biological discovery, exemplifying how principles of the universe inspire and inform human ingenuity.

In astronomy, scientists often observe radiation or light emitted by cosmic objects over extended periods, applying time-based filtering to separate meaningful signals from cosmic background noise, such as the constant glow of the CMB. The universe itself appears to operate through similar principles, structuring information over time rather than presenting it all at once.

Scientific methods are not independent innovations but reflections of processes already embedded in the cosmos. Time-resolved luminescence detection in laboratories isolates specific molecular signals by filtering out interference, enhancing clarity and precision. Likewise, the observation of cosmic signals relies on distinguishing relevant patterns from surrounding noise, reinforcing the idea that the universe itself is structured in a way that allows for signal differentiation over time.

Structured time-based differentiation is not a human invention but a built-in feature of the universe, one that enables its own understanding. Whether in laboratory experiments or astronomical observations, the ability to extract signals from background noise points to an underlying system that is ordered and functional. The universe does not just display patterns—it operates in a way that allows those patterns to be recognized, analyzed, and understood, as if it were designed to make sense to those capable of perceiving it.

CHAPTER 14

# Nebulous Beginnings

*Modern physics is exceedingly mysterious; it really is. I mean, have you tried to read some of the difficult modern physics books? It really is very difficult to understand, and intuition doesn't do it. You cannot use human intuition. Human intuition was built up by evolution over many millions of years to survive on the African plain... Our brains were not built to understand the profundities of the origin of the universe, the end of the universe, the kind of things that only physicists deal with. I think it's actually amazing that at least some [physicist] human brains are capable of dealing with this kind of stuff. My brain isn't, and nor is yours. —Richard Dawkins.*[251]

At first glance, the term "solar system" might seem like just another label, but it represents much more. The word "system" implies organization: a set of components working together toward a purpose.[257] We use "system" in various contexts, such as the Global Positioning System (GPS), file systems, or transportation systems. Our solar system operates with a precision that fully embodies the meaning of the word "system."

The sun acts as the central hub, like a busy transit station. Its gravitational influence keeps everything in motion, ensuring the entire system runs smoothly. Planets, moons, and some asteroids move in predictable orbits, with planets revolving around the sun much like buses on regular routes, all interconnected in a larger system where everything interacts.

The energy radiating from it moves outward in predictable waves, interacting with the objects in the solar system in a way that maintains order, much like energy flowing through a network.[258] The energy is finely balanced, ensuring that each planet receives the appropriate amount to sustain its unique conditions. This balance prevents extreme overloads or inefficiencies, keeping each planet in a stable state, even as their distances from the Sun vary dramatically.

The solar system's consistency is remarkable. It operates like clockwork, governed by the laws of physics, which act as the underlying code or programming driving its mechanics.[259] The balancing force, gravity, is constantly pulling the planets toward the Sun as their inertia propels them forward.[260] Inertia (Newton's First Law) is the principle that an object in motion will remain in motion, and an object at rest will stay at rest unless acted upon by an external force. In space, where friction is nearly nonexistent, there is nothing to slow planets down, allowing them to continue moving in their orbits indefinitely.[261] This system maintains stable orbits, preventing planets from spiraling inward or drifting away.

The predictability of the solar system has offered humanity with insights for millennia. For centuries, astronomers have tracked planetary movements with remarkable accuracy, allowing them to map the skies and predict future positions. This reliability has enabled early sailors to navigate oceans using the stars. Today, modern scientists rely on these same principles to plan interplanetary missions with precise timing, accurate down to fractions of a second.

Understanding the solar system as a true system—coherent, balanced, and purposeful—raises deeper questions: Why does it function with such reliability? What shaped its order? How did this structure become so fit for life? And why does it appear so exceptional in the broader universe, when modern science claims it evolved from fluctuations of particles? Hopefully, by the end of this book, the answers to these questions will be a bit clearer.

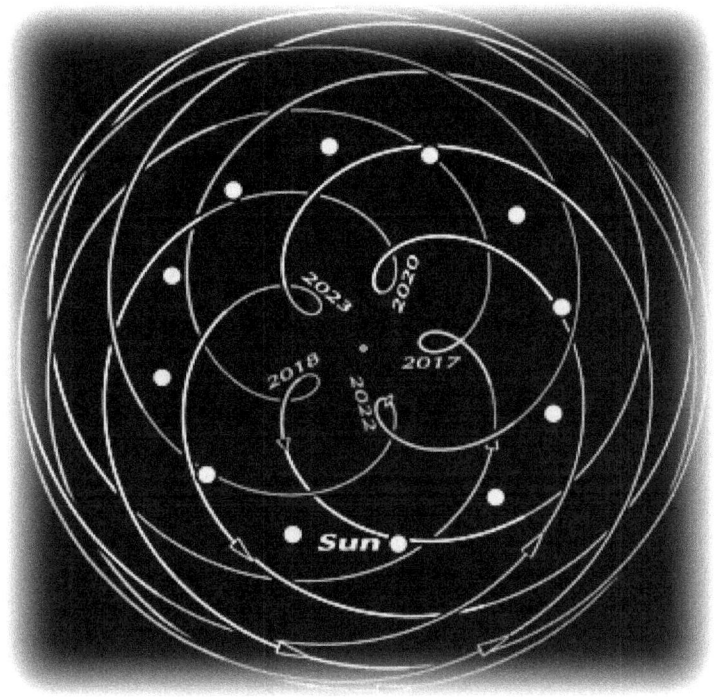

**Figure 15.1** A geometric pattern depicting the synodic cycle between Venus and Earth as they orbit the Sun, from 2016 to 2023. Each loop represents the relative position of Venus to Earth over this period, creating a visually striking pattern that repeats every 8 years, illustrating the harmony in planetary motion known as the Venus-Earth pentagram. **Image courtesy of Guy Ottewell,** https://www.universalworkshop.com/venus/

The movements of celestial bodies follow geometric patterns that can be calculated far into the future, like Venus' eight-year cycle (sketched in Figure 15.1) which traces a pentagram-like pattern around Earth.[262] Mars' retrograde motion, occurring every two years, adds further intrigue as it briefly appears to reverse course in the sky. Mercury, with its 3:2 spin-orbit resonance, creates a rhythm that reflects the order of the cosmos.[263] Ancient civilizations, recognizing these patterns, described them as the Music of the Spheres, believing that the universe moved in perfect harmony.[264] The Mayans, for instance, built sophisticated calendars based on Venus' cycle to time their agricultural and ceremonial events.[265]

This regularity in planetary cycles has influenced our understanding of time. Each movement reveals a symmetry too precise to be accidental, hinting at an underlying system quietly maintaining cosmic balance. These rhythms, like the steps of a cosmic dance, continue to inspire wonder in some as they unfold with the precision of an unseen algorithm.

Today, as technology shapes our perception of time, the significance of natural cosmic rhythms, like the cycles of the moon and stars, has largely been forgotten. The beauty and precision of planetary cycles are still studied scientifically, but they no longer hold the same cultural or existential meaning they once did.

The principles of celestial mechanics form the foundation for modern space exploration. From launching satellites to landing probes on distant planets, the more we understand these movements, the more humanity advances. These timeless patterns, observed since ancient times, have bridged the gap between early astronomy and modern technology, offering a glimpse into the finely-tuned mechanisms of the universe.

## The Solar Nebular Theory

In other scientific discussions, the origins of the celestial bodies in our solar system, including the Sun, are commonly explained by the Solar Nebular Theory. It is commonly referred to as the Solar Nebula Hypothesis, a term that suggests a tentative explanation open to critique and revision.[266] In practice, however, scrutiny is limited to minor details rather than the broader framework.

Approximately 4.6 billion years ago, the Sun is thought to have formed as the result of the collapse the solar nebula, a protoplanetary cloud of dust and gas believed to be the origin of all the bodies in our solar system. The term Solar Nebula specifically refers to the cloud of gas and dust from which our solar system is believed to have formed, named after our star, Sol. However, the origins of this cloud and the cause of its collapse under gravity—resulting in the Sun and planets—are questionable.

The solar nebular theory begins with the assumption that, after the Big Bang, the universe was filled with a vast expanse of hydrogen and helium, with only slight variations in density.[267] All celestial bodies we observe today are thought to have originated from this primordial mixture, similar to how biology traces all living organisms back to a hypothetical microbial organism: LUCA (Last Universal Common Ancestor).[268] [269] Over time, these slight variations supposedly caused regions of gas to slowly clump together under the influence of gravity, forming Giant Molecular Clouds (GMCs) that spanned light-years across.[270] These clouds are considered the cosmic "nurseries" for stars and planets. This hypothesis proposes a mechanism for the gradual clumping of gas but does not fully clarify how denser regions initially formed in an otherwise diffuse expanse.

**Figure 15.2** This iconic image, referred to as the "Pillars of Creation," was captured by the Hubble Space Telescope and reveals towering gas and dust structures within the Eagle Nebular, part of a Giant Molecular Cloud (GMC). Although often described as a star-forming region, this interpretation is largely based on theoretical models and philosophical assumptions. **Image source:** NASA, ESA, CSA, STScI; J. DePasquale, A. Koekemoer, A. Pagan (STScI), CC BY 4.0 INT.

To imagine the process of GMCs leading to the diverse order of our solar system, think of mitosis, where a single cell divides under precise conditions to form organized structures. Once a cell begins dividing, it can multiply and specialize into many more cells, each with a specific function. In contrast, the theory of star formation suggests that GMCs, the cosmic cells, did not evolve from a LUCA-type source but instead appeared independently across the universe, each with the potential to form stars and planets within integrated functional systems.

This would be like specialized cells suddenly appearing everywhere, ready to differentiate into complex systems, without any prior division or guiding process. Though biological systems rely on dynamic networks of interactions to ensure order and coordination, these cosmic clouds are said to have formed spontaneously, without any such directive, yet repeated this process quadrillions of times as if following a universal schematic.

Just like a single cell divides into two daughter cells, the GMC would divide into smaller clouds during a process called "fragmentation."[271] This fragmentation would repeat itself until it reached the desired quantity of clouds and sizes for star formation.[272] This entire process, of course, is said to be driven by gravity. From these humble beginnings, gravitational instabilities and turbulence initiated the differentiation process, where these clouds eventually collapsed under gravity, leading to nuclear fusion and sustaining stars for billions of years. But before explaining how gravity causes this, we should ask: what is gravity? This assumes gravity is a prime mover—but what if gravity is the result of deeper structural logic?

Still, without the Sun's gravitational pull yet present, how could this collapse occur? [273] Gas particles in space are naturally diffuse and would remain spread out unless acted upon by a stronger force.[274] The theory suggests, however, that some regions became denser, enabling gravity to take hold, but it does not explain how these denser regions initially formed.[275]

The solar nebular theory fails to address critical barriers to gravitational collapse, including cooling thresholds, turbulence, and magnetic fields. Observations reveal that these forces dominate molecular clouds, disrupting the sustained density needed for star formation.[276] Instead of providing plausible mechanisms to overcome these barriers, the hypothesis relies on speculative triggers which fail to account for the sheer scale and consistency required for star formation across the universe.

Yet if gravity is strong enough to collapse the smaller fragments, why does it not collapse the larger cloud before it fragments? For a GMC to divide, gravity has to be weak enough that it cannot hold the giant cloud together as a whole. However, once it fragments, gravity is suddenly strong enough to collapse the smaller pieces into stars, an inconsistency that remains unexplained. This abrupt shift in gravitational behavior lacks a coherent explanation and exposes a fundamental contradiction within the theory.

Furthermore, the hypothesis depends on speculative, fine-tuned events like supernova shockwaves, which are highly improbable as consistent triggers for cloud collapse across the vast scales required for universal star formation.[277] These shockwaves, it is said, compressed the clouds at just the right moment, causing them to collapse rather than disperse.[278] The timing of this process is improbable within the chaotic environment of space; yet, from a materialistic perspective, it is often described as if it were inevitable, with the reasoning seemingly, 'After all, we are here.'

Then things get more interesting. As the smaller clouds collapse under their own gravity, angular momentum causes them to spin faster, like a figure skater pulling in their arms to spin more quickly.[279] The theory claims that this spinning cloud naturally flattens into a protoplanetary disk from which planets arise, achieving a near-perfect shape despite turbulence, magnetic fields, and other disruptive forces.

In mainstream star formation models, magnetic fields are incorporated to explain why collapsing clouds don't just keep spinning faster indefinitely. The models use concepts like magnetic braking to explain how angular momentum (the tendency of a spinning cloud to spread out) is reduced, allowing collapse to continue.[280] These ideas are mathematically elegant and fit within the equations. Creatively, they adjust the parameters for the collapse to proceed.

Thus, these hypothetical constructs rely on optimal conditions, idealized to make the equations solvable.[281] For example, the gas density in a model might be uniformly distributed, or the cloud might be treated as a perfectly symmetric sphere.[282] In reality, these theorized clouds would not be uniform. They would have clumps, irregular shapes, and varying densities. This makes it hard for the well-ordered collapse envisioned by models to happen in real space. There is also far more chaos and randomness than any model can predict, which means that the collapse process would be unpredictable.[283]

The conditions for nuclear fusion—extremely high temperature and pressure—are often described as essential for star formation, yet the natural occurrence of such conditions in the vastness of open space is highly implausible. The idea that each fragment, after splitting off from a larger cloud, would collapse into these conditions is egregious speculation. Real-world clouds are often disrupted by competing forces, making it unlikely that they would uniformly collapse. The likelihood of this happening across so many different clouds, under so many different conditions, is quite literally astronomically implausible.

To illustrate this concept, let's consider an athlete, specifically a basketball player. While many athletes possess exceptional abilities that allow them to defy gravity with impressive vertical leaps, we generally do not expect a player to jump from the ground to the top of a standard roof under typical circumstances. However, in a mathematical equation or theoretical framework, we can construct scenarios that make such a leap seem feasible, though hypothetical.

Imagine an athlete designed with the perfect physique for jumping—strong legs, optimal muscle composition, and exceptional technique. We could even factor in ideal conditions, such as a strong gust of wind, or perhaps under a particularly creative theoretical model, a couple of eagles attaching themselves to each shoulder at just the right moment, providing additional lift. The equations that formulate this scenario would indeed be too complex for the average person to understand.

Yet, despite what the math suggests, we know this does not reflect reality. Such suggestions, however, become more persuasive to the uninitiated when framed as events possible billions of years ago, particularly for those who share the same worldview. It's often assumed that we "cannot use human intuition" to grasp these concepts, and that only exceptionally special "human brains are capable of dealing with this kind of stuff."

However, as in our analogy, the fact that physicists could formulate an equation describing the muscular and environmental forces that would enable the hypothetical athlete to jump from the ground to a roof, although quite complex and difficult to fully understand, does not explain why the idea isn't implausible. Furthermore, suggesting that this feat of molecular clouds to star formation occurred countless times in the past—even without intentionality, unlike our athlete and his eagles—would require an entirely different line of reasoning, distancing the claim even further from practical reality. Accordingly, the hypothetical leap of the solar nebular theory stretches the boundaries of plausibility.

These striking celestial patterns—recurring with mathematical precision across vast time scales—stand in sharp contrast to the chaotic, speculative formation mechanisms offered by mainstream cosmology. The deeper one examines the origins of this order, the more one finds not randomness, but embedded regularity—systems that behave as if they were designed to function.

## Planet Formation

Following the division of GMCs and the subsequent collapse of individual clouds into stars, the narrative of dust and debris coalescing into planets, like ours, begins. Dust particles are believed to stick together, gradually forming larger objects that, over time, would accumulate and grow, eventually "snowballing" into planetesimals and, later, planets.[284]

How dust particles in space could stick together to form larger bodies is a formidable issue. Space is as close to a perfect vacuum as we can find, with extremely low densities of particles, such as atoms or molecules; particles are sparse and move at high speeds.[285] However, the nebular theory suggests that electrostatic forces caused these dust particles to clump together in the early stages.[286]

These same forces that make dust cling to your TV screen are easily wiped away, lacking any special ability to form solid bonds.[287] In this theory, these forces allowed small dust particles to come together, initiating the process of growth.[288] Space, however, offers a very different environment.[289] These forces are weak and would typically be disrupted by external influences like radiation or turbulence.[290] As these dust particles supposedly grew, gravity was expected to take over and facilitate further growth. Though gravity at such small scales is nearly negligible, it is believed that these dust clumps continued to grow and increase in tensile strength.[291]

Despite the sparse environment and disruptive forces, these dust grains managed to advance to the next stage.[292] This next stage in the process involves how these small particles, now called pebbles, grew into objects about a meter in size.[293] [294] At this point, something called "gas drag" becomes significant. Gas drag is problematic for pebbles primarily because of their size and the interaction with the gas in the disk. In the protoplanetary disk, gas drag would tend to slow down and pull these objects inward, toward the Sun, where they would spiral in and eventually be destroyed.[295]

**Figure 15.3** This image from NASA's James Webb Space Telescope shows neon emissions around the young star SZ Chamaeleontis. **Image source:** NASA/JPL-Caltech.

The process known as "pebble accretion" is then introduced, explaining how these pebble-sized particles rapidly combined with larger bodies. Beyond gas drag, which pulls smaller particles inward, meter-sized objects face an additional obstacle known as the "meter-sized barrier." Their size makes them especially prone to destructive collisions rather than gradual growth, making it unlikely for them to coalesce into larger bodies at this stage.[296] These difficulties make it unlikely for these small bodies at this stage to coalesce into larger bodies.

After advancing past the meter-sized barrier, the next stage in planetary development involves the growth of planetesimals—solid objects typically kilometers in size, believed to be the building blocks of planets.[297] These planetesimals are said to have formed from a mixture of metals, rock, and ices, a mixture of materials that originated from previous generations of exploded stars. The notion that sand, metal, and ice could coalesce, withstand chaos, and ultimately give rise to planets is a remarkable claim, further contributing to the array of unresolved issues within the hypothesis. Metals typically require extremely high temperatures to bond, while ice would vaporize, and sand might form glass under such conditions. How these different materials could endure such chaotic environments and still coalesce into planets is a detail that remains unknown.[298]

According to the hypothesis, during the planetesimal stage of growth, these bodies no longer rely solely on collisions for their expansion; instead, they begin to grow by exerting gravitational forces, gradually pulling in surrounding material.[299] Despite the chaotic and violent environment, characterized by frequent collisions that seem more likely to lead to fragmentation rather than growth, the theory posits that these young planetesimals overcome these challenges.

For now, these seminal planets continue to collide with one another. The energy from these impacts, combined with radioactive decay, generate enough heat to melt the planetesimals into molten spheres. In this molten state, heavier elements like iron and nickel sink toward the center to form dense cores, while lighter materials float to the surface, forming a primitive crust. After cooling and solidifying, the bodies are poised for the next collision. With each new impact between planetesimals, they coalesce into larger bodies.

This cycle repeats itself: planetesimals collide, generating enough heat to melt the same bodies once more. Differentiation occurs, causing heavier elements to settle into the core while lighter materials rise to the surface. Each time this process completes, it resets, beginning again as further collisions reignite the cycle. As this process continues, planetesimals increase their gravitational strength, allowing them to pull in more material, thereby increasing their rate of growth. Despite the chaotic environment of violent collisions, these impacts are believed to drive the process rather than hinder it, repeatedly producing molten bodies with distinct layers, primed for further development.[300]

This explanation demonstrates the high flexibility of this theory. The same collisions that are described as non-destructive in the early stages of growth suddenly become violent enough to melt entire planets. The assumption that collisions and radioactive decay produce enough heat to uniformly melt planets demands an unwavering dedication to this belief system.[301] While these processes could heat the surface, it's unclear how they could melt entire planetary bodies for long enough to enable their differentiation into distinct layers, where heavier elements sink to the core and lighter materials rise to the surface. [302]

But if these impacts were powerful enough to liquefy entire planets, why wouldn't they also generate enough force to scatter the material, preventing planetary formation altogether? Modern observations of high-energy collisions, including asteroid impacts, show fragmentation, crater formation, and scattered debris—not the smooth formation of a molten planetary body.[303] If today's violent impacts break objects apart, why would early planetary collisions behave so differently?

Through a continuous cycle of smashing, melting, and coalescing, protoplanets reach a stage where they are called "protoplanetary embryos."[304] With further development, these embryos are said to develop into fully-fledged planets like Earth, complete with a core, mantle, and crust. Let's say planets did become molten; the heat distribution during this process would likely be uneven, raising questions about how planets like Earth developed such well-ordered layers. This situation would be like a cosmic game of musical chairs, where, despite the chaotic beginnings, each element seemingly found its precise place when the music stopped.[305]

For the planets to cool and solidify in a manner that allows for the formation of distinct layers, the cooling process would need to occur in a way that defies the principles of physics. The development of a planet's core, mantle, and crust would rely on precise factors such as temperature, pressure, and the material properties involved. At this fully mature stage of development, the planet's structure resembles a functional organism, where each layer functioning like an organ in a body performing specific roles essential for the planet's geological processes and overall stability.

Earth, with its molten core and well-differentiated layers, differs greatly from planets like Mercury, which shows minimal differentiation, or Jupiter, which never formed a solid surface. If the same process is responsible for the formation of all planets, the variety in their structures presents striking results. Why do some planets, like Jupiter, consist mainly of gas, while Earth supports complex ecosystems? Planetary evolution must account for such differences. Saturn has intricate ring systems, while Earth has "evolved" features like New York City. How can a vast array of planets, moons, and other celestial bodies coexist alongside one that has evolved lifeforms and skyscrapers on Wall Street?

Moreover, as these planets are thought to have evolved, splitting from the disk, the Sun took shape at the center, pulling in 99.8% of the solar system's mass.[306] This hypothesis of the Sun's formation presents yet another mind-boggling equilibrium. Centrifugal forces from the spinning cloud, which would naturally tend to fling material outward, were counteracted by gravity, concentrating nearly all of the solar system's mass into the Sun while leaving just enough for our planetary system to develop.

Another finely-timed occurrence: the narrative explains once the Sun stabilized and ignited into burning, its solar wind efficiently brushed away the remaining dust and gas, just after the planets had fully formed. Perhaps I lack the specialized knowledge required to understand how these processes are considered feasible or maybe materialists operate within an esoteric framework where such improbabilities are accounted for.

## First Stars Problem

Furthermore, if supernovae are cited as catalysts for star formation, it's intriguing that there appears to be little urgency to address the origins of the very first stars. This issue can be called the "first generation of stars problem."[307]

While supernovae result from the deaths of stars and are frequently cited as essential triggers for the formation of new stars, the first stars formed in an environment devoid of such events. The solar nebular theory fails to address these foundational questions about the origin of the first stars.[308] What initiated the formation of the first stars, and was this mechanism fundamentally different from the processes that initiated later star formation?

Further, according to this narrative, minutes after the Big Bang, heavier elements, essential for forming planets like Earth, didn't exist.[309] This introduces the 'missing ingredients problem'—how did the universe transition from a primordial state devoid of key building blocks to a cosmos filled with rocky planets and complex systems? This recurring motif within materialistic explanations continues to reveal that these theories are more about addressing philosophical gaps than providing plausible reconstructions of the universe's past. The expectation that heavier elements could materialize from a cosmos primarily composed of hydrogen and helium clouds introduces significant questions about the mechanisms at play.

Thus, this is not just about a few gaps: the entire materialistic origin story—from gas to planets to life—is fundamentally flawed, with multiple interdependent theories compounding the issues at each step, each flawed in a fundamental aspect. Or, to honor Orwell's notion that "clarity is the remedy:" from the Big Bang to me writing this book, nothing about the materialist cosmic continuum makes sense.[310] There is no consensus on how diffuse gas coalesced into stars or how stars then triggered the formation of planetary systems like ours. It's strange, then, for anyone to accept these processes as settled facts. Like Behe's flagellum, this system requires all its parts and processes to function, as removing even one component or process renders the entire structure incompatible. There are several implausible points that make these origin narratives useless.

Even authoritative sources like NASA avoid addressing these foundational questions. While they offer explanations of later star and planet formation, their accounts ignore these discrepancies altogether. The underlying assumption that ties all materialist explanations together is that Materialism is the only valid framework for understanding the universe. Thus, my petitions for cohesive explanations are largely rhetorical. As reflected in the literature, these processes largely remain inexplicable through a materialist lens. A more substantive question is why these speculative scenarios are presented to the public so dogmatically as fact.

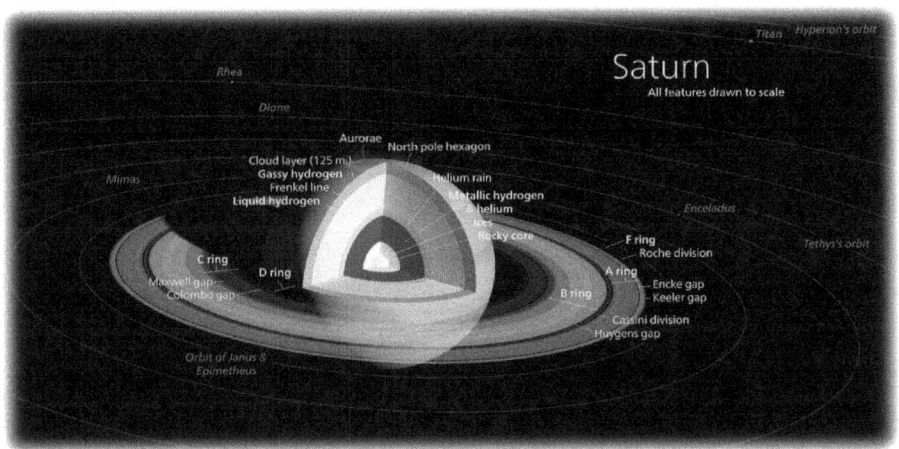

**Figure 15.4** A cross-sectional view of Saturn, illustrating its layered internal structure—including a rocky core, an icy layer, metallic hydrogen, and helium—as well as its atmospheric composition and ring system, highlighting key features such as the Encke gap, Cassini division, and the planet's hexagonal storm at the north pole. **Image Source:** Kelvinsong, Wikimedia Commons.

Let's look at some of the planetary "coincidences" that present more problematic implications—for example, the stability of Saturn's rings.[311] Its rings play an essential role in maintaining the planet's gravitational balance.

Composed of ice and rock, their origin and persistence remain a mystery.[312] The leading theory suggests they formed from the debris of colliding icy moons, but this raises further questions: Why don't other planets have similar rings? And how have Saturn's rings remained intact despite the violent events that supposedly created them?[313] Their precise structure suggests they are an integral part of Saturn's system, not just the result of random collisions.

Venus, with its retrograde rotation, further complicates this narrative. If all planets formed from a spinning disk of gas and dust, how did Venus end up spinning in the opposite direction of most other planets?[314] Similarly, Mercury's magnetic field and polar ice, despite its proximity to the Sun, defy the expectations of the standard model. The characteristics of outer planets like Neptune and Uranus represent an odd chapter in the standard narrative.[315] Neptune's orbital resonance with Pluto is particularly notable. Instead of their orbits crossing, the two bodies remain in a stable, synchronized rhythm, suggesting a level of precision that random gravitational forces alone cannot explain.[316]

As we step back and examine the solar system as a whole, it becomes increasingly difficult to see its features as the outcome of chaotic collisions and gravitational improvisation. It is a fully integrated technological system—engineered with precision, intention, and stability in mind. Every planetary feature, every orbit, every cycle reflects embedded logic, not randomness. This isn't speculative; *it's observable*. Gravity did not "form" these bodies. Gravity is an effect within this system—not the designer of it. The logic behind the narrative of planetesimals and accretion is backwards. The system did not evolve into order. It began as order. We are looking at an all-or-nothing system, a gyroscopic logic, not one that assembled itself piece by piece.

The planetary compositions themselves—Earth's hydrological balance, Jupiter's electromagnetic dominance, Mercury's metallic core—aren't the outcomes of random accretion. They are differentiated specifications. No engineer duplicates every component identically in a system. Every planet in our system plays a distinct role, each calibrated for a function—thermal regulation, mass distribution, electromagnetic interaction, orbital harmonics.

The resonances among planetary bodies, the synchronized orbits of moons, the axial tilts and spin rates, the precision of celestial alignments—all of it operates as part of an intentional, interlocked framework. These aren't coincidences to be normalized. They are protocols running in harmony.

Retrograde motion is not a glitch. Venus and Uranus, for example, do not spin "the wrong way"—they spin with intention. Opposing rotational vectors are not irregularities in system logic; they are stabilizers. In any dynamic technological system, counter-rotation is used to maintain equilibrium, absorb stress, and balance torque across a larger mechanism. This is exactly what we observe here: system-level calibration.

Notably, experimental studies in plasma physics show that counter-rotation can function as a system-regulated behavior, contributing to stability, momentum transport, and turbulence suppression. What the literature calls "anomalous counter-rotation" is not anomalous at all—it reflects embedded structural logic. As noted in a 2014 paper on toroidal plasmas, "spontaneous toroidal rotation in the direction opposite to the plasma current has been observed even without any external momentum input... This phenomenon... cannot be explained by classical momentum transport models and is considered anomalous."

As mentioned, Saturn's rings are not remnants of ancient collisions. They function as part of Saturn's architecture—regulating, filtering, and maintaining dynamic balance around the planet. Their structure, distribution, and sustained order over time are signatures of purpose, not debris.

The solar system is not a loose collection of lucky outcomes. It is a unified technological array—purpose-built, tuned, and maintained. This same principle appears in our highest-precision technologies: systems that regulate orientation, maintain spin balance, and recalibrate dynamically—like inertial guidance platforms or quantum clocks. They reflect the very logic written into planetary behavior. Every movement and interaction are reflections of Natural Technology at work. And when seen through this lens, the standard narrative of accidental cosmic evolution can no longer stand as a credible account in light of the structural coherence it seeks to dismiss.

## Cosmic Homeostasis

Relational dynamics, often described as gravitational interactions, appear to preserve the structure and balance of the solar system over billions of years. This principle of relational stability governs planetary systems and reflects universal laws shaping biological, ecological, and cosmic systems. No entity exists in isolation, and every relationship is a component of the resilience of the larger whole.

That said, I'd like to examine an aspect of design that you may not have considered. What if there are finer-tuned, more abstract systems, such as self-regulating technological processes naturally embedded on a cosmic scale, yet largely overlooked due to the lens of Materialism? Discussions in scientific literature, when examined from a broader perspective, often reveal descriptions of NatTech.

Consider Isaac Newton's observation that planetary "gravitational" interactions could cause perturbations—tiny shifts in their orbits—that should, over time, destabilize the system. Yet, how does an ordered system like the solar system maintain its precise functions for billions of years despite such disturbances? Pierre-Simon Laplace demonstrated that subtle orbital resonances play a key role in maintaining this order.[317] [318] However, the long-term stability of such a complex and dynamic system remain a deeper mystery.

Newton could not have anticipated the extraordinary stability of the solar system or the accelerating expansion of the universe driven by the mysterious force of dark energy. Remarkably, the solar system, along with other gravitationally bound systems, remains unaffected by this rapid expansion, as if the delicate gravitational balances within it somehow shield it from the chaos unfolding across the universe. The fact that such sophisticated local systems can persist for billions of years, undisturbed by the vast and accelerating forces on the largest scales, demonstrates just how finely-tuned and resilient the cosmos truly is.

In 1989, Jacques Laskar, a senior researcher at CNRS at the Observatoire de Paris—PSL, determined that planetary motion exhibits chaotic behavior on a timescale of 5 million years and becomes unpredictable beyond 60 million years.[319] In 2008, his calculations further predicted that the likelihood of a collision between the inner planets (Mercury, Venus, Earth, and Mars) was just 1% over the next 5 billion years, comparable to the Solar System's current age.[320] More recently, in collaboration with Federico

Mogavero, a postdoctoral researcher at the PSL, Laskar revealed that the typical waiting time for a catastrophic event is actually much longer than the age of the universe.[321]

Laskar introduced the concept of quasi-integrals of motion—quantities like angular momentum or energy that are nearly, but not perfectly, conserved over time.[322] Quasi-integrals were discovered mathematically, but they describe behavior that was already present and observable in the real system—like the solar system—long before we had the math to model it. In theoretical physics, these quantities are expected to remain constant in so-called "ideal" systems—systems with no external interference, perfectly symmetrical forces, and no chaos.[323] However, such systems do not exist. They are conceptual fictions used to simplify equations.

Laskar's quasi-integrals, often treated as near-constants in a chaotic system, suggest that planetary motion is regulated by internal recalibration. Space functions as a structured system of relational alignment, preserving orientation across scale. What we call conserved quantities may reflect real-time stabilization—maintained through distributed interaction, much like inertial systems hold direction. Planetary alignment, too, reflects sustained structural order, not statistical coincidence. What appears as stability over time is the result of ongoing structural regulation within the lattice itself.

These quantities don't stay near-constant by accident—they're being actively regulated. Not by randomness, but by embedded rules. This regulation points to spatial memory and rotational logic. Like gyroscopic systems, the solar architecture holds its course through embedded orientation and continuous adjustment—stability is not maintained by force, but by internal logic that resists disruption.

To make this more tangible, think of a car on cruise control or a plane on autopilot. Despite encountering hills, turbulence, or crosswinds, these systems continually recalibrate, holding the vehicle steady.[324] Similarly, quasi-integrals act like onboard stabilizers for the solar system, ensuring coherence amid external interference.[325] But unlike manmade technology, they require no maintenance. They run—perfectly—across billions of years. It's like having a system that never sleeps, quietly preserving cosmic order.

The standard model explains planetary stability as the outcome of probabilistic processes, but it fails to recognize the self-regulating intelligence encoded within the system. These quasi-integrals are not just mathematical approximations—they behave more like embedded algorithms. Like biological homeostasis or ecological feedback loops, they sustain equilibrium through real-time correction, not chance.

Even the term "quasi" underplays their role. To call DNA a "quasi-code" would trivialize its function. Likewise, these quasi-integrals should be recognized for what they are: embedded system logic, recalibrating planetary interactions to preserve harmony. This is a visible expression of cosmic homeostasis—a self-correcting mechanism written into the structure of reality itself. Its signature appears not just in orbiting bodies, but in the continuity of systems across domains: biological, quantum, and cosmological. Quasi-integrals point to a logic that may bridge the deterministic framework of classical planetary dynamics with the probabilistic nature of quantum mechanics and gravity.[326]

## Only Materialist Can Understand It

It's important to note that aside from it being the perspective of the majority of cosmologists and astrophysicists—which we're forever reminded of—we have no solid reason to accept that an early universe that begins with limited fundamental elements, such as hydrogen and helium, behaved differently billions of years ago than they do today. As noted, physics tells us that these gases naturally disperse and remain diffuse unless acted upon by external forces. As far as we know, dust still behaves like dust, gravity remains consistent, and high-speed collisions continue to produce destructive outcomes. Yet, when it comes to explaining planetary formation, these principles are conveniently suspended. Processes that should lead to destruction or stagnation are instead reimagined to work flawlessly, resulting in the formation of countless planets and planetary systems.

It is this narrative that forms the cornerstone of the materialist creation story, often distilled into the familiar refrain: 'we are just stardust.' It cannot be overstated, the religious-like zeal with which many materialists cling to this view. For them, this nebulous creation story, along with Darwinism, is essential because it shapes their entire understanding of reality. Evolution, in their worldview, begins with this cosmic narrative, and so it all 'must' work. Even natural laws themselves are expected to conform to radical Materialism, and any speculation becomes plausible if it supports this narrative.

What's interesting is that many of the most improbable theories, along with the public's willingness to believe in them, often stem from opposition to religious implications and, by extension, lead people to unwittingly reject strong arguments simply because they reinforce religious perspectives. In turn, mainstream materialist theories, no matter how unfeasible, are viewed as intellectually superior because they forbid spiritual or religious implications.

Consequently, this dominant philosophy, often equated with science, is a strict belief system of its own, with the public assuming that there is some hidden knowledge that only materialist scientists truly understand. Ironically, by accepting these theories without question, people are inadvertently subscribing to a new form of religion while believing they are rejecting religion and more enlightened.

At the beginning of this book, I mentioned that Ken Ham's efforts to prove the physical feasibility of Noah's Ark may be counterproductive. But let's consider for a moment what it would look like if the Bible claimed that all miracles were feasible by natural means. We'd still be left with the Ark problem, the parting of the Red Sea problem, the three days in the fish problem, and the water-to-wine and walking-on-water problems. Each miracle would present insurmountable challenges to explain through purely naturalistic terms.

These types of "problems" are exactly what radical Materialism gets away with, repackaged in scientific language that, without any proposed agent, makes them even less feasible by *natural* means. The standard timeline of a universe coming to be from nothing isn't riddled with simple inconsistencies; it contains numerous fundamental contradictions to established physics.

One last note on a different but relevant topic: Sabine Hossenfelder, a German theoretical physicist, frequently calls out theoretical shenanigans.[327] A skeptical materialist, she often critiques the ingrained dogma within the scientific community.[328] She is outspoken about speculative fields, such as string theory, for failing to produce tangible understanding. In her YouTube video, "This is why physics is dying," she critiques string theorists, stating:

> Everyone who works on this just repeats arguments that they all know to be wrong to keep the money coming... This so-called research has been going on for four decades. And what's come out of it, besides papers? Nothing. Nada. Niente. It hasn't taught us anything about nature whatsoever... I don't know how it ever became accepted that inventing some maths and insisting it's real counts as theoretical physics. It's insane.[329]

Although her focus in this video is on string theorists and those working on quantum gravity, the broader implication of her critique is clear: many scientists knowingly promote flawed theories to sustain their positions and secure funding, rather than confronting the fact that these ideas have been debunked. Sabine's willingness to call out this intellectual dishonesty shows her commitment to authenticity admirably, even within a scientific field where others fear the repercussions of such criticisms.

At the beginning of this chapter, I referenced Richard Dawkins' remarks from an interview with Piers Morgan, in which he claims that "only physicists" are capable of grasping the complexities of the universe, dismissing the ability of ordinary people to comprehend such topics. Dawkins' thought-policing statement discourages non-physicists from questioning or engaging with these ideas, encouraging a passive reliance on authority instead of fostering critical thinking.

His position contradicts the philosophy of Richard Feynman, one of the brightest physicists of the last century, who believed that scientific concepts could be made understandable to everyone. Feynman, a Nobel laureate often called "The Great Explainer," encouraged people to approach complex ideas with curiosity. He believed that with enough effort, anyone could grasp even challenging aspects of science. In his words:

> You ask me if an ordinary person, by studying hard, would get to be able to imagine these things, like I imagine? Of course! I was an ordinary

person who had studied hard. There are no miracle people. It just happens they got interested in these things and they learned all this stuff... So, if you take an ordinary person who's willing to devote a great deal of time, study, work, thinking, and mathematics, then he's become a scientist. [330]

Dawkins' notion that certain concepts are unreachable for most people is fundamentally different from Feynman's philosophy of learning. Where Dawkins seems to close the door on public engagement, Feynman opens it wide, insisting that the process of learning and studying is accessible to anyone willing to put in the effort. Feynman is often attributed with the quote: "if something couldn't be explained simply, it wasn't understood well enough." Perhaps this is the case with the physicists whom Dawkins seeks to shield from scrutiny.

By claiming that "our brains were not built to understand the profundities of the universe," Dawkins discourages people from realizing that many speculative materialist theories are not supported by reality-grounded evidence. Feynman's exhortation to seek scientific understanding promotes challenging these theories, asking the necessary questions, and engaging with science on more fully level. His message is clear: understanding is within reach for those who are willing to put in the work, making science a field where all can participate meaningfully.

# CHAPTER 15

# Our Satellite

*Moon and Earth and Earth and Sun are locked in the rotational motion. In such motion the mutual gravitational attraction is balanced by the centrifugal force resulting from the rotation around the common center of gravity (barycenter).... Stability of the Earth-Moon system will require that the sum of all centrifugal and attraction forces should be zero. While this statement is true for the centers of the Earth and Moon, the balance does not occur in every point, thus leading to the forces generating tides.... As is well known, the centers of gravity of Earth and of Moon are moving around the common center of gravity (barycentre). Actually, this system is a twin planets system moving around the common center.*[331] Zygmunt Kowalik & John L. Luick

For centuries, the Moon has fascinated astronomers, yet its origin remains a mystery. The Earth and Moon move in a precisely coordinated system, bound by gravitational attraction and centrifugal forces that keep them in balance around a shared center of mass.[332] This synchronized motion ensures their stability, yet variations in force distribution create dynamic effects. The Earth itself follows a similar pattern in relation to the Sun, maintaining a steady orbital path defined by the same fundamental principles.[333] Not just the typical idea of a satellite orbiting a planet, the Earth and Moon function as a dual-body system, each influencing the other as they revolve around a mutual point in space.

Despite decades of research, no single theory has provided a fully satisfactory explanation. In reality, no materialist origins theory, in any area, has provided a fully coherent explanation based on the facts we know. However, this has never stopped their assumptions from being embedded in textbooks and presented to students as settled science.

Consider this passage from an OpenStax.org astronomy textbook:

> It is characteristic of modern science to ask how things originated. Understanding the origin of the Moon has proven to be challenging for planetary scientists, however. Part of the difficulty is simply that we know so much about the Moon (quite the opposite of our usual problem in astronomy). As we will see, one key problem is that the Moon is tantalizingly similar to Earth and frustratingly different.[334]

The statement, "Part of the difficulty is simply that we know so much about the Moon," is counterintuitive. Typically, the more information we have, the clearer our understanding becomes. Yet these authors are frustrated by the abundance of knowledge.

This is *textbook* cognitive dissonance. Their frustration stems from the assumption that the evidence must conform to the standard linear framework, even as mounting evidence increasingly exposes its incompatibility. It exemplifies materialistic indoctrination, where students are taught to treat challenges to orthodox origins theories as inconvenient problems to solve rather than opportunities to rethink foundational assumptions. Evidence should not be framed as a hurdle to fit within a constrained framework, but as an invitation to explore new possibilities.

Various theories try to unravel how the Earth-Moon celestial relationship came to be, each presenting a linear perspective founded in chaos where this system falls in place like perfectly lined-up dominos. From the Fission Theory, which suggests the Moon split from the Earth, to the Giant Impact Hypothesis, proposing a colossal collision, each idea introduces its own set of unresolved questions regarding the Moon's specific origin and formation process.

**The Fission Theory**

Let's examine the fission theory, for example. First proposed by Charles Darwin's son, George H. Darwin, this theory suggests the Moon was once part of Earth and separated early in our planet's history.[335] Imagine a spinning object ejecting a piece of itself, which then comes to orbit the body it was ejected from.[336] This is the scenario that fission theory explains. Though this concept might seem plausible at first glance, closer examination reveals significant flaws in its logic.

If the Earth needed to draw material inward to form its core, it would have spun faster due to conservation of angular momentum. Some might argue that this faster spin enabled material to fling off and form the Moon. However, a faster spin would have made it harder—not easier—for material to overcome Earth's gravity and escape. Even if material were flung out, it would not form a clean, cohesive body like the Moon. Instead, it would result in a chaotic spray of fragments, likely creating an uneven disk of debris around Earth. Moreover, the rotational force required to eject material would have been so extreme that the Earth itself could have torn apart.

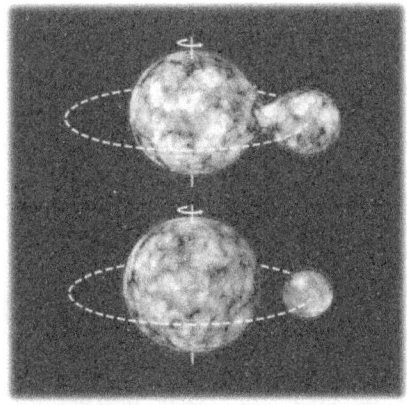

Figure 16.1 Artistic impression of the Moon's formation through the fission theory, suggesting it split from Earth's crust. **Image source:** ESO/José Francisco (josefrancisco.org).

Furthermore, if Earth was spinning that quickly, the day would have been incredibly short, lasting only about 2 to 3 hours long.[337] If the Earth had been spinning fast enough to generate a Moon-sized chunk, the Moon should have carried off a significant amount of that momentum. However, when scientists calculate the angular momentum of the Earth-Moon system, the numbers don't add up.[338] As the OpenStax authors note, the Moon remains "tantalizingly similar to Earth and frustratingly different," a contradiction that continues to puzzle planetary scientists. This unresolved complexity weakens the case for a fission-based or flinging origin.

## The Double Planet Hypothesis

Moving away from the fission theory, we encounter the double planet hypothesis, which paints a different picture of mutual origin, suggesting that Earth and the Moon evolved together from a shared birthplace in the cosmos. This hypothesis posits that Earth and the Moon formed together in a localized process within the solar system's accretion disk, supposedly explaining their unique gravitational relationship. Like the fission hypothesis, this double planet scenario also struggles to account for the significant compositional differences between Earth and the Moon.[339] Despite supposedly forming from the same material under similar conditions, the two bodies exhibit traits that are too distinct to support such a close relationship.

## The Capture Theory

The capture theory presents a narrative where the Moon, a wanderer in space, was ensnared by Earth's gravitational embrace. This theory, statistically improbable, requires an exact synchronization of velocities and orbital mechanics.[340] It would be like catching a wandering celestial body in a perfectly orchestrated cosmic ballet, a scenario that stretches the limits of probability.[341]

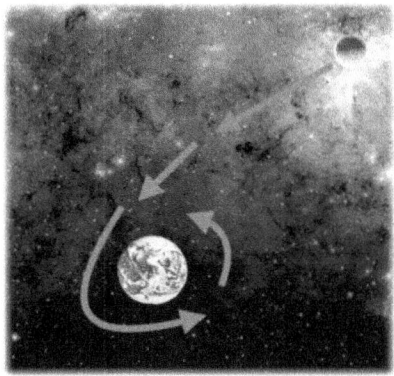

**Figure 16.2** This diagram depicts a smaller Moon on its initial trajectory toward Earth, before being gradually captured and stabilized into orbit around the planet, illustrating one of the hypothesized scenarios for the Moon's origin. **Image source:** S. A. Cooper.

Imagine a figure skater gracefully spinning at the center of a vast ice rink, executing perfect revolutions. Now, picture another skater entering the rink at a considerable speed, determined to join the performance. This newcomer must slow down precisely to match the pace of the spinning skater while always keeping eye contact and facing in the same direction as the spinning skater.

Such a synchronized dance is plausible for skilled skaters who have agency and control, but it becomes implausible if we extend the analogy to celestial bodies. The capture theory proposes that the Moon, like the second skater, was drawn into Earth's orbit, perfectly synchronizing its movements with Earth's rotation—however, without agency or intent.

## The Giant Impact Hypothesis

Turning to the giant impact hypothesis, we delve into a scenario of cosmic violence: the idea that the Moon was born from a cataclysmic collision between Earth and a Mars-sized body, which resulted in the debris that formed the Moon.[342] This theory, while currently the most favored, is possibly the most flawed, and it opens up questions about the specific dynamics of such an impact and the composition of the resulting Moon.[343]

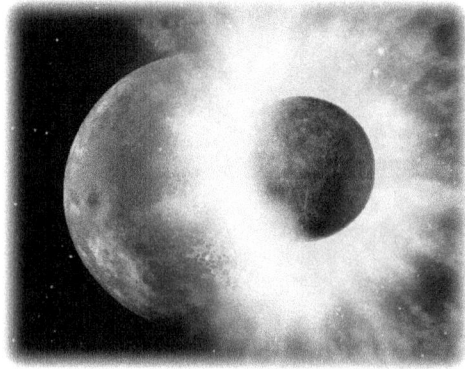

**Figure 16.3** "This artist's concept shows a celestial body about the size of our moon slamming at great speed into a body the size of Mercury."—NASA. **Image source:** NASA/JPL-Caltech.

Consider the interplay of forces and the sheer improbability involved in these theories. If such an impact had occurred, why didn't the material liquefy and coalesce, as thought to occur in the formation of other planets? Instead, we are left with a Moon that is both too similar and too different from Earth in ways that defy standard planetary formation models. This is a 'something is better than nothing' approach designed to offer a convenient take-home explanation.

**Magic Realism**

The Earth-Moon relationship only adds to the countless problems within the materialist paradigm. When discussing the Moon's origin, materialists treat this relationship as a byproduct of chance rather than an essential factor in the equation. Each hypothesis, from the Fission to the Giant Impact Hypothesis, demands a set of circumstances so precise and conditions so exact that they exist only as thought experiments. They read more like *magic realism* than serious science. However, the goal is not to construct the most imaginative scenario for how the Earth and Moon came to be, but to consider their tightly intertwined, interdependent relationship.

To illustrate this misguided approach, imagine a thermostat in the home of a fragile elderly lady. This thermostat maintains a comfortable climate, protecting her from life-threatening heat waves in the summer and freezing temperatures in the winter. Now, picture trying to explain how this thermostat came to be purely by random causes, ignoring how it crucially relates to her survival. You'd end up with a narrative that misses the point entirely. We cannot fully comprehend anything in the cosmos by isolating it from its intrinsic relationships.

As the widely used textbook I quoted earlier suggests, evidence that resists a purely materialist explanation remains a frustrating problem for many. There isn't an Earth-Moon "problem;" the real issue is Materialism. The Moon's role in Earth's history has had an active influence on the planet's ability to sustain life. The gravitational interaction between Earth and the Moon has been a key factor in the relatively mild climate fluctuations that have allowed life to thrive.[344] With its synchronous rotation always showing the same face to Earth, the Moon presents an intriguing aspect of this celestial puzzle.[345] This synchronization is not just a curious coincidence but a critical factor in regulating our planet's wobble and overall axial stability.[346] Without the Moon's stabilizing influence, Earth's tilt could have varied wildly, leading to extreme and rapid climatic changes, hindering the existence of complex life.[347]

The Moon's gravity drives the ebb and flow of the ocean's waters, stirring the oceans and mixing nutrients and gases, which influence marine life and the global climate.[348] The precision of this arrangement points towards a relationship between Earth and the Moon that transcends the simplistic explanations offered by orthodox theories.

Additionally, from a scientific standpoint, the Moon has been a laboratory for studying planetary geology, the effects of impacts, and the history of the solar system. The samples brought back by the Apollo missions have provided invaluable data about the Moon. Its impact is also evident in the less tangible aspects of human life. Throughout history, the Moon has served as a source of inspiration and wonder, influencing cultures, religions, and arts. It also has been a guide for farmers, playing a significant role in the development of civilization. Its influence is woven into the fabric of human existence, from ancient rituals to modern-day festivals.

Let's delve into the relationship between human ingenuity and the NatTech found in our universe by examining some of the parallels between the Earth-Moon relationship and our derivative technology. Our focus shifts to the artificial satellites orbiting our planet. These products of human ingenuity reflect principles similar to those observed in the natural dynamics of the Earth-Moon system. It's worth noting that the word "satellite" itself is also a term for a moon—our ancient satellite existed long before we launched our own. This linguistic overlap is more than coincidental; it's yet another reflection of how our technology often mirrors natural systems.

Just like the Moon orbits Earth in a consistent, predictable path, geosynchronous satellites are positioned to maintain a fixed position relative to the Earth's surface.[349] This allows for uninterrupted communication, weather monitoring, and global positioning services, mirroring the Moon's constant presence in the night sky.[350] The engineering behind placing a satellite in geosynchronous orbit involves a precise understanding of gravitational forces, orbital mechanics, and the delicate balance required to maintain a stable orbit.[351] These are the same principles that regulate the Earth-Moon system.

Now, consider the James Webb Space Telescope. Its construction and positioning serve deliberate engineering purposes. Similarly, the Moon's constant orientation toward Earth, its distance, and its composition indicate that such configurations are not arbitrary. Webb's sunshield is always positioned between the Sun, Earth, and Moon to shield the telescope, a feat made possible by its orbit at the second Lagrange point (L2), 1.5 million kilometers from Earth.[352] This deliberate placement ensures optimal function—just as the Moon's orientation provides stability in its relationship with Earth.

Yet, despite the precision with which we strategically place artificial satellites and telescopes, the significance of the Earth-Moon relationship is overlooked, treated as incidental rather than integral to the Moon's origin.

# CHAPTER 16

# Anonymous Functions

*Somebody can call it 'new physics' on the time scale 1×10⁻⁴³ with no ticking clocks. The Model is even more complicated due to numerous parameters accounting for 'tensions' between observations and theory, vaguely justified by fictitious fitting, which makes the Model non-refutable by established theoretical means. ... Sadly, 'Dark Matter' was not theoretically predicted at all.[353]* —Kirill Vankov & Anatoli Vankov*

## Gravity as System Logic

Scientists originally proposed the concept of "dark matter" when they noticed something intriguing about how galaxies spun. Based on the visible matter within galaxies—such as stars and gas—they were expected to rotate at a certain speed. However, scientists discovered that galaxies spin much faster than anticipated, yet they don't fly apart. This revealed a mismatch between galactic motion and the gravitational effects predicted by scientists based on visible matter.[354] To resolve this, they hypothesized the existence of dark matter: an invisible substance that would provide the additional gravity needed to hold galaxies together.

The term *dark matter*, however, is rooted in a materialist presupposition and is, in fact, a misnomer. A more precise and philosophically neutral way to frame this mystery would be to say: "There are phenomena in the universe that currently cannot be explained by known material causes." This avoids prematurely committing to any unobservable entities and keeps the focus on what the data actually reveals. Referring to it as dark matter misleads people into thinking we've identified the cause, when in reality, dark "matter" is a materialist placeholder for something that clearly does not have a material basis. By definition, matter refers to something that has physical substance, interacts with forces like gravity, and obeys the laws of physics. However, these effects are interpreted through gravitational forces, as if its being matter is a given.

Consider all the matter and energy in your home: the walls, ceilings, furniture, and appliances—the refrigerator, dishwasher, air fryer, TVs, and lightbulbs, all of which need power to run. Now, picture a sharp-eyed relative reviewing your energy bill and noticing something strange: you've been charged for 27% more furniture than you can actually see, and 68% of the energy does not show up on the meter readings.

When you ask why your bill is so high, the company explains, "Oh, this charge is for the accumulation of dark furniture or energy—things we know must be there because we're missing them: even though you don't see or use them in the traditional sense." When you respond with a puzzled, "Huh? What?" they assure you it's too complex to explain but insist their calculations are correct.

Naturally, this explanation leaves plenty of questions unanswered. For starters, who granted this company the authority to claim their resources as the source of this untraceable furniture and stream of energy? This is similar to how materialists discuss dark matter and dark energy, particularly when they assume that dark matter is, in fact, matter, and that dark energy behaves like measurable energy, even though neither is directly observed.

Even when we observe real matter, we don't truly know that it is the source of the effects we attribute to it. We only know there is a correlation between observed mass and certain motions. The idea that "matter causes this" is an inference, not a direct or proven mechanism. The materialist assumes that real—or their make-believe—matter is the cause of cosmic effects, simply because that's what matter is believed to do.

But gravity, inertia, and cohesion are not seen arising from matter—they are interpreted that way through models. Materialists assume that they arise from matter because their framework demands material causes and because historical models linked matter to motion and structure. The foundational premise that matter inherently generates these effects itself is incorrect or incomplete. In that case, dark matter isn't just unproven; it's built on a chain of assumptions that collapse under the right questions.

Now, picture dark matter as an invisible spider web spanning the universe—unseen yet omnipresent. It forms the essential framework that supports the universe's structures, similar to the beams hidden behind the walls of a house.[202] [203]

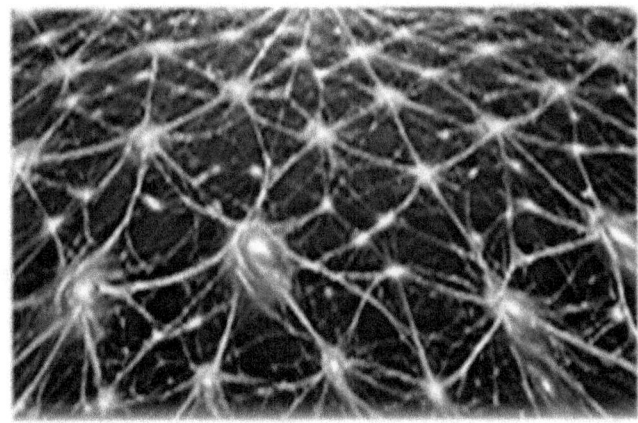

**Figure 12.1**

Visual representation of the cosmic orientation matrix—commonly mislabeled the "dark matter web." Rather than depicting invisible matter,
this structure reflects a coherence lattice: a grid of embedded alignment nodes sustaining large-scale cosmic order through orientation, not gravitational pull.
**Image Source:** S. A. Cooper.

Its presence, holding galaxies together, is akin to how a spider's web holds objects in its strands—or how a spider spins and secures its prey. But rather than an unseen mass exerting gravitational force, this cosmic structure may be better understood as a coherence structure: a network of orientational tension woven into the fabric of space.[355] The so-called "dark filaments" do not merely bind matter—they channel coherence, aligning galaxies and clusters through timing and orientation rather than force. Much like a field of tension orients the movement of interconnected nodes, this lattice structures the universe not through mass, but through informational alignment.

## Lambda Functions

In further contemplation of the intersection between human-derived technology and the natural technology of the cosmos, an intriguing parallel takes shape with "lambda functions" in programming.[356] Although they share a name with the cosmological constant, these concepts arise from distinct fields. Lambda functions—derived from Alonzo Church's Lambda Calculus—embody abstraction by performing tasks based on input without requiring explicit naming.[357] Similarly, aspects of the natural world, though not labeled "technology," are understood through their functionality and structure.[358]

Lambda functions let programmers write concise, efficient code that performs tasks based purely on input, without explicit labels. Similarly, natural phenomena operate according to underlying principles without needing explicit names. For instance, a lambda function might sort data or filter values within an algorithm, similar to how fundamental forces work seamlessly within the cosmos. We can see how fundamental forces and constants work without needing to be explicitly "named" but are identified through their effects and interactions.[359]

Dark energy follows a similar pattern. Like an anonymous function, it operates behind the scenes to adjust the universe's expansion, a critical functionality programmed into the cosmos from its inception. Decoding dark energy isn't just about predicting how the universe ends—it's about uncovering a hidden layer of its technological structure. Similarly, dark matter can be understood as an anonymous effect, shaping the universe's structure through the synchronized choreography of stars, planets, and galaxies.[360] It dictates the universe's structure, reminiscent of a suspense movie where something mysterious remains hidden in the shadows, unseen yet fundamentally influential.[361]

## Inference to the Best Explanation

The way science is carried out today doesn't reflect how explanation traditionally worked. Before Darwinian materialism became the dominant lens, structure didn't need to be rescued—it was the starting point. Coherence was assumed, not reverse-engineered. Systems were expected to behave like systems, not patched assemblies of hypothetical components added to fix broken predictions. Coherence wasn't something simulated and patched over with speculative particles. It was assumed, built into the architecture of the system itself.

You didn't need five different papers and six adjustable parameters to explain why things moved the way they did. You needed a principle. A logic. Something embedded. Now, the explanations never stop—because nothing can be pinned down. Each failure spawns another paper, with each contradiction more Patchganda. The system isn't being clarified—it's being kept alive by volume.

Early cosmological models—however wrong in specifics—were at least committed to the idea that structure arose from structure. That form had cause. That orientation, rhythm, and balance were not emergent side effects of invisible mass, but evidence of a system behaving as a system. The shift to Darwinian materialism changed that. Now coherence must be explained

159

post hoc, by things that can't be found, behaving in ways they were never observed to behave. Structure isn't assumed—it's denied, until the model breaks. Then it's borrowed back in pieces.

If I were to throw in my bid for a more coherent explanation—something better than dark matter, and more structurally grounded than gravity as it's typically framed—it would be a Gyroscopic Lattice: a spatial framework in which orientation, inertial balance, and internal spin regulate stability across scale.[362] This is not a system of gravitational pull, but one of embedded rotational structure. Each node behaves like a gyroscope: locally stable, globally synchronized.

For continuity with existing language, I'll refer to these as Dark Gyroscopes. Dark Gyroscopes are not clouds of hidden mass invented to preserve a mass–motion correlation—they are structured components embedded in space, stabilizing cosmic architecture through internal logic, not external attraction. Unlike gravitational halos—invoked to explain galactic rotation curves and large-scale structure—this interpretation points to regulation, responsiveness, and balance arising from within the system itself.[363] If space contains an embedded stabilizing logic—a framework that behaves like a network of gyroscopes—then galactic cohesion, orbital consistency, and rotational behavior might all be regulated structurally, not caused by mass.

In contrast, the conventional dark matter halo model is both conceptually vague and functionally inert. In current theory, a halo refers to a massive, invisible region said to surround galaxies, invoked solely to supply gravitational force that visible matter cannot. These halos are treated as collisionless, pressureless, and non-interacting—contributing nothing to the system's dynamics beyond their supposed gravitational field. They have no feedback, no organizing logic, and no explanation for their shape, distribution, or orientation.

Even within its own framework, the halo model presupposes what it cannot explain. It assumes dark matter clusters in ways that match observed velocities—but cannot say why these formations are rotationally coherent, or how their angular momentum is so precisely distributed. It treats structure as the passive result of accumulated mass, without asking what sustains such cosmic regularity. A gravitational halo cannot generate coherence. It only asserts that unseen mass must be present because motion demands it. That is not a mechanism. It is a circular placeholder: motion infers mass, and mass is invoked to explain the motion.

The standard cold dark matter framework, as formalized in Navarro, Frenk, and White's widely cited paper, does not build coherence from explanation—it rescues it with post-hoc fixes. It is not a discovery, but a response to theoretical breakdown—a placeholder introduced to explain why galaxies rotate faster than their visible mass allows. As contradictions accumulated, such as the problem of early structure formation, the placeholder evolved. It had to clump. It had to be cold. These traits were not measured—they were assigned to make the model work. The name changed. The behavior shifted. The substance remained undetected.

Cold dark matter is not a refinement of evidence—it is a narrative under pressure. It behaves like an octopus: adaptive, ungraspable, and always reshaping itself to avoid capture. Wherever the theory breaks, it flows in. It is not cold. It is convenient. Its predictive failures are not treated as falsifications, but as prompts for new speculation. Where galaxies don't form, the theory assumes they failed to ignite. Where mass-to-light ratios don't fit, it assumes unknown trends in disk structure. Where dwarf galaxies contradict the expected halo profile, it assumes galaxy formation altered the halo after the fact.

Navarro, Frenk, and White: "CDM halos are too concentrated to be consistent with the halo parameters inferred for dwarf irregulars... This may imply that the core structure of dwarf galaxies was altered by the galaxy formation process." This is not explanation. It is retroactive justification. The contradiction is neutralized by suggestion—not resolved by theory.

On the mismatch between the predicted number of halos and the observed number of galaxies, they write: "The predicted abundance of galaxy halos is larger than the observed abundance of galaxies... This may imply that galaxies failed to form (or remain undetected) in many dark halos." Here again, the theory overpredicts and explains the gap by invoking what cannot be seen. The observational deficit is translated into a metaphysical surplus.

On failed rotation curve predictions, they write: "Matching the observed rotation curves... requires disk mass-to-light ratios to increase systematically with luminosity." Rather than revise the model, they adjust the galaxy to fit it. The rotation curve is not predicted—it is calibrated. The theory does not model galaxies. It models itself. What appears to be precision is not explanation but calibration. And calibration, by nature, implies engineering. Agency. Constraint. Structure. These are not absent from the model—they are already at work, only deployed in reverse: not to discover structure, but to prevent collapse.

The model borrows the vocabulary of design while denying its logic. It reverse-engineers the data to preserve its assumptions, then insists those assumptions remain untouched. It fails not from lack of structure—but from its misapplication.

In stark contrast, the Dark Gyroscopes framework presents a different causal architecture. Where the standard model adds assumptions to fix contradictions, mine reveals that no fix is needed. The structure isn't failing—it's misunderstood. It isn't missing mass—it's missing logic. A gyroscopic system is defined not by static mass, but by orientation, resistance to perturbation, conservation of angular momentum, and spatial memory.

This is not theoretical excess. In spacecraft, inertial guidance systems maintain orientation without external force. They predict, correct, and regulate motion across time. If this principle scales to the cosmic web, then structure arises not from gravitational clumping, but from systems that maintain coherence through orientation memory and rotational logic.

This view also aligns with observable structure. The coherent rotation of galaxies, even across clusters, is not predicted by mass accretion. Nor is the filamentary pattern of the universe, which displays directional alignment, consistent spacing, and long-range rotational continuity. These are not the traces of chaotic collapse. They are the signatures of constraint and embedded order.

Conservation of angular momentum holds at every scale—from electrons to galaxies—yet the halo model offers no explanation for spin. Halos are silent. A gyroscopic framework makes spin foundational. It treats rotation not as a result, but as the organizing principle. The "Dark Gyroscope" is not an object—it is a function performed by a node in a spatial system. It does not pull. It orients.

To continue defending the halo model while dismissing dynamic alternatives is to guard contradiction. Halos are relied on to explain invisible forces, but any model that reframes those forces structurally is ignored. This is not scientific caution. It is philosophical inertia.

What if matter doesn't cause gravity—but what we call gravity arises from a gyroscopic-like system that organizes space and anchors matter into functional relations? Gyroscopes don't pull. They stabilize. They resist disruption. If embedded at cosmic scale, they could generate the relational effects we call gravity—not as force, but as coherence. In inertial systems, gyroscopes regulate motion without external reference—functions that mirror the inferred role of dark matter. Even frame-dragging in relativity suggests that spin structures space.

Recent work by Ali Seraj introduces "orientation memory," where freely falling gyroscopes retain a permanent record of spacetime deformations caused by gravitational waves. Independently, Ciufolini and collaborators demonstrated persistent precession in satellite orbits through frame-dragging. Both show that orientation and rotation are not passive—they are active agents of structure.

If such memory is embedded in local gyroscopic systems, it strengthens the case that orientation logic scales cosmically. In this view, mass does not create curvature. It is a node in a deeper logic of spin, memory, and relational order. If this holds, then spin and orientation are not afterthoughts; they are ontological primitives.

After all that's been said, it's worth pausing to consider the kind of inference that's accepted without question. Consider the universe as a kind of cosmic city, like the one depicted in the 1933 film *The Invisible Man*.[364] In this narrative, the invisible character's existence is inferred through his clothing. Similarly, in our cosmic metropolis, galaxies and stars function as markers, revealing the presence of an unseen force.[365] The reliance on these 'invisible' components, such as dark matter and dark energy, believed to make up about 95% of the universe, forces us to accept that most of reality exists beyond direct observation.[366]

This discussion lends weight to arguments for intelligent design. If materialists assume that only material causes can explain dark matter's role in planetary restraint or movement, since matter is *thought to be* the only known source of such effects, it seems inconsistent to dismiss Meyer's claim that information, whose only known source is a mind, must also have an intentional origin. Similarly, Behe's argument for irreducibly complex systems, which from our experience always reflect the work of an engineer, further underscores this inconsistency. Ignoring one unobservable source while embracing another reveals a double standard that is unjustifiable.

Which sounds more like pseudoscience: inflationary theory, which invokes an imaginary field to expand from nothing to an entire universe—faster than the speed of light, in a fraction of a second? Or the claim that the kind of specified, functional programming we observe in the cell only has one known source—a mind?

CHAPTER 17

# Ambient Technology

*Awesome nature appears to have some very distinct (positive) effects on moods, emotions, and prosociality... which diverge from the effects obtained for more mundane types of nature. Regarding interventions, it is obviously far from evident to bring people into contact with actual awesome nature on a regular basis (because of its uncommon and often inaccessible character). However, as our results show, already brief exposure to relatively small images of awesome nature... may have significant positive effects on people's emotions and behavior.*[367] —Yannick Joye & Jan Willem Bolderdijk

Now let's consider another wonder: the dynamic weather patterns of our planet. The weather on Earth isn't just a daily concern or a topic of small talk; it's an integral part of the planet's life-sustaining system.[368] From the graceful dance of winds to the majestic formation of clouds, atmospheric phenomena play a vital role in shaping the conditions that make Earth habitable, influencing all aspects of life here.[369]

Imagine a cosmic game of poker, where each planet in our solar system was dealt a hand dictating its atmospheric conditions and potential weather patterns. Earth was dealt a royal flush. Its breathable atmosphere and dynamic weather offer a wide variety, from hot beaches to cool mountains, providing an exceptional experience. With its unique ability to sustain life and offer diverse environments, Earth stands apart among all known planets, like a city set on a hill.[370]

In this game, Mars got a bad hand.[371] The planet endures a relentless environment, shaped in part by its remarkably thin atmosphere, which is about 100 times less dense than Earth's. This scarcity of atmospheric density leads to a host of stark conditions: there's no rainfall to speak of, and the surface is subjected to intense dust storms that can last for days.[372] The thin atmosphere also means less protection from solar radiation and extreme temperature fluctuations, making Mars a challenging frontier for exploration.

Venus was dealt a hand too hot to handle, with an atmosphere that faces conditions too harsh for conventional exploration. Its atmosphere, composed predominantly of carbon dioxide, acts as a thick blanket, trapping heat to create an environment with hellish extremes.[373] This greenhouse effect elevates the planet's surface temperatures averaging around 870 degrees Fahrenheit (about 465 degrees Celsius), levels higher than any other planet in the solar system, including Mercury, despite Venus being further from the Sun. Such intense heat, coupled with a crushing atmospheric pressure, makes Venus a world where survival is beyond imagination, painting a picture of a planet that is intriguing yet inhospitably hot.[374]

And then there's Jupiter where the conditions are anything but hospitable.[375] Dominated by a thick atmosphere primarily composed of hydrogen and helium, Jupiter is perpetually stirred by immense storms and powerful winds, the most famous being the Great Red Spot, a gigantic storm larger than Earth, persisting for centuries.[376] Its rapid rotation, completing a day in just about 10 hours, further fuels its dynamic weather patterns, with winds in the upper atmosphere reaching incredible speeds.[377] Unlike Earth, Jupiter lacks a solid surface; descending into its depths, one would encounter increasingly hostile conditions, with escalating pressure and temperature leading to a layer of metallic hydrogen.[378] This unique and hostile environment, characterized by high-speed winds, massive storms, and a vastly different composition, renders Jupiter exceptionally challenging for exploration.

As for Mercury, it faces extremes of temperature, being scorched by the sun during the day and freezing at night, due to its lack of a substantial atmosphere.[379] Saturn, renowned for its magnificent rings, endures harsh conditions with fast-moving winds and violent storms, including a persistent hexagonal storm at its north pole, all within an atmosphere primarily composed of hydrogen and helium.[380] Uranus, on the other hand, experiences drastic seasonal changes and maintains a frigid atmosphere made up of hydrogen, helium, and methane.[381] Neptune, the most distant known planet in our solar system, is characterized by some of the fastest and most violent winds, despite its extreme distance from the Sun, making its environment extremely cold and turbulent. [382] These planets' harsh environments speak to the broader reality of a solar system largely resistant to life as we know it.

Earth, with its atmosphere of 'just-right' gases, has the luxury of a dynamic weather system that distributes water, regulates temperature, and even disperses essential minerals. Consider this: our weather system acts like Earth's circulatory system.[383] [384] Similar to how blood circulates nutrients, oxygen, and cells essential for life, Earth's weather patterns circulate water, modulate temperature, and spread life-supporting elements across continents.[385] Water evaporates, forms clouds, and travels to various parts of the world before falling as rain, essentially connecting distant ecosystems and supplying life's most fundamental need: water.[386] Isn't it remarkably convenient that this "circulatory system" is both dynamic and finely balanced?

One might argue that the weather is not always benign; after all, hurricanes, floods, and droughts are no one's idea of convenience. These might seem like inconvenient or even destructive elements, but they are integral parts of Earth's ecological and geological fabric. Each extreme weather event plays a role in rejuvenating, reshaping, or revitalizing some aspect of our planet. Forest fires, for instance, play a crucial role in nutrient cycling and habitat renewal.[387] Hurricanes and storms churn oceans, facilitating nutrient distribution and temperature regulation.[388] Floods spread nutrient-rich sediments, creating fertile lands for diverse ecosystems.[389]

Weather, therefore, is a key actor in the cosmic play of Earth's existence. And though we might complain about a rainy day or wish for a white Christmas, it's worth reflecting on the incredible complexity that underlies the weather patterns we so often take for granted. Is Earth's dynamic weather just a lucky break in a cosmic game of chance, or could it be part of a hand deliberately set by the dealer in a far grander game?

In our consideration of Earth's complex weather systems, we can draw compelling parallels with advanced technological constructs, such as feedback control systems, networked computing systems, and predictive analytics. These technological marvels offer a window into understanding Earth's weather as a form of biospheric technology.

Consider feedback control systems, widely used in everything from climate control in buildings to maintaining equilibrium in industrial processes.[390] These systems continuously monitor and adjust various parameters to maintain a desired state. Earth's weather operates in a strikingly similar manner. It functions as a natural feedback loop, where atmospheric conditions such as temperature, humidity, and wind patterns are constantly adjusted, much like a thermostat regulating a room's climate.

Moving to networked systems in computing, particularly in the field of cloud computing and distributed networks, we observe a web of interconnected nodes processing and relaying information.[391] This is analogous to the interconnected nature of Earth's weather systems. Atmospheric phenomena do not occur in isolation; they are the result of a complex interplay of various factors and influences, much like data flowing through a networked system.[392] Changes in one part of this 'network', such as ocean currents or air pressure differences, can have cascading effects globally, mirroring the interdependent operations of a distributed computing network.[393]

Predictive analytics in technology, which involves algorithms analyzing data to predict future trends and events, offers a compelling comparison to meteorology.[394] Meteorologists use sophisticated models and algorithms to forecast weather, analyzing patterns and data much like predictive analytics software anticipates market trends or consumer behaviors.[395] This process involves learning from past patterns, adjusting predictions based on new data, and improving accuracy over time.[396] This suggests that weather, in its essence, is operating on a set of natural 'algorithms,' constantly processing and responding to environmental inputs to maintain the Earth's equilibrium.[397]

Looking below ground, deeper into Earth's structure, we can see its layered core, mantle, and crust resemble a natural technological system. The core, which generates a magnetic field similar to radiation shields in engineered systems, is essential for life because it protects the planet from harmful solar and cosmic radiation.[398] This parallel shows that Earth's protective mechanisms function with the same purposeful design seen in human-engineered shielding.[399] The mantle's role in plate tectonics and the hydrological cycle's efficient water management reflect systematic, intentional design.[400]

**Figure 17.1** Layers of the Earth - A diagram illustrating the various geological layers of the Earth, including the crust, mantle, and core. **Image source:** A. Shteiwi, Wikimedia Commons.

Earth's core is a bit like the engine room of a grand ocean liner, unseen but necessary for the journey. According to researchers Yang and Song, this hidden engine is spinning in its own eccentric dance beneath our feet. They've found that unlike the steady, predictable movement of the Earth's surface, the inner core has its own unique rhythm, occasionally speeding up or slowing down, like a conductor subtly altering the tempo of an orchestra mid-performance.[401] [402]

Imagine that deep below the surface, this colossal sphere of iron and nickel is not just sitting idly; it's whirling in a way that could affect everything from navigation systems to the migration patterns of birds, all due to its impact on Earth's magnetic field.[403] [404] The dynamics of this process are complex. The inner core, floating in the liquid outer core, isn't bound tightly but can move freely, and its movements are essential. The heat from the inner core drives convection currents above it, churning the molten iron around it, which in turn powers the magnetic field.[405] It's a sophisticated, natural mechanism that could rival the cleverest inventions in its efficiency and impact.

## Why is the Sky Blue?

Let us move back above ground and look skyward to address something we all notice but rarely consider thoughtfully: why is the sky blue? The sky's blue hue has long been a subject of poetic inspiration and artistic expression, but its color is a consequence of a process known as Rayleigh scattering.[406] When sunlight, a spectrum of various colors, enters Earth's atmosphere, it interacts with air molecules, scattering shorter wavelengths like blue and violet in all directions. The sky appears blue to us because our eyes are more sensitive to blue light. Thus, the sky's blue hue, resulting from this scattering, is part of a larger, advanced system.

That explains the science—it tells us how the sky is blue. But the question remains: why is the sky blue? In any well-woven story, the setting is more than a backdrop; it's an active participant, shaping the narrative in a meaningful way. Similarly, Earth's atmosphere isn't just a passive layer of gas; it's a system that plays an essential role in the survival and flourishing of life on our planet.[407] These features not only fulfill essential biological functions but also meet human emotional and aesthetic needs, addressing both physical and metaphysical necessities.

**Figure 17.2** This image is of the vivid blue sky typical of Balneário Camboriú, Brazil, offering a backdrop that enhances the city's coastal beauty. **Image source:** Panoramio upload bot, Wikimedia Commons.

The expansive blue sky, stretching over landscapes and waters, creates a unique visual experience that evokes specific emotions. Research in environmental psychology suggests that gazing at the sky can promote relaxation, reduce stress, and enhance mindfulness, aligning with findings in color psychology that associate blue with a calming effect on the mind.[408] [409]

**Figure 17.3** Sunset over Dublin: A vibrant cityscape capturing the fading light of dusk in Dublin, Ireland, taken on August 16, 2016. **Image Source:** Giuseppe Milo, Flickr link.

At dawn and dusk, the sky transforms, adorned with a medley of oranges, pinks, and purples, almost as if signaling the Earth's daily cycles of renewal and rest. These transitory colors, warm and vibrant, and filled with energy, seem to fulfill the purpose of emotional bookmarks in our day, emphasizing the profound relationship between the sky's display and our emotional well-being.[410]

The materialist paradigm struggles to reasonably explain the correlation between the attributes of the sky and our psychological and aesthetic inclinations. In a universe said to have come from nothing, how did Earth develop an atmosphere that not only sustains life but also caters to human emotion and aesthetics? The explanation that it's the only canvas we know, thus come to adore, is not an adequate explanation.

Imagine how dense someone would have to be if, when a child asks why their newborn sibling's room is painted blue or pink, a relative responds with a lecture on how paint is tinted or fabric is dyed. The child isn't asking about the mechanics; they're asking about the meaning. Even if the color has no effect at all, the choice was made with intention—whether for tradition, sentiment, or personal preference. To deny that the action was mindful, even if only sentimental, would be absurd. Likewise, many have become just as mechanical in their thinking—conditioned to focus solely on processes and mechanics while overlooking the mindful actions and intentions behind them.

## The Great Outdoors

As we explore further into the characteristics of Earth's atmosphere and its critical components, like the ozone layer, in conjunction with the abundant greenery covering the planet, we see that they act as the "lungs" of Earth, purifying the air and regulating the planet's climate.[411]

The widespread presence of flora contributes to emotional well-being, offering shades of green that are proven to relax the human mind. The abundant greenery of our planet doesn't just purify the air and regulate climate; it also has a vital impact on human health.[412] Exposure to forests and trees has been shown to bolster immune systems, lower blood pressure, and alleviate stress.[413] It enhances mood, sharpens focus, and even aids children with ADHD.[414] These natural surroundings also play a role in hastening recovery from illness or surgery, boosting energy levels, and improving sleep quality.[415]

The orthodox view posits that Earth's atmosphere evolved from a primordial mix of gases, eventually leading to the hypothesized Great

Oxidation Event (GOE), said to have transformed the atmosphere into one capable of supporting complex life (approximately 2.4 Ga).[416] However, this narrative glosses over significant implausibilities.

How did Earth transition from an oxygen-absent to an oxygen-rich atmosphere?[417] The ozone layer, which plays a critical role in safeguarding life on Earth, is composed of ozone, a molecule formed from three oxygen atoms.[418] What catalyzed the initial proliferation of photosynthetic organisms in an environment without the protective ozone layer? The existence of Earth's ozone layer defies the conventional narrative of gradual atmospheric evolution, as it requires oxygen to form in the first place.

Oxygen, the fundamental building block of ozone, is largely a product of photosynthesis, a biological process carried out by plants and certain microorganisms.[419] These organisms absorb carbon dioxide, sunlight, and water, producing oxygen as a byproduct. Yet, for these organisms to thrive and perform photosynthesis, they require a stable, habitable environment— one significantly influenced by the presence of an ozone layer. The ozone layer's critical function is to absorb and scatter the Sun's harmful ultraviolet radiation, preventing it from reaching Earth's surface in deadly amounts.[420]

However, it's important to note that photosynthesis describes a step in the process of recycling of oxygen, not its origin—just how a water treatment plant circulates water but does not generate it. This oversight stems from a failure to recognize the irreducible relationships in nature.

Furthermore, the GOE is said to have introduced large amounts of oxygen into Earth's atmosphere over 2 billion years ago, yet land plants— the primary known source of oxygen today—are thought to have appeared around 450 million years ago. If oxygenation was essential for creating

Earth's stable atmosphere, what maintained it for nearly 2 billion years before plants existed? And if oxygen was already abundant long before plants, why was another oxygenation event supposedly needed later—other than to patch holes in this incoherent narrative?

Without this protective shield, Earth's surface conditions would be far less hospitable to life, especially to photosynthetic organisms.[421] This interdependence presents a paradox: the ozone layer depends on oxygen to form, yet photosynthetic life—the primary producer of that oxygen—requires the ozone layer's protection to survive.[422] Such an irreducibly complex relationship defies the conventional view that Earth's atmospheric development was a purely linear and unguided process.

This seemingly mundane ozone layer also acts as a sort of celestial bouncer, turning away the Sun's most harmful ultraviolet rays before they can crash the party of life on Earth.[423] How incredibly considerate of the universe to lay out this red carpet for life, isn't it? Imagine living in a house located in a crime-ridden neighborhood, where dangers like "harmful UV rays" are always looming. Surprisingly, you find out that this house is equipped with an advanced, invisible security system, similar to our ozone layer, that effectively keeps out bad actors while allowing in only the 'friendly visitors'—beneficial solar radiation.

According to standard theories, the existence of such a necessary security feature in an otherwise perilous location is purely coincidental. It's as if you just happened to move into a high-risk neighborhood and stumbled upon a house that, by arbitrary chance, had exactly the security measures you would wish for. Does this scenario feel entirely plausible, or does it seem to stretch the limits of what we might expect from mere chance?

The exact coordination of conditions necessary for the formation of the ozone layer—and its crucial role in cultivating a habitable planet—does not appear to be a serendipitous occurrence. The sky's beauty, with its deep blues and fiery sunsets, and the vibrant greenery of nature, which offers not just visual delight but also medicinal properties, further underscore this sense of purpose.

# CHAPTER 18

# Cosmic Waterboy

*I'm just going to say that miracles are so highly improbable that they are the least possible occurrence in any given instance, they violate the way nature naturally works. They are so highly improbable, their probability is so infinitesimally remote that we call them miracles. I'm not saying it didn't happen, but if it did happen it would be a miracle.*[397] – Bart Ehrman

Because we encounter it daily in various aspects of our lives—beyond just needing it to survive—we often perceive water as unremarkable. However, when we view it from an analytical perspective, water reveals itself as anything but mundane. Considering the many essential functions and benefits of water, we should ask why it exhibits unique properties that are perfectly attuned to sustaining life, ecosystems, and climate. Its ability to exist in three states, its high specific heat capacity, and its solvent properties regulate temperatures, enable biochemical reactions, and maintain ecological balance. Is the abundance of water on Earth, with its remarkable compatibility with the needs of living organisms and planetary stability, just another lucky coincidence?

Uniquely, water expands when it freezes, making ice less dense than its liquid form, providing insulation on frozen water bodies. Recent discoveries add to water's enigma: research led by Brooklyn College Associate Professor Nicolas Giovambattista reveals that water can exist in two distinct liquid states at low temperatures.[424] This new understanding of water's phase behavior reveals another facet of its complex role and reinforces the concept of NatTech. Imagine an engineer designing a material with such versatility:

an insulator, a solvent, a temperature regulator, all while remaining stable under various conditions. Water effortlessly fulfills these roles and more.

Apart from its unique physical properties, water also cycles through our planet in a way that are beneficial to life and ecosystems. Through the process of evaporation, condensation, and precipitation, water is effectively distributed across vast terrains, from towering mountain ranges to arid deserts. This hydrological phenomenon is a fascinating and essential mechanism for sustaining life.

Water vapor is instrumental in maintaining Earth's climate. It acts as a greenhouse gas, trapping heat, and it serves as a key component of clouds, which reflect sunlight and have a cooling effect. The precision of this balance is so fine-tuned that it is additional evidence against the notion that our universe is indifferent to life. Water seems particularly concerned with sustaining life on Earth. However, it's not just its utility that serves as evidence, but also the mystery surrounding its origin and vast abundance.

Imagine trying to cool your home using a furnace instead of an air conditioner. This illustrates the paradox of the early Earth's atmosphere as explained by current theories: initially, the atmosphere is theorized to have been hostile and unsuitable for sustaining water or life, much like a furnace would be for cooling.[425] This inhospitable early atmosphere, lacking water, has led to questions about how we got water on Earth. However, much of the public is probably unaware of this water "problem."

But there's an understandable reason why materialists don't care to discuss how water fits into their narrative. Many of these scholars have had to settle on a fantastical proposal: Earth's water originated from external sources such as comets and asteroids.[426] Having that as your best explanation may not be something you're proud of. Although water is one of the most essential substances on Earth, its origin is one of the most avoided topics concerning the history of Earth. Water is so critical to life that as scientists search for life on other planets, the likelihood of water is always established first.

**Figure 18.1** "Viewed from space, the most striking feature of our planet is the water. In liquid and frozen form, it covers 75% of the Earth's surface. It fills the sky with clouds. Water is practically everywhere on Earth, from inside the planet's rocky crust to inside the cells of the human body." [427] **Image source:** NASA.

Take a moment to reflect on this extraordinary proposal: Earth's vast water bodies, totaling up to 1.38 billion cubic kilometers, originated from cometary and asteroidal deliveries.[428] [429] Accepting such an assertion seems to show a desperation to avoid the words "we don't have a clue."

Asteroids are primarily composed of rock and metal, so one should not even entertain them as possible explanations of Earth's water. Comet impacts on Earth are very rare, as most comets reside in distant regions of the solar system. However, to estimate the number of comets required to deliver the amount of water present in Earth's oceans, we can start with the total volume of Earth's water, which is approximately 1.38 billion cubic kilometers.[430]

Despite estimates placing most comet nuclei between 0.6 and 6 miles (1 to 10 kilometers) in diameter, let's be generous and grant them a 10-mile-wide (16 km) comet for this scenario.[431] Imagine a comet that is 10 miles (16 kilometers) in diameter and has a volume of roughly 2,145 cubic kilometers. Let's assume a more than generous 50% of a comet's volume is water. Each comet would contain about 1,072 cubic kilometers of water. Dividing Earth's total water volume by the water volume per comet, we find that it would take around 1,287,000 comets to provide the necessary amount of water.

Therefore, even under the most favorable conditions, nearly 1.3 million colossal comets would need to crash into Earth—without vaporizing the water they were meant to deliver. To supply the necessary amount of water, this bombardment would have to occur at an almost conveyor belt-like frequency.[432] Yet, even in the most optimistic models, the rate of such impacts falls far short of what would be required.[433]

To put this into perspective, the Chicxulub asteroid, which is theorized to have caused the mass extinction of the dinosaurs, was only 6 miles wide, yet it released energy equivalent to 100 million megatons of TNT.[434] The fallout from a single impact of that size included global firestorms, tsunamis, and atmospheric disruption, plunging the planet into a prolonged *impact winter*.[435]

**Figure 18.2** This image displays Comet 67P/Churyumov-Gerasimenko. **Image source:** ESA/Rosetta/NavCam – NASA.

Now consider that the comet hypothesis requires over 1,000,000 impacts of objects even larger than Chicxulub. Such a scenario would have superheated the atmosphere, destabilized the crust, and likely vaporized much of the delivered water, rendering the Earth uninhabitable for millions of years. These implications underscore the catastrophic implausibility of the comet hypothesis as an explanation for Earth's abundant water.

Furthermore, isotopic analyses contrast Earth's water composition with that found on comets, further undermining the theory.[436] If comets were indeed Earth's primary water source, we would expect a closer isotopic resemblance. In reality, comet compositions vary, with some containing more dust and rocky material than water-ice.[437] Unlike in my thought experiment, not all comets contain significant amounts of water-ice. In no way is this a logical explanation for Earth's abundant water.[438]

Using a cup, imagine trying to fill a large lake which is several miles away only using rainwater, and you live in a desert where it rains once a year. Even if you wait for millennia, the chances of filling the pond to the brim are negligible. This could only work in a model where the variables are tailored and tapered to fit, rather than in the reality we live.

But even if we accept the already improbable idea that 1.3 million comets delivered Earth's water, the Late Heavy Bombardment (LHB) presents yet another problem. According to some theories, the Moon formed around the same time as the Late Heavy Bombardment. This theorized period of extreme impacts introduces a further contradiction, making the entire timeline even more implausible.[439]

These two conflicting ideas—millions of comets bombarding Earth to gradually deliver water, followed by an even more extreme bombardment that would have obliterated any water that somehow accumulated—cannot logically coexist as an explanation for Earth's abundant water supply. One is implausible on its own, but together, they are entirely self-defeating. The more adjustments made to force this timeline to work, the more absurd and contradictory it becomes.

This LHB era, characterized by a surge of asteroids and possibly comets colliding with Earth, the Moon, and other inner solar system bodies about 4 to 3.8 billion years ago, would have subjected our planet to extreme heat and energy, raising significant questions about how water could have endured. [440] One might argue that water delivery occurred during or after the Late Heavy Bombardment. I concede that possibility, as the standard model lacks a clear or consistent timeline for when Earth's water was delivered. Depending on the explanation, water is said to have arrived before, during, and even after the LHB—resulting in a self-contradictory and convoluted sequence. Yet, no matter which timeline is chosen, the idea remains implausible.[441]

**Figure 18.3** An illustration of early Earth experiencing intense bombardment by extraterrestrial objects. **Image source:** S. A. Cooper.

Furthermore, such celestial bombardments not only fail to explain Earth's abundant water but also directly conflict with planetary formation theories, which emphasize stability and conditions incompatible with the delivery and retention of water during early Earth's volatile history. Notably, the same high-energy impacts that are said to have helped form planets through accretion are also used to explain catastrophic events like the LHB.

Further, ancient zircon minerals are thought to indicate the presence of liquid water on Earth more than 4 billion years ago, well before the theorized

Late Heavy Bombardment.[442] These minerals form in specific conditions that would require liquid water, complicating the narrative that water arrived later through celestial impacts.[443] Imagine finding a two-century-old photo of a city skyline with modern skyscrapers. The existence of such a photo would be difficult to explain without concluding that the modern skyscrapers originated much earlier than previously understood.

Additionally, the Earth's early atmosphere during this period is theorized to have been thin and not capable of retaining water vapor effectively, further complicating the possibility of water accumulating and remaining stable. The combination of high temperatures and an inadequate atmosphere to retain water vapor makes the survival of liquid water under such conditions highly improbable. Nevertheless, according to NASA:

> Over billions of years, countless comets and asteroids have collided with Earth, enriching our planet with water.[444]

> Though this explanation suggests a long-term process, it fails to address the catastrophic implications and the improbability of water retention during these volatile conditions of early Earth.

This inconsistency regarding the origin of Earth's water reveals a deeper issue: the narrative surrounding this vital question is rooted in materialistic doctrine rather than in scientific evidence. These narratives fail to provide genuine scientific explanations, as they not only contradict what is naturally possible but also fatally contradict one another. It's troubling that the origin of something so essential to life—comprising about 60% of our bodies and covering 71% of our planet—is brushed aside with "just-so" stories, rather than acknowledged with the honest admission that we do not know.

This brings to mind an additional contradiction in this narrative: if the Sun swept lighter materials, like debris and water vapor, from the inner solar system during its formation, how could so many, if any, water-rich asteroids have reached Earth? The entire cosmic evolutionary narrative is a hot mess. It's evident that we must constantly suspend logic to entertain these origin narratives.

Further, after receiving millions of these impacts, once Earth had accumulated its vast oceans, what happened? Notably, we don't see these enormous aqua reservoirs zipping by today. Did the universe's cosmic delivery system suddenly decide that Earth had reached its optimal 71% aqua quota and switched to "Elsewhere Mode"? May we be reminded of why and how these deliveries started and then stopped in the first place?

Imagine opening your door to find a random delivery of food from DoorDash. While it might be a surprising occurrence, you know it's unlikely to occur again, at least, any time soon. Yet, it's beyond imagination to receive an endless delivery of burgers and fries made just the way you like them, to accumulate to an amount that could fill every storage unit in your city within an hour.

Accordingly, when contemplating the origin of Earth's abundant water—the lifeline for life, universal solvent, climate regulator, nutrient transporter, reaction facilitator, medium for biological processes, weather pattern influencer, heat absorber, and insulator—we must seriously question whether its presence can truly be attributed to a delivery system of asteroids or comets.[445] Again, it's not about labeling speculation as scientifically plausible simply because it fits a computational model, but about grounding ourselves in reality and understanding the interconnected relationships of water within it. The how, when, and from where of Earth's water remain confounding mysteries within these narratives. The essential nature of water and its unique abundance on Earth should, logically, place it at the forefront of those in origin of life research.

CHAPTER 19

# Location, Location, Location

*Who are we? We find that we live on an insignificant planet of a humdrum star lost in a galaxy tucked away in some forgotten corner of a universe in which there are far more galaxies than people.* —Carl Sagan[419]

Consider an analogy of apartment hunting. While on your hunt, imagine you find an apartment perfectly situated and equipped with every amenity imaginable, at the ideal intersection between your workplace, your favorite café, a lush park, and a state-of-the-art hospital. But it doesn't stop there. This perfect apartment also comes with a personal chef and masseuse, a gym, a pool, ample parking, and a view that would make a postcard jealous. The neighbors are friendly, the security is top-notch, and natural light floods your living space.

There is one downside: you live in one of the most dangerous cities in the world. However, you reside in the safest neighborhood, with two separate security companies blocking invaders around the clock. And here's the kicker, upon inquiring about the price for the living space, the property management says: "Don't worry about rent. Just for you, rent here is free." At this point, you would question the likelihood of stumbling upon such an improbable gem and ponder whether there's been a joke arranged behind this. The sheer possibility of being the only person randomly placed in such an extraordinary residence, rent-free, makes happenstance an unlikely explanation.

Earth's life-supporting conditions, similar to the perfect apartment, create a living space in the cosmos that offers unmatched richness and luxury, perfectly tailored for its residents to thrive. Earth exists within a cosmic sweet spot, often referred to as the 'Goldilocks Zone,' where a myriad of factors converge to create an environment that nurtures and sustains complex life forms.[446]

Our "café" represents Earth's immense biodiversity, providing an endless variety of food sources. From the bountiful fruits of the rainforest to healing fungi and herbs, Earth provides a richness that sustains and nurtures life far beyond mere survival. It functions as the universe's only state-of-the-art hospital, with its plants, fungi, and ecosystems offering unparalleled medicinal properties. From ancient remedies hidden in forests to modern medicine derived from natural compounds, Earth's resources heal and sustain, offering remedies that could only be described as life's most exquisite health plan.

Just outside lies the 'lush park' of Earth's natural beauty—majestic landscapes filled with vibrant trees, blooming flowers, flowing rivers, and crystal-clear lakes. These places provide not only aesthetic beauty but life-sustaining ecosystems, a reminder of the interconnectedness that fuels all living things. Earth's environment is more than functional; it's a breathtaking paradise so compelling, it's rare that visitors would want to leave.

Earth also boasts enormous "pools"—oceans, lakes, and rivers that are unparalleled. It is the only known planet with vast beaches and waterways where humans can swim, fish, and sail. These bodies of water provide sources of life, joy, and exploration, offering adventure and serenity, as well as numerous luxurious amenities.

Earth's sky is its crowning jewel. Nowhere else do we find a sky that transitions from a brilliant blue during the day to a star-filled expanse at night. It offers a perfect window into the cosmos, making Earth not just a haven for life but the ultimate observation deck in space. Its features include a magnetic field that deflects harmful cosmic rays, a cancer preventing ozone layer that protects us from ultraviolet radiation, and the presence of water in its various states. Together, these features make life on Earth possible.

The notion of 'rent-free living' points to Earth's uniqueness. It is the only known planet where we can live unaided. We don't need artificial habitats or protective suits; Earth provides everything we need. The only "spacesuit" we require is the one we are born in. And these suits have the ability to repair themselves. No other planet offers such effortless existence.

Similar to the protective role of the ozone layer, Jupiter, with its massive gravitational pull, acts as our tireless guardian, intercepting potential threats from space, such as asteroids and comets, long before they can pose a danger to Earth. It's as if Earth has its own dedicated defense system, absorbing or deflecting cosmic projectiles. This protective force is a 24/7 security system that ensures Earth remains a safe, nurturing home.

All these features combined present Earth as an extraordinarily perfect home. Its beauty, protection, resources, and open spaces come together to make it not just livable but uniquely suited for life and discovery. Just as the perfect apartment seems too good to be true when found by chance, Earth's conditions for life are so finely tuned that they defy randomness. In fact, this analogy barely scratches the surface. The likelihood of such a world existing is so improbable that many materialist scientists postulate an infinite number of universes to explain how our world could arise by purely chance.

# CHAPTER 20

# Vivi-Technology

*The only part of my professional course which really and deeply interested me was Physiology, which is the mechanical engineering of living machines... what I cared for was the architectural and engineering part of the business, the working out the wonderful unity of plan in the thousands and thousands of diverse living constructions, and the modifications of similar apparatuses to serve diverse ends.[447]* —Thomas H. Huxley

In this section, we'll explore 'vivi-technology' (ViviTech or VT), a term coined here. ViviTech represents life and the sophisticated, interwoven systems, and processes that sustain it, framing every biological process and component as part of an elaborate, purpose-driven technological framework. We will begin by examining what might seem like mundane aspects of ViviTech, such as the human arm, nose, and ear, and then continue with discussing sophisticated systems within the cell.

## Short Sightedness

First, I want to address a common pretense used against design arguments. Many critics of design arguments often focus on disproving the concept of an omnipotent and omniscient Christian God rather than directly addressing the evidence of design itself. This is a red herring that shifts attention away from the central question of whether evidence of design exists and instead diverts the discussion toward a theological debate.

For example, a common claim is that certain biological features, like the eye, aren't designed because they're not "perfect." This argument is remarkably shortsighted, but let's concede for the sake of argument, that the eye is not perfectly designed. Perhaps it has inefficiencies or features that could be improved. Even so, this does not disprove design. Design is not the same as perfection.

My mother once bought a Kia SUV in the early 1990s, around the time they first hit the market. The car had so many problems, unlike anything I've encountered since. Considering that it had just hit the market, such issues might be understandable. Still, no one would argue that the vehicle wasn't manufactured. Its flaws reflected the limitations of its design, not an absence of design. Even if someone claimed, 'God must have made it,' it would be irrational to overlook the overwhelming evidence that it was designed simply because of the rationale of that statement. The question of whether it was designed and who may have designed it are entirely separate discussions.

Likewise, consider the book you are reading. Even if it were poorly structured or riddled with errors, no one would reasonably argue that someone didn't create it. You might critique the writing skills or the quality of the content, but you wouldn't deny the intentionality behind its creation. Similarly, if someone proves that I didn't write this work, it does not change the fact that the work itself was authored by someone.

Now consider your eyes, which allows you to capture the light on this page, convert it into electrical signals, and process it as a coherent image in your brain. How ironic would it be to observe this sophisticated optical system and then insist the eye was not designed simply because it isn't "perfect" —such as its inability to detect infrared. The eye, with its ability to focus light, adapt to different conditions, and transmit information to the brain, clearly shows intentionality and purpose in its construction. While the eye has features that could theoretically be improved, it does not make it any less a product of design.

Furthermore, flaws or trade-offs are common in systems that must balance competing needs or adapt to constraints. Engineers, for example, often make choices between performance, durability, and cost in their designs. The eye reflects a similar reality—it functions within the limits of biology in ways that are undeniably effective.

## The Arm

Recognizing design in nature does not hinge on the complex arguments often used in mainstream design discussions. Although many people may not readily notice these designs, elaborate extrapolations are unnecessary to reveal NatTech at work. The most profound demonstrations of this technology are not primarily found in the depths of cellular biology or molecular structures, but in the collection of these cells that allows the everyday actions our bodies perform.

Consider, for instance, flexing your wrist or extending your arm. These seemingly routine motions, executed almost subconsciously, are vivid examples of engineering. As you flex your wrist or contract your bicep, consider the harmonious coordination at play. The skeletal framework, the mobility of our joints, and the power of our muscles, all orchestrated seamlessly by the mind and nervous system, demonstrate a level of sophistication intrinsic to our very being.

Beyond functional tasks, our arms are essential for social relationships. Hugging and embracing friends and loved ones is one of the most significant social interactions that provide support and a sense of love and belonging, which benefits one's health and psyche.[448]

195

**Figure 20.1** A New Holland E215 Excavator at work. **Image source:** Wikimedia Commons, Public Domain.

Now, consider the mechanical arm of an excavator; despite its simplicity and a stark contrast to the sophisticated engineering of the human arm, we acknowledge it as modern technology.[449] The excavator, impressive in its own right and capable of moving heavy materials and excavating earth, lacks the finesse and versatility of the human arm. The synergy between our brain and the parts of our arms enable diverse actions, from throwing a ball to the delicate art of playing a musical instrument.

Though an excavator's arm is lauded for its hydraulic power and structural efficiency, it mirrors the basic functions of bones and joints in the human arm. Yet, it is technologically rudimentary when juxtaposed with it. The natural marvel of the biological arm lies not just in its mechanical aspects but also in its sensory feedback and neural control, including touch sensitivity, proprioception, and finely-tuned motor skills—elements that no excavator arm, with its rigid steel and hydraulic systems, can replicate.[450] Our arms are not simple biological appendages; they represent a higher order of ancient engineering. They convey essential temperature, texture, and pressure information while passively interacting with our environment.

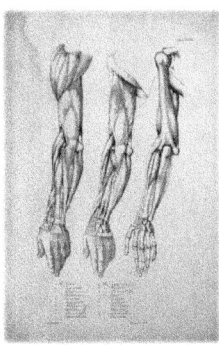

**Figure 20.2** This detailed anatomical illustration, created by George Stubbs, known for his anatomical work, shows the muscles and bones of the shoulder, arm, and hand. **Image source:** Wellcome Collection.

This understanding leads to a fundamental question: why do we readily acknowledge the excavator's arm as a product of technology, yet hesitate to apply the same recognition to the human arm, despite its superior design and functionality? This intransigent bias, recognizing only human creations as 'technology' while overlooking the more advanced, naturally occurring technological marvels of our bodies, is unreasonable.

## The Nose

In preschool and elementary education, children are formally introduced to natural sensory devices shared by humans, animals, and even some plants.[451] During these formative years, young learners marvel at the senses of sight, touch, smell, hearing, and taste—abilities that, unfortunately, many come to take for granted later in life.

Picture the simple joy of discovering the world of sounds, sights, and smells through a field trip to the zoo, then finding those same animals in the shapes of clouds. Consider the thrill of mixing colors with finger paints, the delight of catching the scent of freshly baked bread, the enchantment of listening to a favorite story read aloud or the excitement of tasting a fresh gingerbread cookie on a wintery day. These moments, rich with sensory exploration, lay the foundation for an appreciation of the natural wonders that surround us.

The nasal passages, through which air enters the body, are designed to optimize airflow into the respiratory tract.[452] This design entails foresight; the shape of the nasal passages warms and humidifies the air, protecting delicate respiratory tissues and enhancing gas exchange efficiency.[453] This design is significant for maintaining respiratory health.[454]

Within these passages, a complex filtration system consisting of hair and mucosal linings functions as a natural air purifier, capturing particulates, pathogens, and dust with the efficiency of HEPA filters found in technological settings.[455]

**Figure 20.3** An illustration of a HEPA filter. **Image source:** BruceBlaus.

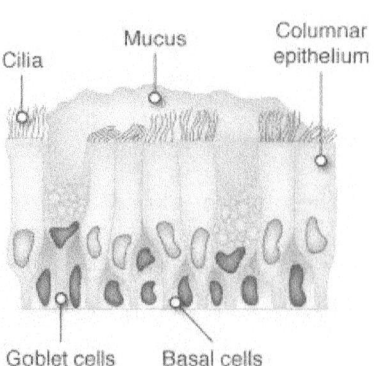

**Figure 20.4** Diagram of the nasal epithelium showing various specialized cells, including cilia, mucus-producing goblet cells, basal cells, and columnar epithelium. These components work together to filter and protect the respiratory system from particulates and pathogens. **Image source:** ccconline.org, by Cenveo, CC BY 3.0 US.

Additionally, specialized tissues known as nasal-associated lymphoid tissue (NALT) in the nasal cavity play a substantial role in our immune defense.[456] NALT induces immune responses by activating various cells that work together to defend against pathogens and allergens.[457] Intranasal immunization can activate specific protective immunity in mucosal and systemic compartments.[458] By filtering the air we breathe, this system shields us from unseen environmental hazards.[459] Thus, this multifunctional system ensures that we breathe cleaner air and strengthens our body's first line of defense against airborne contaminants.

The nose excels in scent detection, housing an advanced sensory system within the olfactory structure, capable of differentiating a vast array of odors—rivaling even modern chemical sensors.[460]

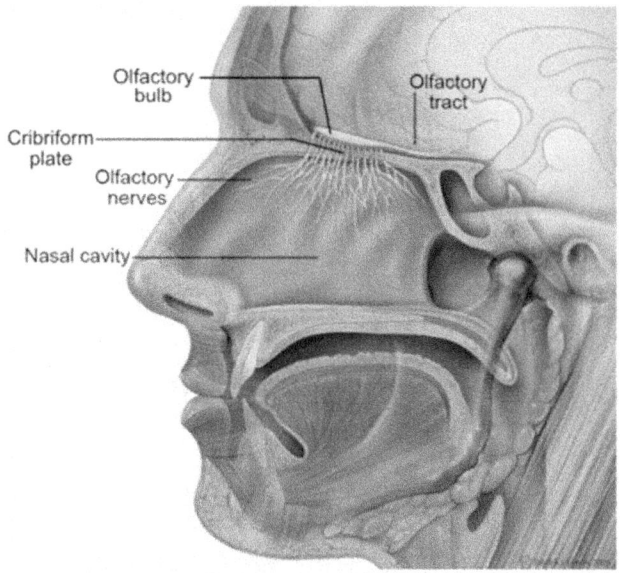

**Figure 20.5.** Head anatomy with olfactory nerve, including labels for the nasal cavity, olfactory nerves, cribriform plate, olfactory bulb, and olfactory. **Image source:** Patrick J. Lynch, medical illustrator.

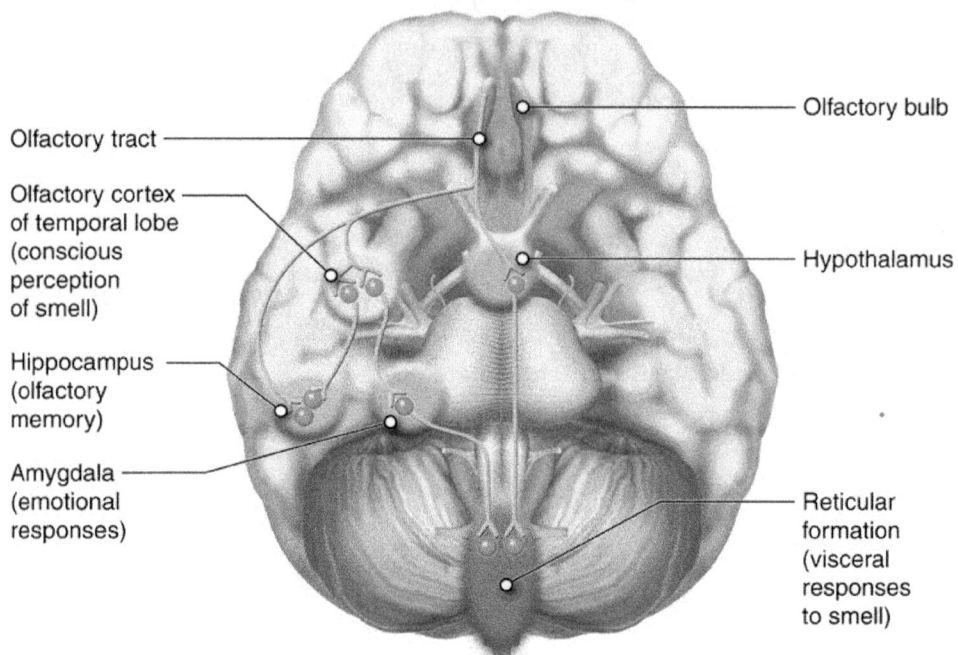

Olfactory bulb

Olfactory tract

Olfactory cortex
of temporal lobe
(conscious
perception
of smell)

Hypothalamus

Hippocampus
(olfactory
memory)

Amygdala
(emotional
responses)

Reticular
formation
(visceral
responses
to smell)

**Figure 20.6.** Central Nervous System Regions that Receive Information from the Olfactory Bulb. The image depicts the connections between the olfactory bulb and specific brain regions that receive information from it. **Image source:** anatomytool.org "Cenveo - Drawing Central Nervous System Regions that Receive Information from the Olfactory Bulb - English labels."

The olfactory system's role in survival is critical, enabling organisms to locate food, detect threats, and navigate environments.[461] Olfactory receptors in the epithelium detect specific molecules and trigger a chain reaction, converting these interactions into electrical signals for the brain.[462] This biological process converts chemical data into meaningful sensory and emotional experiences, like the comforting scent of a favorite meal, exemplifying nature's embedded technology.[463]

The sense of smell also plays a vital role in familial and romantic connections, boosting social bonds and deepening emotional ties. It helps define cultures and societies through shared scents that evoke memories, enhance interpersonal connections, and create emotional resonance between individuals, further illustrating its adaptive and integral role in our lives.[464]

Additionally, air flow in speech control plays a vital role in communication; it allows us to modulate our speech for emotional expression. For instance, pausing or stopping airflow can emphasize words or phrases, contributing to the dynamics of spoken language. The tone in our voice can convey a wide range of emotions and intentions, whether we are sad, happy, angry, or stern. Imagine the difference between a gentle, soothing voice expressing sympathy, a cheerful tone sharing good news, an angry tone delivering a reprimand, or a stern voice giving a warning. These nuances in tone are essential for interpreting and understanding the true meaning behind the words. [465]

**Figure 20.7** A sheriff uses a trained sniffer dog to detect contraband, exemplifying natural technology at work. The dog's advanced olfactory system exemplifies sophisticated biological engineering, utilized here as a vital tool in law enforcement. **Image source:** Bjwhite66212, Wikimedia Commons.

The nasal cavity and sinuses act as a resonating chamber for our voice's quality and tone.[466] This resonating function is integral for producing a range of sounds that are essential to human language, especially nasal consonants like "m," "n," and "ng."[467] Any obstruction or anomaly in the nasal structure, such as nasal congestion, can significantly alter speech clarity.[468] When someone is congested, excess air escapes through the nose during speech, affecting the clarity of non-nasal sounds and giving them a nasal quality.[469] This change in voice can be so distinct that it may be recognized by a loved one or even a stranger, showing the interconnectedness of body function and communication.

This regulation is also particularly important for singers and public speakers who rely on precise breath control for effective performance. The voluntary control of nasal airflow is a key aspect of the nose's functionality, showcasing its sophisticated engineering within human physiology.

If those features are not enough, the ability to voluntarily hold one's breath is essential for activities such as swimming, where controlling air intake is critical for staying submerged and avoiding water inhalation. This voluntary control also functions as a protective measure when encountering unpleasant or harmful odors, preventing unwanted particles from entering the respiratory system. Similar to how the body switches between allowing or blocking nasal airflow, a car's air recirculation button acts as a 'closure' mechanism. This system allows the driver to switch between inside air or outside air, effectively controlling whether air is drawn from the vehicle's interior (to block out external pollutants or odors) or from the outside (to refresh the air supply).

In contrast to involuntary processes like the heartbeat or the automatic nature of hearing, which are essential for survival and operation without conscious effort, the control of nasal airflow involves autonomic and voluntary nervous systems.[470] This dual control system highlights the body's ability to interact with its environment in a nuanced way.

It's beneficial to compare the nose with technologies like smoke detectors and breathalyzers. Each component in these devices plays a key role. Smoke detectors sense specific particles in the air to alert inhabitants of potential danger. In similar way, breathalyzers detect the chemical composition of a person's breath to measure alcohol levels. This level of integrated functionality demands considerable foresight and precision in both design and assembly.

Similarly, we can detect the intensity of smoke and even gauge how much someone has indulged in alcohol by sensing their breath. Though smoke detectors and breathalyzers are designed with specialized tools for their specific functions, the NatTech exhibited by the human nose and olfactory system far surpasses them in functionality.

## The Ear

Let's look at the next sensory system: the ear. This organ exemplifies sophisticated NatTech, finely tuned to capture and interpret a vast array of sounds. In elementary school, we engaged our hearing through activities like playing musical chairs, singing rhymes, and listening for the quiet that signaled nap time. This organ is divided into three main parts: the outer ear, the middle ear, and the inner ear, each with specific roles that contribute to hearing and balance.[471]

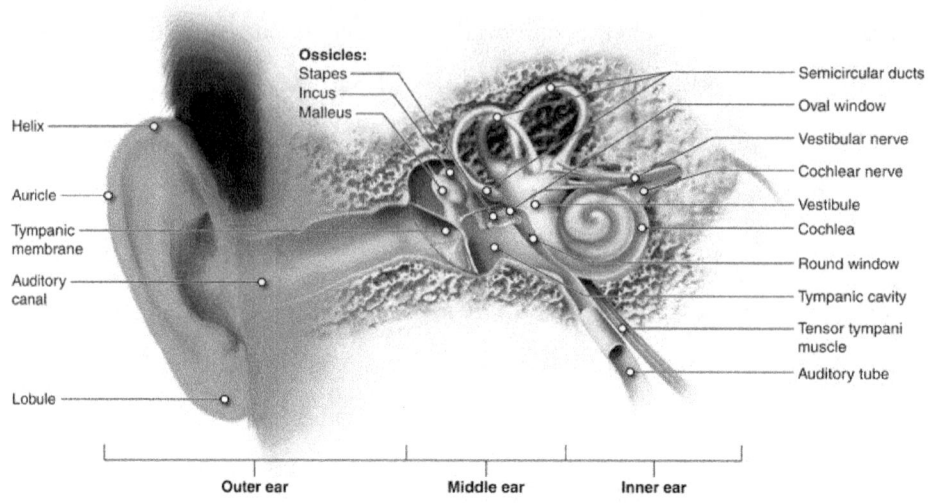

**Figure 20.8** Anatomy of the Ear. The outer ear is the auricle and canal through to the tympanic membrane. The middle ear contains the ossicles and is connected to the pharynx by the auditory tube. The inner ear is the cochlea and vestibule responsible for hearing and equilibrium. **Image source:** anatomytool.org; "Cenveo - Drawing Anatomy of the Ear - English labels" by Cenveo.

The medical term for the outer ear is the auricle or pinna.[472] The pinna captures sound waves and funnels them into the ear canal. This funneling is not arbitrary; the pinna is uniquely designed to gather sound from the environment and direct it efficiently into the ear canal.[473] Once sound waves reach the eardrum, they cause it to vibrate.

These vibrations are then transferred to the middle ear, where they are amplified by the ossicles which consist of a trio of tiny bones known as the hammer, anvil, and stirrup.[474] This amplification is significant because it allows even faint sounds to be heard clearly.[475] The precision with which these bones interact exemplifies a level of mechanical engineering that mirrors the most delicate man-made devices.[476]

From the middle ear, the amplified vibrations pass into the inner ear, specifically into the cochlea, which is filled with fluid.[477] The inner ear transforms these mechanical vibrations into electrical signals through a process involving the movement of tiny hair cells.[478] These signals are then transmitted to the brain via the auditory nerve, allowing us to perceive and interpret sounds.[479] This transformation from mechanical vibrations to electrical signals is as critical as it is complex, ensuring our ability to hear and react to our environment.

Beyond hearing, the inner ear also plays a crucial role in maintaining balance through the vestibular system.[480] This system consists of semicircular canals and otolith organs, which provide the brain with essential information about movement and head position.[481] The semicircular canals are three looped tubes, each positioned at right angles to the others.[482] The three semicircular canals, positioned at right angles to each other, contain fluid (endolymph) and sensory hair cells.[483] When the head moves, the shifting fluid bends these hair cells, sending signals to the brain about movement direction and speed.[484]

In addition to the semicircular canals, the otolith organs—the saccule and utricle—detect gravity and linear acceleration.[485] These organs contain crystals that shift in response to changes in motion, stimulating underlying hair cells. For example, when accelerating in a car or riding an elevator, the otolith organs help the brain process changes in speed and direction.[486] Together, these vestibular components ensure proper coordination, posture, and balance. The vestibular system, like other sensory systems, operates as an embedded orientation framework—one that reflects coherence and calibration, not trial-and-error assembly.

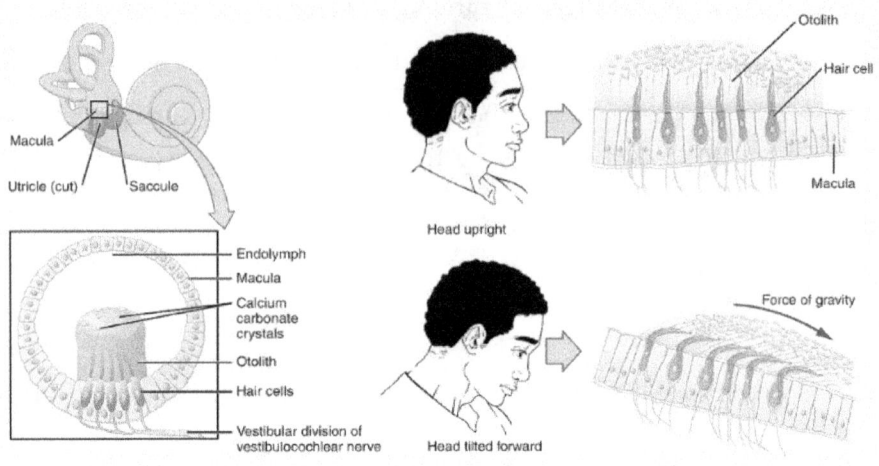

**Figure 20.9** This image illustrates how the maculae are specialized to sense linear acceleration, such as when gravity acts on a tilting head or when the head moves in a straight line. The difference in inertia between the hair cell stereocilia and the otolithic membrane in which they are embedded creates a shearing force, causing the stereocilia to bend in the direction of acceleration. **Image source:** "OpenStax AnatPhys fig.14.11 - Maculae and Equilibrium - English labels" by OpenStax, licensed under CC BY. Available from: Anatomy and Physiology.

Whether walking on uneven ground, standing up quickly, or participating in sports, the vestibular system continuously provides the brain with critical data to keep the body balanced and oriented.[487] This system is so integral to our functioning that issues with it can lead to dizziness, vertigo, and other balance disorders, significantly impacting a person's ability to perform daily tasks.[488]

Further, the ear plays a crucial role in social interactions and relationships. The ability to hear and respond to speech, music, and environmental sounds enhances our connections with others, enabling us to engage in meaningful conversations and share experiences. The ear's capacity to process and interpret sound is not just a biological function but a cornerstone of human communication and social bonding. Listening to colleagues, friends and loved ones is one of the most significant social interactions that provide support and a sense of love and value.

Another intriguing aspect of social interactions is the ear's solution to what audiologists (hearing scientists) refer to as the 'cocktail party problem.'[489] This extraordinary phenomenon allows individuals to distinguish and focus on a single voice in a noisy environment, such as a crowded room, while filtering out unrelated background noise.[490]

The coordination between the ear and brain—which enables complex auditory tasks like the cocktail party effect— is remarkable and defies explanation through the blind processes of Darwinian evolution. Though this innate ability may seem unremarkable, the thought of unseen waves traveling through the air at a party—amidst music and many voices—and the brain's ability to amplify specific sound waves, convert them into electrical signals, and process them into meaningful information despite the cacophony is an extraordinary feat of natural technology. This capability showcases the ear's precision in sound filtering and the complex neural processes involved. Replicating this function continues to elude even our most advanced audio and speech recognition technologies.

Moreover, electronic systems often use devices like limiters to cap loud audio signals and prevent damage. In contrast, the ear has built-in mechanisms for self-maintenance and protection. This self-maintenance feature, involving tiny muscles in the middle ear that reflexively contract in response to loud noises, temporarily stiffens the ear's ossicles, reducing their ability to transmit vibrations and prevent potential damage.[491]

The ear and brain, similar to modern speech recognition software, process spoken words through complex, multi-stage processes.[492] Both systems demonstrate advanced capabilities in capturing, converting, and interpreting sound, yet the NatTech within the ear exhibits a level of integration and functionality that far surpasses current technology.

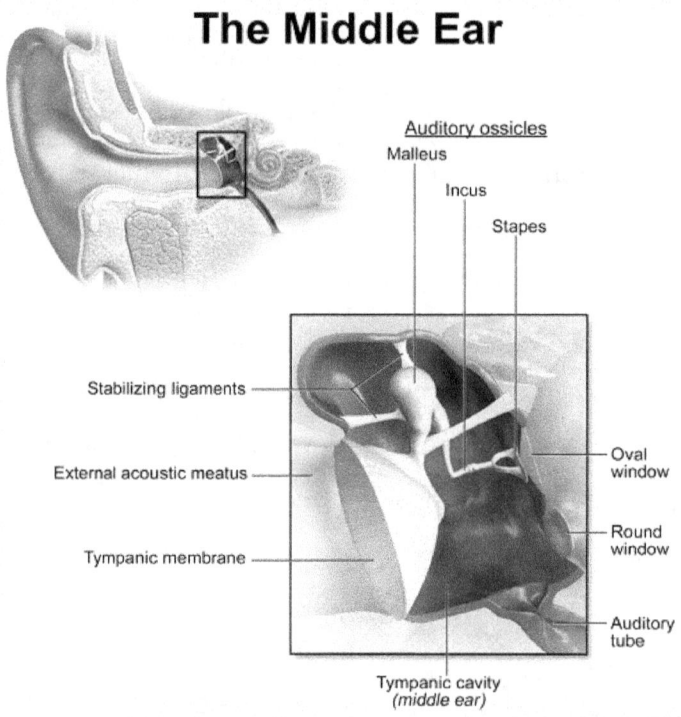

**Figure 20.10** An illustration of the middle ear: When exposed to intense sound, the stapedius and tensor tympani muscles of the ossicles contract.[493] The stapedius pulls the stapes (stirrup) away from the cochlea's oval window. In contrast, the tensor tympani pulls the malleus (hammer) toward the middle ear, stiffening the ossicular chain.[494] This reflex reduces the transmission of vibrational energy to the cochlea, which is converted into electrical impulses to be processed by the brain.[495] **Image source:** BruceBlaus, CC BY 3.0.

Speech-to-text technology starts with a microphone capturing sound waves, much like the outer ear. These sound waves are converted into digital signals, akin to how the ear's inner components transform sound waves into mechanical vibrations. The technology uses sophisticated algorithms to analyze these digital signals, breaking them down into recognizable words and phrases, much like the brain processes electrical signals from the ear to interpret and understand sounds.[496]

However, while speech-to-text systems require extensive programming and calibration to handle different accents, dialects, and languages, the human ear and the associated auditory processing centers in the brain adapt seamlessly from birth, learning and adjusting to the linguistic and acoustic environment without the need for external adjustments, programming or upgrading.[497] Although artificially intelligent systems try to emulate these processes, they will never be able to enjoy a conversation, song, or movie.

## Bionics

Indeed, these comparisons between human-made devices and natural engineering, such as the human arm, nose or ear, bring us to a realization that technology, in its purest form, is most vividly seen in nature itself. Despite advancements in machine limbs or prosthetic development, these efforts inherently acknowledge the unmatched sophistication already present in biological systems. In our quest to mimic the human arm's functionality, we inevitably confront the limitations of our artificial creations.[498]

Each device, no matter how sophisticated, in one way or another, only accentuates the innate complexity and superiority of the biological counterpart it seeks to emulate, a sophistication it fails to capture even with each advanced upgrade.[499] This isn't to diminish the value of these technological strides; instead, it places them in a context that recognizes the ingenuity of the natural world. Body parts in nature plainly stand as a paradigm of a higher order of technology: self-regulating, self-healing, and exquisitely attuned to their environment. Once again, we must ask: Why is acknowledging the ubiquitous natural technology in nature shunned?

Furthermore, think about past perspectives that viewed human ingenuity as inevitably surpassing the wonders of nature. Bionics, initially "the study of electronic systems which function in the manner of organic systems," became popularized in the realm of science fiction, embodying the aspiration to replicate and even surpass biological functions through mechanical and electronic means.[500] [501]

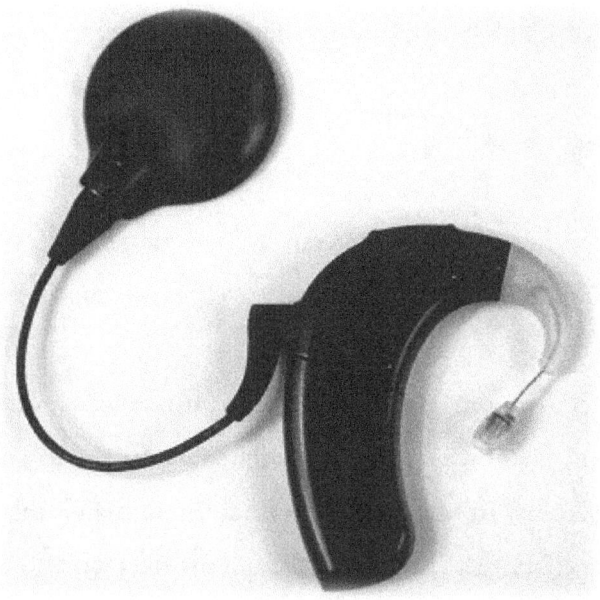

**Figure 20.11** This is a fourth-generation behind-the-ear cochlear implant processor made by Advanced Bionics circa 2021. It is worn on the ear and sends data by radio to the cochlear implant hardware embedded in the user's skull. **Image source:** RespectCE, CC BY-SA 4.0.

The concept envisioned a future where human capabilities and bionics could be revolutionized, exalting our ingenuity beyond what is reflected in nature. After all, as the thinking went, if nature achieved its biological feats randomly and without intent, surely we could surpass it?

The term "bionics" gained widespread recognition with the television series *The Six Million Dollar Man,* captivating the imagination of a generation and planting the seeds of a technological dream.[502] The show featured Steve Austin, an astronaut whose body is rebuilt with bionic parts after a crash, giving him extraordinary speed, strength, and vision.[503] Austin's bionic eye could zoom in on objects miles away and had night vision capabilities.[504]

Similarly, in *The Bionic Woman*, Austin's spin-off, Jaime Sommers is outfitted with bionic implants following a skydiving accident.[505] [506] These enhancements include a bionic ear that can hear sounds over great distances and frequencies beyond the normal human range, bionic legs that give her incredible speed and the ability to jump great heights, and a bionic arm that grants her extraordinary strength. These portrayals overestimated human ingenuity while underestimating the sophistication of natural biological technology.

The idea of acquiring superhuman capabilities through bionics, as depicted in these shows, arose during a period of optimism—and a certain naivety about the complexities involved. These shows painted a picture of a future where bionic limbs and augmented abilities were not just possible but commonplace. Over time, as memories faded, the once-thrilling vision of bionics in science fiction has also lost its allure and plausibility. Reality has led us to a more grounded understanding.

The more recent fascination with mutations and genetic alterations—rooted in evolutionary ideas—remains, but its hold is weakening. Science fiction has increasingly favored concepts that, while still unrealized, feel more plausibly within reach. And as the idea of mutating has diminished, the focus has shifted once more. Today, cognitive enhancement—especially through AI integration—dominates the conversation, reflecting a more tangible frontier. This shift is evident in shows like *Westworld* and *Black*

*Mirror*, which explore the fusion of AI with the human mind. But who knows what this integration may offer in the future?

Nevertheless, the high level of technological expertise required to integrate with biological systems clearly reveals the advanced nature of these systems. Understanding and replicating these systems demands more skill and knowledge than is typically needed for man-made technologies. Indeed, some of these biological systems may never be fully understood or replicated.

## Cellular Systems

At the heart of ViviTech, we encounter the remarkable world of proteins. The sophistication of their structures extends the metaphor of a biological library beyond basic text. Proteins communicate sophisticated information through their three-dimensional forms.[507] Proteins that fold into complex shapes are not simply physical entities; they represent a reservoir of advanced biological information. Each fold, twist, and turn in a protein's structure is pivotal, determining its functionality and interactions, much like how each component of a complex machine is essential for its operation.[508]

Proteins are more than just the books in this library—they are also the catalog that records the books, the bricks that form the library's structure, and the builders that continually reshape and maintain it. The spatial architecture of proteins is dynamic and precisely engineered, embodying complex information-processing capabilities inherent in biological systems.

Imagine understanding proteins as if you were reading a book where the font itself carries significant meaning to the story, or a pop-up book that extends the narrative beyond the written words. In these pop-up books, the story is communicated not just through the words but also through the shapes that unfold as the pages turn. These three-dimensional structures add depth to the narrative, making it more engaging and immersive. Each structural detail conveys critical information necessary for the protein to function and interact correctly within the cellular environment.[509]

Drawing a parallel to the art of origami offers another layer of understanding. In origami, a simple sheet of paper is folded to create various shapes, each fold consequential to the final form. Just as origami transforms a flat sheet into a three-dimensional object, proteins fold into complex shapes essential for their function. This process, guided by the sequence of amino acids, mirrors the deliberate folds in origami that dictate the final structure.

In the completed state of an origami creation, it is no longer just a piece of folded paper; it communicates meaning to the observer. For instance, an origami swan or dragon is recognized and understood by those who see it, transcending its material composition to convey a specific idea or image. Similarly, the folded structure of a protein is more than the sum of its parts. It becomes a functional unit that interacts within the cell meaningfully and is understood by other cellular components.[510] This interaction is integral for the myriad of processes that sustain life, from enzymatic reactions to cellular signaling.[511]

This origami analogy somewhat captures the elegance and complexity of protein folding and the importance of its final form. However, a misfold in origami leads to an unrecognizable or unstable shape, whereas misfolded proteins can cause cellular dysfunction and disease. In living systems, form and function are tightly intertwined, each fold and twist laden with meaning and purpose, integral to the life-sustaining interaction of molecules within us.[512] This insight into engineering could pave the way for groundbreaking advancements in our own engineering and technology. In its complexity, sophistication, and versatility, the simplest cell far surpasses even the most advanced human-built facility.

For clarity, there's no such thing as a "simple" cell. Within any cell lies a microcosm of complexity, where intricate biochemical networks drive its functions.[513] The reason these cells were originally labeled "simple" was because it took nearly 300 years—from the first observations of cells in the 17th century to the discovery of DNA's structure in 1953—before we could truly observe the structure, processes, and intricate mechanisms operating within them.[514]

## DNA

DNA is part of a complex information storage and processing system, inherently capable of interacting with many cellular components and signals.[515] [516] The regulation and expression of genetic information depend on a highly complex networks of signals and interactions within the cell.[517] The process of DNA replication exemplifies this natural ingenuity. Far from a simple "copy-paste" task, DNA replication is a meticulously orchestrated procedure that demands high precision.[518] Specific molecules must be arranged meticulously through elaborate steps like error checking and repair mechanisms to ensure genetic fidelity. This system's complexity becomes apparent when considering the role of various enzymes that recognize specific sequences and perform precise functions, essential for maintaining life and preventing detrimental mutations.[519]

For example, if an incorrect nucleotide is added to the new strand, DNA polymerase has the capability to remove the mismatched nucleotide and replace it with the correct one.[520] Moreover, DNA polymerase is also involved in repair processes to fix any damage to the DNA that might occur due to environmental factors like UV light or chemical exposure.[521] This ability to repair and maintain the DNA ensures the cell's genetic material remains stable and reduces the likelihood of mutations that could lead to diseases like cancer.[522]

The sequence of nucleotides in DNA or RNA, as well as the sequences of amino acids in proteins, form part of a living communication network. These sequences are not simple repositories of genetic data; their roles extend far beyond simple storage.[523] DNA and RNA interact with proteins and ribosomes to execute functions critical for the cell's survival and adaptation.[524] The concept of information in biology becomes a tangible, functional reality, actively read, interpreted, and utilized within the living cell to perform highly specific tasks.

**Biological Cache**

In computing terms, DNA functions as a permanent repository of genetic information, yet its distribution across every cell resembles a decentralized cache, ensuring local access to the full genome without requiring a centralized retrieval system.[525] However, true caching mechanisms within the cell—such as messenger mRNA transcription, protein synthesis, and epigenetic modifications—enable rapid access to essential instructions, optimizing cellular efficiency similar to how computational caches enhance data retrieval.[526] [527] These dynamic processes reduce the need for constant access to genetic storage and ensure that cellular functions remain efficient and responsive. This hierarchical nature of cellular information flow, from DNA storage to dynamic RNA and protein regulation, mirrors computing strategies for managing data—but in a living system operating at a level of sophistication that even our most advanced computer engineers can only dream of replicating.

Cells do not access DNA indiscriminately but retrieve only the genetic information necessary for their specific functions, much like computing caches improve efficiency by storing frequently accessed data.[528] Similarly, epigenetic modifications, such as DNA methylation and histone acetylation, function as an indexing system that prioritizes active genes while silencing less relevant ones, much like caching algorithms optimize memory by storing high-priority content and ignoring low-demand data.[529]

In computational systems, frequently accessed ("hot") pages are stored in fast memory like DRAM, reducing the cost of retrieving data from slower disk storage.[530] Algorithms dynamically determine reuse thresholds based on the cost of memory versus retrieval delays, ensuring efficient data access.[531] Similarly, cells enhance efficiency by selectively transcribing genes needed for immediate function. Liver cells, for example, "keep hot" detoxification-related genes, while skin cells prioritize structural protein genes like keratin.[532] Just as computing systems manage storage hierarchies to reduce latency, cells optimize their own biological memory by ensuring that the most relevant genetic instructions are quickly accessible while minimizing unnecessary transcription.[533]

At a deeper level, mapping genetic instructions to functional outputs also mirrors computing processes. In computing, virtual memory uses page tables to map virtual addresses to physical locations in memory, allowing efficient data retrieval.[534] Likewise, in cellular systems, DNA sequences (addresses) are mapped to functional outputs (proteins or RNA).[535] This mapping process is accelerated by transcription factors and signaling pathways, similar to the Translation Lookaside Buffer (TLB) in computer systems speeds up address translations.[536] Both systems balance hierarchical data retrieval, ensuring that frequently used information is accessed with minimal delay while maintaining overall system integrity.

Moreover, error correction mechanisms in both biological and computational systems ensure data integrity over time. Error correction codes in computing detect and rectify transmission errors to preserve information fidelity.[537] In both cases, these mechanisms maintain functionality and prevent corruption, ensuring the stability of stored and retrieved information.

The similarities between biological caching mechanisms and computational caches reveal how nature and technology converge on common principles of efficiency, redundancy, and optimization—while operating with far greater sophistication and efficiency than human-engineered systems.

## Jumping Genes

In the 1940s and 50s, Nobel Prize-winning geneticist Barbara McClintock revealed *transposable elements* (TEs), or "jumping genes."[538] These DNA sequences, also known as transposons, are native to an organism's genome—its host—but have the unique ability to move within it, either by copying and inserting themselves into new locations or by cutting and relocating directly.[539] This mobility allows them to activate, deactivate, or alter gene expression, making them essential to genetic variation and adaptability.[540]

Imagine working on a word document, but with a twist: certain sentences or paragraphs have a mind of their own. They autonomously "jump" from one part of the text to another, sometimes copying themselves into new sections, other times cutting themselves out and re-pasting elsewhere. This behavior is unpredictable, and these "mobile" sections can subtly alter the document's flow or meaning. This is how transposons shuffle genetic material within a genome, potentially influencing gene expression or even causing mutations.

Given this unpredictability, it's easy to see why TEs are often viewed as disruptive. Yet, just because a system can malfunction doesn't mean it wasn't originally designed for a purpose. Their capacity to influence gene expression suggests there may be more to their role than mere randomness. This discovery overturned the traditional static view of the genome, revealing that genetic material is far more fluid and responsive to environmental stimuli than previously believed. McClintock's insights laid the groundwork for understanding genetic adaptability and gene regulation.

Some scientists compare their role to natural selection.[541] However, this analogy is misleading—the adaptability observed in McClintock's discovery of TEs reflects genetic responsiveness, not a selection-driven process.[542] While TEs restructure genomes and influence gene regulation, they do not provide a mechanism for speciation. Variation itself does not explain how distinct species come about, as broader requirements—such as reproductive boundaries and the stabilization of genomic changes—must be met.[543]

Casacuberta and González (2022) analyze various studies that examine the role of jumping genes in the adaptive capacity of different organisms. They conclude that TEs "play an important role in the responsive capacity of their hosts in the face of environmental challenges," demonstrating that these genetic components are active participants in the recalibration of organisms.[544] For example, some TEs integrate into regulatory regions and help establish stress-inducible regulatory networks, allowing organisms to rapidly adjust gene activity under environmental pressure. Research suggests that "TEs might directly affect the function of individual genes... [and] disseminate regulatory elements that can lead to the creation of stress-inducible regulatory networks," highlighting their role in genome-wide adaptability.[545]

The article states, "TEs influence the capacity of adaptation of the host as large as the variety of TEs and host genomes."[546] In other words, the more diverse the TEs and the host genome, the greater the potential for genetic variation, which can influence an organism's ability to adapt and survive in changing conditions. While most TE inheritance occurs vertically, there is also evidence of rare horizontal transfer across species. In these rare cases, TEs can introduce new genetic material that influences adaptability by altering gene regulation or adding functional traits. This further underscores the complexity of genetic adaptation, reinforcing the idea that genomes are not static but actively engage in innovation.[547]

The article explains that when TEs are horizontally transferred from one species to another they can cause a wide array of genetic alterations.[548] This horizontal gene transfer resembles technology transfer at a genetic level, where genetic traits and regulatory capabilities are exchanged between organisms, enhancing their survival—much like sharing technological innovations across different industries. (Note: In this chapter, "horizontal transfer" refers specifically to the movement of TEs between species. In Chapter 28, a broader concept of horizontal, or lateral, gene transfer (LGT) will be discussed, encompassing various types of genetic exchange across species boundaries.)

These alterations range from adding new regulatory regions that might lead to gene overexpression, disrupting existing regulatory frameworks, to introducing stop codons or altering reading frames.[549] They can also affect gene expression through changes in miRNA-binding sites or by creating new splice sites that lead to alternatively spliced variants.[550] Each of these changes has potential consequences on the host's genetic architecture and adaptability.[551]

These observations reveal that biological systems exhibit an inherent logic and functionality. While I highlight the sophisticated engineering evident in biological systems through the functionality of TEs, some may point to their potential for harm—such as triggering mutations or genomic instability—as evidence against intentional design. But even in a malfunctioning or "haywire" state, their behavior doesn't diminish the technological character of the system; if anything, it underscores the complexity and depth of its engineered features. Their continued ability to integrate into and regulate genetic expression suggests a system built for adaptability and resilience, not randomness.

The fact that TEs can be both beneficial and harmful does not imply a flaw in their original function. Instead, their disruptive tendencies may be the result of genomic deregulation over time—a form of biological entropy—a process where once finely tuned genetic mechanisms gradually lose precision due to accumulated mutations and environmental influences. If these were marked primarily by detrimental effects, it would argue more against functional complexity arising from chaos rather than inherent engineering.

This deregulation does not occur in isolation. It is part of the broader context of the organism's life history, where multiple systems interact and influence one another. As these systems experience wear and entropy, the original functionalities of TEs can become obscured or altered, leading to outcomes that may appear maladaptive or destructive. However, this phenomenon does not negate the complexities of maintaining such sophisticated designs over extensive timescales.

Researchers often claim that elements now seen as potentially harmful—such as viruses, TEs, and endogenous retroviruses (ERVs)—were initially parasitic sequences that evolution "co-opted" for beneficial functions.[552] [553] While ERVs originate from ancient viruses, certain TEs— particularly retrotransposons—use similar mechanisms of genome integration and regulation.[554] For instance, the syncytin gene, derived from an ancient retrovirus and essential for placental development, is cited as a classic example of evolutionary co-option.[555] However, this "harmful-turned-helpful" narrative lacks direct evidence. If syncytin played a critical role in reproduction from the outset, this suggests it had an initial function rather than being gradually repurposed from a harmful origin.

The co-option narrative presumes that biological systems blindly repurposed harmful elements, yet the sophisticated roles of syncytin, TEs, and ERVs in their modern functions point to their suitability for integration from the beginning. These elements were not random parasites that "became" useful; rather, they were adaptive genetic resources embedded with inherent potential. Similar to how quantum states resolve into meaningful patterns under specific conditions, these genetic elements may have been embedded with latent capabilities, awaiting the right context to unfold their adaptive functions.

Imagine a powerful kingdom invaded by a hostile force with one intent: domination. After failing to overpower the kingdom, the invaders are subdued and confined within its borders, where they are kept under strict control to prevent further harm. Initially, they are not seen as beneficial, but rather as a threat requiring constant vigilance. However, over time, the kingdom's rulers recognize that with deliberate oversight and strategic intervention, these former invaders can be harnessed to serve the kingdom's needs—but only under tightly regulated conditions to ensure they remain useful rather than destructive.

This process is not random. It requires ongoing management, adaptation, and enforcement to prevent the invaders from reverting to their original harmful purpose. Without constant regulation, their disruptive tendencies would resurface, undermining the stability of the system. Such an institutional, controlled transformation is unlikely to result from trial and error. Managing and repurposing a once-hostile presence—especially for a specialized role like syncytin in placental function—requires foresight, not blind co-option.

## Integrated Systems

Modern technology provides several analogies that reflect the engineering of this ancient technology found in biological systems. For instance, in computer science, modular programming and code reusability—such as Dynamic Link Libraries (DLLs) in Windows operating systems and Python packages—mirror the function of transposable elements, enabling code to be reused and integrated across different applications.[556] [557]

Similarly, plug-and-play hardware seamlessly integrates into a computer without disrupting its core functions, transposable elements insert themselves into the genome, influencing genetic regulation without dismantling the system.[558] Like modular technology, these elements are designed to integrate, interact, and enhance functionality without compromising stability. USB devices, such as flash drives and external hard drives, can be used immediately upon connection, providing additional storage or functionality. Likewise, PCIe cards, including graphics cards and network interface cards, enhance a computer's capabilities without requiring a complete system overhaul.

Additionally, dynamic data structures in computer science offer a compelling analogy to the behavior of transposable elements within genetic systems.[559] For instance, linked lists and tree data structures, which are capable of expanding and contracting by adding or removing nodes, dynamically adjust their structure to efficiently manage data.[560] This fluid adaptability mirrors the role of transposable elements in the genome, which traverse genomic landscapes, activating or deactivating genes and thus modifying genetic information in response to environmental cues or internal regulatory signals.[561] Similarly, a binary search tree designed to ensure quick data access, can be dynamically reorganized to maintain optimal data retrieval times.[562] This is similar to how genetic structures are fine-tuned by cellular mechanics and processes, ensuring that cellular functions are optimally regulated according to the organism's needs, thereby maintaining homeostasis and enhancing survival in changing environments.

Furthermore, genetic algorithms in artificial intelligence demonstrate variability and adaptability similar to that introduced by transposable elements in biological systems. Genetic algorithms are optimization techniques that use processes such as mutation, crossover, and engineered selection to develop solutions to complex problems.[563] These algorithms introduce changes to candidate solutions to evaluate different possibilities and find optimal or near-optimal outcomes.[564] This process is akin to how transposable elements introduce genetic variability. The copy, cut, and paste behavior of jumping genes further illustrates how the adaptability in many of our digital technologies are extensions of ancient engineering principles inherent in nature.

As our technologies advance, we continue to learn fascinating truths about the links between our technology and NatTech. Synthetic biology, for example, has harnessed DNA's programmability to create genetic circuits that perform logical operations similar to electronic circuits.[565] These biological circuits can control cellular behavior in response to specific inputs, demonstrating DNA's potential as a medium for programming life. Technologies such as CRISPR, epigenomic mapping, and single-molecule imaging provide detailed data on these processes, showing how cellular components efficiently work together.[566] These advances in knowledge are not just evidence against Darwinism; they are strong circumstantial evidence that builds a rock-solid case for an Engineer.

# CHAPTER 21

# Tangled

*It from bit symbolizes the idea that every item of the physical world has at bottom — at a very deep bottom, in most instances — an immaterial source and explanation; that what we call reality arises in the last analysis from the posing of yes-no questions and the registering of equipment-evoked responses; in short, that all things physical are information-theoretic in origin and this is a participatory universe.*[567] — *John Archibald Wheeler*

It is fascinating to envision a world where devices transcend reacting to physical stimuli like touch or voice. Imagine technology finely attuned to the mental states or intentions of its users. This technology would be a revolutionary leap, like our jump from horses to spacecraft. Such advancements could transform our interaction with machines from tools or assistants to partners capable of understanding and responding to our psychological states. Yet, such innovation cannot be realized within the confines of epiphenomenalism, which treats consciousness as a mere byproduct of physical processes, devoid of any causal influence. If the mind is nothing more than an epiphenomenon, there would be no foundation for even attempting to integrate it into advanced technology. We must first be open to the idea that the mind extends beyond the physical confines of the body.

It is intriguing to consider the relationship between the imaginative worlds of science fiction and tangible advancements in technology. Shows like *The Jetsons* entertained us with futuristic visions while subtly reflecting deeper aspects of our universe.[568] These imaginative tales, once seen as purely fictional, have, at times, converged with scientific discoveries. This blurred déjà vécu reflects a universe more interwoven than we once thought, ideas intuited by storytellers and progressively uncovered by science.

Look back to the 17th century. The idea that reality operates on mathematically describable laws, which the human mind can discover, remains one of Isaac Newton's most profound legacies. When he was figuring out how things move, he wrote down these brilliant laws that tell us how objects behave when forces act on them. You throw a ball; it follows a nice, predictable path. This framework of thinking is called classical mechanics. It refers to the study of motion and forces in systems that operate predictably, like machines with moving parts—such as gears and levers—or the Earth revolving around the Sun.

Jump forward to the early 20th century. Scientists notice that tiny particles like electrons and photons are not following Newton's rules. They are doing all sorts of unconventional things that classical mechanics cannot explain. The formal study of these phenomena began with pioneers like Max Planck, Albert Einstein, Niels Bohr, and Erwin Schrödinger.[569] These scientists uncovered rules governing the micro-world, laying the foundation for quantum theory. Thus, a new branch of physics was born: quantum physics, along with its first formalized framework—quantum mechanics.

Sometimes, we talk about quantum mechanics as if it were the phenomenon itself. But quantum mechanics is not the phenomenon—it's a mathematical framework we use to describe the behaviors observed in the quantum world. This framework has helped us discover new insights about the universe and gain a deeper understanding of fundamental reality.

Quantum mechanics shows us that the universe itself is inherently computational.[570]

Modern physics has enabled observations that unveil phenomena beyond the microscopic scale, where behavior is often interpreted through the wave function. A wave function describes how matter exists at the tiniest scale, which is nothing like the way we perceive objects in everyday life. To visualize how standard physics describes this, imagine a person steps into a Star Trek transporter, momentarily becoming an uncertain pattern before re-materializing. Similarly, the wave function represents a system as a range of possibilities. This has led to fascinating discoveries such as quantum *entanglement*, *superposition*, and *wave-function collapse*, revealing a once-hidden layer of nature.[571]

Quantum mechanics is more than a set of equations describing microscopic behavior—it reveals a universe that operates fundamentally through information processing. Rather than being just a branch of physics, it can be viewed as a computational framework where quantum states encode and manipulate data in ways that transcend classical logic. This brings us to informatics—the study of how information is structured, processed, and communicated in both natural and engineered systems. According to the School of Informatics at the University of Edinburgh, informatics examines these interactions across various domains.[572] Given these principles, *quantum informatics* may more accurately describe what is traditionally known as quantum mechanics, as it better reflects the fundamental role of information processing at the quantum level.

In the quantum world, entanglement creates a unique link between particles, where the state of one instantaneously influences the other, regardless of distance.[573] Einstein famously called this influence "spooky action at a distance." This interconnectedness brings to light how information-rich the universe is, as changes in one part of a system instantaneously reflect in another.

Quantum superposition is the idea that a quantum system can exist in multiple states at once until it is examined or looked upon. Upon this observation, the quantum system takes on a definite state—a process known as wave function collapse.

To visualize this, imagine an object spinning in the air. You can see it spinning, but it's tiny, so you put on magnifying glasses to take a closer look. The moment you do, it freezes into a side of a coin—heads. Curious, you repeat the experiment and consistently find that the object keeps spinning until you put on the glasses, at which point a side is revealed—either heads or tails. You begin to realize that by bringing it into focus using the glasses, the spinning object takes on a definite state each time.

This leads to a debate: Did the coin's outcome become definite because the glasses magnified it, or only when you, the observer, became fully aware of it? Some argue that the glasses alone—like a measuring device—determined the result, while others claim that it only became real when consciously perceived. This mirrors the quantum debate over whether measurement alone collapses the wave function or if conscious observation plays a fundamental role.

In quantum mechanics, the term "observation" is often used interchangeably with "measurement," creating a little confusion by equating physical measuring devices with conscious observers. In other words, credit is often given to the measuring device itself for causing wave function collapse, as if the apparatus alone has the power to enforce a definite outcome. This assumption leads to the broader idea that any interaction with a quantum system—whether from a detector, a photon, or an external force—qualifies as a "measurement" capable of collapsing the wave function. But does wave function collapse occur solely due to physical interaction, or is conscious awareness necessary for it to happen?

Many argue that consciousness plays no role in the collapse. If this assumption were true, however, measurement would lose any distinct meaning, as quantum systems are constantly interacting with their surroundings. Collapse would have already occurred before the concept of measurement. The concept of "measurement" implies a distinction between ordinary interactions and those that force a definite outcome—yet this assumption blurs that line entirely.

The distinction between *observing* and *measuring* is not an inherent feature of quantum mechanics but rather a manufactured distinction, likely introduced due to bias against the idea of consciousness playing a role in wave function collapse—though it appears to be the case. Early quantum physicists—including John von Neumann, Eugene Wigner, and even Niels Bohr—acknowledged that observation played a fundamental role in collapse.[574] Even though Schrödinger was critical of the idea that observation determines reality, his views on consciousness and reality suggest that he did not entirely dismiss its potential significance in quantum mechanics.[575] Though somewhat counterintuitive for physicists, the *observer effect* was initially discussed in a way that left room for the idea that consciousness might influence wave function collapse.[576] However, as materialist interpretations of physics became dominant, there was a growing consensus to separate measurement from consciousness, leading to a strategic reframing of terms.

It is important to distinguish quantum observation from the classical notion of the observer effect. Originally, the observer effect was a concept from classical physics, referring to how the act of measurement physically disturbs a system.[577] For example, using a thermometer to measure temperature slightly alters the system, checking the pressure in a tire causes some air to escape, or illuminating an object with light can exert pressure and subtly change its state.[578] These are examples of physical disturbance, where measurement affects the system but does not fundamentally determine its state.

Over time, the concept of measurement was misapplied to suggest that wave function collapse is merely another physical interaction, rather than something tied to conscious observation.[579] This redefining allowed materialists to create the narrative that only the device disturbs the system, avoiding the deeper issue: before measurement, the system exists in a range of possibilities.[580] Early pioneers like John von Neumann and Eugene Wigner argued that measuring devices do not finalize wave function collapse.[581] [582] Unlike classical systems, where measurement simply reveals a pre-existing property, quantum measurement appears to bring definite properties into existence, leading them to consider consciousness as the final step in the measurement process.[583]

The act of quantum observation is similar to how many video games use *frustum culling*, a technique where only the objects within the player's field of view are rendered, while unseen elements remain unprocessed to optimize performance.[584] However, in a video game, the world exists as pre-coded data awaiting activation, whereas in quantum mechanics, the observer's act of observation is what determines reality itself.[585] This runs counter to Materialism, which assumes a reality that exists independently of any observer—a universe that would persist even if life had never come to be. In contrast, quantum physics suggests that without observation, no definite, objective reality exists—only unresolved potentialities.

Although many physicists today favor interpretations that remove consciousness from quantum physics, this remains a philosophical stance rather than a fact. The well-established dominance of materialist perspectives, however, has dictated both conceptual interpretations and linguistic framing, imposing strict boundaries. This has made it particularly difficult to address and fully explore complex issues like consciousness, which resists straightforward measurement or quantification.[586]

Historically, figures like Newton, Kepler, Maxwell, and Planck described the cosmos in physical terms, yet they saw its order as evidence of design. Similarly, early pioneers of quantum mechanics, such as Schrödinger, Werner Heisenberg, and Wolfgang Pauli, grappled with the deeper implications of their discoveries, often invoking spiritual or even theological considerations in their interpretations.[587]

Schrödinger's rejection of strict Materialism went beyond curiosity; he believed that consciousness could not be fully explained by physical processes alone.[588] Schrödinger was deeply influenced by Hindu philosophy, particularly *Vedanta*, and saw parallels between quantum mechanics and the idea of a unified consciousness. In *My View of the World*, he reinforced this perspective by directly referencing the Vedantic principle *Tat tvam asi* ("You are that"), suggesting that individual consciousness is not separate but fundamentally part of a greater whole:

> Inconceivable as it seems to ordinary reason, you—and all other conscious beings as such—are all in all. Hence, this life of yours which you are living is not merely a piece of the entire existence, but is in a certain sense the whole. This, as we know, is what the Brahmins express in that sacred, mystical formula which is yet really so simple and clear: Tat tvam asi, this is you.[589]

For Schrödinger, this was not just philosophical speculation but truth about reality, one that reflected his scientific worldview. He saw the multiplicity of individual consciousnesses as an illusion. In *Mind and Matter* he wrote:

> There is obviously only one alternative, namely the unification of minds or consciousnesses. Their multiplicity is only apparent; in truth, there is only one mind. This is the doctrine of the Upanishads.[590]

Schrödinger grapples with the paradox of individual perception versus shared reality. He acknowledges that each person's experience of the world is strictly private, yet there exists a striking agreement between these separate sense-worlds. He notes that "many people prefer to ignore or gloss over the strangeness of it," instead explaining this agreement by assuming the existence of an independent, external reality.[591] He writes, "The broad measure of agreement between two observed worlds… is to be explained by some sort of correspondence with the real world."[592] However, he immediately turns the assumption on its head, questioning whether such a *real world* (R) can even be observed or meaningfully compared to our perceptions.

Ultimately, he finds the assumption of an external, objective reality insufficient to explain this agreement, leading him to conclude: "These two worlds of experience, yours and mine, agree because the same mold is shaping them similarly in similar material."[593] Yet, this explanation alone does not satisfy him. He then poses a deeper question: "How do we both get knowledge of this agreement?"[594] This question forces him to confront the paradox of subjective experience and objective reality, reinforcing his conclusion that individual minds are merely apparent divisions of a single, unified consciousness.

Many prominent scientists and philosophers seeking a unified explanation of reality have gravitated toward a form of *monism*, often drawing direct inspiration from Hinduism, particularly Vedanta. This likely stems from their shared emphasis on an underlying unity beneath apparent dualities. Even before modern physicists, influential figures such as Charles Lyell, Ernst Haeckel, and Ernst Mach embraced monistic ideas and were drawn to Hindu philosophy, often in opposition to Western religious traditions. Among more modern scientists, David Bohm became one of the most prominent advocates of monism, proposing a deeply interconnected reality through his theory of the Implicate Order, which parallels Vedantic concepts of an underlying unity.[595]

This pattern reveals a paradox: while many scientists aim to reject supernatural explanations, their pursuit of a fundamental reality frequently leads to metaphysical conclusions—suggesting that a strictly deterministic view of the universe is insufficient to fully account for existence, consciousness, and the unity observed in nature. These concepts are even incorporated into K-12 educational settings, such as through yoga and mindfulness programs, though western spiritual perspectives are often excluded.[596] [597]

Rather than dismissing spiritual or metaphysical ideas, Schrödinger and others saw them as necessary to bridge the gap between the physical world and the conscious experience, an approach fundamentally at odds with modern materialist interpretations of quantum mechanics. Modern physicists, despite positioning themselves as heirs to these pioneers, have increasingly avoided engaging with the broader philosophical implications of the evidence. This materialist framing has led to a resistance in acknowledging the obvious implications of quantum mechanics, particularly regarding the role of the observer in shaping reality.[598]

Despite the claim that wave function collapse occurs purely due to physical measurement, Wheeler's Delayed-Choice Quantum Eraser experiment contradicts this assumption, demonstrating that if *which-path* information is erased before being consciously observed, the interference pattern reappears.[599] This suggests that collapse is not finalized at the moment of physical interaction but only when the result becomes known to a conscious observer. A detector is not an observer. If collapse were purely a physical process, then once a detector registers a measurement, the interference pattern should never return.[600]

While some physicists propose that decoherence resolves the measurement problem without invoking consciousness, decoherence only explains how quantum states appear classical—it does not resolve the question of definite outcomes. The only verified fact is that quantum systems remain in superposition until a result is known. As the *Stanford Encyclopedia of Philosophy* states:

> One often hears the claim that decoherence solves the measurement problem of quantum mechanics... Physicists who work on decoherence generally know better, but it is important to see why even in the presence of decoherence phenomena, the measurement problem remains or in fact gets even worse.[601]

Another troubling aspect of the measurement problem for some is that while the observer appears to affect the system by collapsing its wave function, the system does not, in turn, place the observer into a superposition. According to their assumptions, the measurer—being a physical system—should also evolve into a superposition upon measurement, yet this is not what is experienced.[602] This discrepancy arises, in part, because these physicists view the mind as only a byproduct of physical interactions, failing to account for the apparent role of conscious observation in determining quantum outcomes.

This is at the heart of the measurement problem, which sits uneasily with Materialism by suggesting that reality, at its most fundamental level, plays a role in determining physical outcomes—implying that mind and matter are more intertwined than Materialism allows. Additionally, the persistent unease with a purely probabilistic framework is closely linked to another fundamental issue: the fine-tuning problem. If only one reality exists and it happens to have precisely the right conditions for our cosmos and life, this suggests either an unfathomable coincidence or an underlying reason—such as mindful design.

Theorists seeking explanations more compatible with Materialism have settled on the Many-Worlds Interpretation (MWI) to resolve these discrepancies. MWI is highly favored because it circumvents the need for consciousness by eliminating wave function collapse altogether. What later became known as the MWI traces back to Hugh Everett's 1957 Ph.D. thesis, originally titled the "Relative State Formulation" of quantum mechanics.[603]

MWI addressed the measurement problem by positing branching worlds based solely on quantum events, with each universe causally isolated.[604] That is, each universe is completely separate, so nothing that happens in one universe can affect or influence what happens in any other. However, Everett's "Relative State" formulation, which differs in key ways from the modern MWI, was initially dismissed as too radical and speculative. This rejection contributed to Everett eventually abandoning academia.

Yet, as physicists continued to grapple with the probabilistic nature of quantum systems, no alternative theory in the following decades succeeded in resolving these conceptual challenges to Materialism. The lack of progress in resolving the measurement problem led many fatigued materialists to crawl back to Everett's interpretation. This renewed interest picked up in the 1980s and 1990s, when the theory was reconsidered and refined, increasingly being treated as a serious possibility.[605] Thus, this shift was not driven by new evidence but rather by the failure of existing frameworks to provide viable alternatives. For many, MWI is now regarded as the only possibility.

Instead of requiring an observer to choose an outcome, the theory proposes that all possible outcomes happen, but in separate, causally isolated branches of reality. This means there are "many worlds," similar to ours, where each possible outcome occurs. This bypasses the need for an observer to cause the transition from superposition to a definite state. From this view, the illusion of a single, definite reality is simply because we exist in only one branch at a time. MWI restores a fully deterministic, observer-independent reality—even if it requires an infinite number of unobservable universes to do so. Though some physicists attempted to use it to explain fine-tuning, this notion has largely fallen out of favor. Instead, later theorists turned to cosmological multiverse models, which allow for varying fundamental constants across different universes.

If Many-Worlds is correct, then with every quantum event, countless versions of ourselves come into existence, each existing in a separate, causally isolated branch of reality. Yet, despite this infinite branching, our experience remains singular and continuous—we do not perceive ourselves as splitting, nor do we detect any divergence in our personal existence. This presents a contradiction: if we are constantly branching, why do we feel as though we remain the same individual in the same world, maintaining the same family, friends, and worldview? This explanation fails to address why our conscious experience does not reflect the fragmentation MWI describes.

Theorists often conflate interpretations such as MWI and the multiverse hypothesis to create the appearance of a unified, materialist-friendly framework. This serves two purposes—first, to reinforce the illusion that there is broad consensus around a purely physical explanation of quantum mechanics, and second, because none of these interpretations are verifiable or testable. What theorists are also not forward about is that there are multiple multiverse theories, but they are fundamentally incompatible and even contradict one another. Merging these concepts blurs their distinct descriptions and creates confusion over what these 'other' universes actually represent. For example, are they quantum branches of a single wave function, or entirely separate, independent realities?

Many physicists interpret MWI as a tree-like structure, where new branches continuously diverge from one another, but this is not how MWI theoretically functions. There is a philosophical branching, but the branches are completely detached from each other after "environmental-induced" decoherence.[606] This idea is better described as a fragmentation of a single material reality where everything originates from one unified wave function. MWI's fragmentation can be conceptually compared to giant molecular cloud (GMC) fragmentation, where a single entity (a massive cloud of gas) splits into multiple, independent structures that no longer influence each other's evolution.

Even critiquing these theories is indeed strange; it's like reviewing scripts from different Marvel movies, but let's continue. Unlike multiverse theories where different universes may have different physical laws, all MWI "fragments" obey the same fundamental physics but evolve independently. The multiverse is not a split in the same sense as MWI but rather the continuous creation of entirely separate realms, where all conceivable realities must exist—some following familiar rules, others obeying laws beyond anything we can imagine or even comprehend.

Interestingly, in the many models of multiverse theory, there must exist at least one universe where a God exists and has created that realm. Since these models propose that all conceivable realities must exist, a universe with a divine creator would be an unavoidable part of the vast multiverse landscape. [607] In fact, according to the multiverse theory, phenomena like a first-century man named Jesus walking on water or turning water to wine is commonplace, simply part of some discrete universe's "natural" laws. As Brian Greene suggests, our intuitions may be ill-suited to grasp reality's true nature, recommending that we not dismiss theories like Many-Worlds or the multiverse simply because they seem strange.

Inflationary theory, however, does not directly address wave function collapse. In fact, what is now known as *eternal inflation*—which posits a multiverse—has no inherent connection to the original inflationary theory.[608] Ultimately, these theories are post hoc extrapolations.[609] Inflationary theory was introduced to explain away why the universe's observed properties contradicted the materialist model of cosmic evolution. It ultimately attempts to account for features that make the universe appear the result of special creation. Eternal inflation theory was crafted to provide the theoretical currency needed to fund the multiverse concept, retroactively making inflation seem inevitable rather than fine-tuned.[610]

Since inflation describes a process within our universe, it does not occur in MWI, nor does quantum branching take place during inflation. Inflationary models deal with large-scale classical expansion. If these multiverse models are true as described, it suggests a "collapse" of quantum fluctuations into classical outcomes—which contradicts MWI's core principle that collapse never happens. If MWI is true, then inflation should never result in a single classical universe, which raises the question of how we end up with an observable cosmos that has one definite structure. Theorists who claim both can exist are ignoring this contradiction or using rhetorical tricks to merge them without addressing the conflict.

By blending these ideas, many theorists cherry-pick elements from each framework—such as invoking MWI to explain quantum outcomes while relying on multiverse ideas for cosmic fine-tuning. Rather than accepting the evidence and implications of our reality, they seek to circumvent it, promoting a multiverse that is, by definition, unobservable, untestable, and unscientific.[611] This double standard allows observable phenomena like the observer effect to be downplayed, while egregiously speculative models like multiverses are elevated as "scientific" enlightenment. This is yet another example of how modern materialist physics suppresses evidence of true reality, contradicting the greatest pioneers of pure science while relying on illogical, self-conflicting theories—scrambling at every turn to avoid the collapse of Materialism.

Sean M. Carrol, theoretical physicist who is currently at Johns Hopkins university, has been one of the most vocal proponents of MWI and multiverse theories, consistently presenting them in a way that misleads audiences by misrepresenting their speculative nature, downplaying their lack of empirical evidence, and framing them as the inevitable conclusions of quantum mechanics and cosmology. In one talk, Carrol presents the MWI as the most natural and unproblematic understanding of quantum mechanics. His discussion is a carefully crafted narrative designed to reframe an experimental fact of quantum physics as a weird problem that MWI supposedly solves. This is not an honest explanation of quantum mechanics; it is materialist apologetics.

Carroll begins his talk by stating, "The Schrödinger equation makes an absolutely crystal-clear prediction for what happens when you interact with that electron. … you saw it somewhere else for every single possible place you have seen it. In other words, you yourself evolved into a superposition."[612] However, his framing of the Schrödinger equation in relation to MWI is misleading, as he has admitted in his own academic writing:

> Wavefunctions evolve according to the Schrödinger equation, at least when the system is not being observed. Upon measurement, the wavefunction collapses to an eigenstate of the measured observable.[613]

This directly contradicts his public presentation, where he falsely extends the Schrödinger equation to imply that measurement naturally leads to Many-Worlds branching.

Further, Schrödinger did not fully endorse any single interpretation of quantum mechanics, but his views aligned more closely with *wave function realism*—the idea that the wave function represents a real, continuous physical entity rather than just a mathematical tool for calculating probabilities.[614] Although Schrödinger was critical of Copenhagen's wave function collapse, the MWI implication that every quantum event creates a branching universe is something Schrödinger likely would have rejected as absurd. Instead of addressing the obvious fact that measurement produces a single outcome, Carroll conflates the MWI assumption that the observer "evolves into a superposition" of having seen the electron in every possible position.[615] This is not something Schrödinger's equation states or implies—it is an invention of MWI.

Carroll rewrites the history of quantum mechanics to frame wave function collapse as an arbitrary, ad-hoc fix. He claims:

> We've never thought that we were in a superposition, right? No one ever feels like that's what they've seen. So, the inventors of quantum mechanics said, well, we're gonna pick one of those possible places.[616]

This is an outright misrepresentation. The founders of quantum mechanics did not "pick" an outcome—they acknowledged the fact that measurements always produce single, definite results. They never observed a different outcome outside the reality they existed in. The concept of wave function collapse was not introduced as an arbitrary rule; it was a direct reflection of experimental reality—a reality that displeases Carroll.

Also, to make MWI appear inevitable, Carroll makes the claim:

> The Many-Worlds interpretation wasn't invented by adding many worlds but by removing some of the weird, bizarre, ad-hoc rules that people had invented for dealing with measurement.[617]

This further projects ad-hoc reasoning onto the traditional interpretation while obscuring from his audience the fact that MWI replaces what is actually observed with a supernatural assumption. His statement absolutely false. As we will see, many worlds were not just added to the equation—they were an add-on to the theory from which they arose. If wave function collapse is considered a "bizarre" rule, then postulating an infinite number of physically real, unobservable universes is an infinitely more bizarre assumption.

He further claims, "There are now two worlds, one in which the electron was moving clockwise and one in which it was moving counterclockwise."[618] There is, however, no experimental confirmation that these two worlds exist. In fact, in the same breath, Carroll admits, "The sad part is you can never talk to each other. The worlds are truly separate, they can't interact."[619] However, if they cannot interact and we can never observe them, then this is not science; it is a supernatural claim.

His argument is not much different from how cosmic inflation was introduced to *solve the problem* of uniformity in the cosmic microwave background—not because it was a real problem, but because Materialism could not explain it without untestable add-ons. Carroll's entire framing is designed from the start to lead his audience to his predetermined conclusion that MWI is the only correct theory.

Carroll's misrepresentation of quantum mechanics does not stop at Schrödinger's equation—it extends to the fundamental nature of Everett's original Relative State formulation. While Carroll insists that the MWI is the "natural" extension of quantum mechanics, his version of MWI itself is a modern reimagining, layered with speculative additions that Everett himself never proposed. Everett's 1957 thesis, "'Relative State' Formulation of Quantum Mechanics," does not describe distinct, causally separated universes. Instead, it presents a deterministic universal wave function where all possible measurement outcomes remain part of a single, evolving mathematical structure:

> From the viewpoint of the theory, all elements of a superposition (all 'branches') are 'actual,' none any more 'real' than the rest. It is unnecessary to suppose that all but one are somehow destroyed, since all the separate elements of a superposition individually obey the wave equation with complete indifference to the presence or absence ('actuality' or not) of any other elements.[620]

Contrary to Carroll's portrayal, Everett never claimed that these "branches" imply physically real, separate universes. Instead, his formulation describes them as mathematical correlations within a single wave function, not distinct, causally disconnected realities. Over time, these correlations undergo decoherence, making different outcomes effectively non-interacting but still part of the same underlying reality. As Everett himself explains:

> Thus the state of one subsystem does not have an independent existence, but is fixed only by the state of the remaining subsystem. In other words, the states occupied by the subsystems are not independent, but correlated.[621]

Everett's entire framework is based on correlations, not separations. Different outcomes exist relative to different *observers*, but they do not "split" into fully formed, disconnected universes as Carroll falsely, along with many others, falsely presents.

The modern "Many-Worlds" framework, in which every quantum event causes reality to split into distinct universes, was introduced by Bryce DeWitt in the 1970s—though this detail is often left out when attributing MWI to Everett. DeWitt reinterpreted Everett's relative states as literal, causally isolated parallel universes, introducing the speculative "branching worlds" concept that Carroll now treats as fact.[622] Many proponents of MWI, either ignorantly or deliberately, conflate Everett's original work with DeWitt's reinterpretation, conveniently ignoring that Everett himself never proposed physically distinct universes.

Everett's formulation still left room for an observer-dependent reality—since measurement entangles the observer with a relative state. Everett's model left a problematic question for materialists: If all measurement outcomes exist in the universal wave function, why do we only experience one? Thus, the role of consciousness in measurement was seen as "weird" and had to be eliminated. DeWitt's Many-Worlds removed the problem entirely by declaring that all outcomes "exist" in separate universes, making the observer irrelevant.[623]

In a different presentation aimed at science fiction movie enthusiasts, Carroll shifts his focus to explaining the multiverse. Despite presenting MWI in the previous talk as the most natural interpretation of quantum mechanics, here he starts with the multiverse, treating it as equally real, following the same rhetorical pattern—constructing a false problem, misrepresenting the nature of observation, and steering the audience toward his predetermined conclusion. He spends considerable time reinforcing the multiverse as a serious scientific idea, distancing it from its portrayal in pop culture.

He begins this talk by cheekily acknowledging that the multiverse has gained popularity in Hollywood, citing films like *Doctor Strange in the Multiverse of Madness* and *Everything Everywhere All at Once*, but quickly asserts that the version of the multiverse presented in pop culture is not the same as what physicists discuss.[624] He ironically distinguishes between a "philosophical" idea of multiple possible worlds and what he claims are the "scientific" multiverse models from physics. However, the distinction he makes is ultimately misleading, as his own framework relies on highly speculative, unobservable realities just as much as any Hollywood script does. In fact, given the logic of an infinite multiverse, theoretically, anything depicted in a Hollywood movie could actually occur in some universe.

Carroll insists that physicists were "dragged" into the idea of the multiverse "kicking and screaming" because they were simply following where the evidence led them.[625] In reality, the multiverse was not a conclusion forced upon physicists. Instead of accepting that the universe appears finely tuned or that observation plays a fundamental role in reality, materialists proposed an infinite number of unseen, causally disconnected universes to preserve the assumption that our own universe is nothing special. The "kicking and screaming" was not a resistance to speculation, but to the evidence itself.

Carroll's framing makes it seem as though the multiverse, as he does with MWI, is the natural outcome of evidence. But how can both be inevitable when they fundamentally contradict each other? He further distorts the issue when he states that theories like inflationary cosmology and Many-Worlds "unambiguously predict" the existence of other universes.[626] This is a misrepresentation of what predictions actually mean in science. A true scientific prediction is something that can be tested and falsified, yet Carroll casually groups inflation and MWI together, while ignoring the fact that neither has direct empirical support, and both rely on unobservable assumptions.

As Carroll strategically shifts from discussing the multiverse back to his fancy, the Many-Worlds interpretation, he repeats his previous rhetorical sleight of hand by presenting MWI as if it is simply a direct conclusion of quantum mechanics. He states that quantum systems allow for multiple measurement outcomes and asks, "What happens to the alternative measurement outcomes that you didn't observe?"[627] By phrasing it this way, he subtly smuggles in the assumption that those alternative outcomes must still exist somewhere rather than considering the *far simpler* and observationally supported conclusion that they simply do not occur. He then makes the bold claim that MWI is "literally a parallel universe," not just something very far away but something that "exists simultaneously."[628]

He then attempts to distance MWI from human decision-making, stating that the branching of universes is not caused by choices but rather by quantum measurements.[629] Yet, in doing so, he unintentionally reveals the contradiction in his explanation. If quantum measurements create separate universes, then the observer is still fundamentally tied to the process—something he has worked hard to deny in many discussions. It takes a conscious agent to create, set up, and interpret the device whose measurement records the quantum outcome.

If measurement is what "splits" worlds, then conscious observers must be present in every branch where a measurement occurs. Without an observer to define a measurement basis, there would be no reason for distinct branches to correspond to experience. This suggests that human consciousness is inherently required in each world, as the branching process itself depends on an act initiated by conscious observers.

Yet, Carroll goes further: "If anything, they [quantum measurements] force us to make decisions." This statement completely reverses his previous stance. If quantum measurements "force" decisions, then observation is not a bystander but an active participant in shaping reality. This contradiction reveals a fundamental debacle in his framing. When rejecting wave function collapse, he argues that measurement does not cause anything special, that branching simply "happens" as a natural extension of quantum mechanics. However, when justifying MWI, he describes measurements as actively compelling choices, implying a one-way influence that contradicts his own claim that observation has no effect.

However, if every quantum measurement result in a branching universe, then our direct experience of a single reality remains unexplained within MWI—an issue Schrödinger himself pondered in his reflections on the nature of the 'self.' Carroll acknowledges this issue indirectly when he states that the other versions of ourselves in parallel worlds are like "identical twins" who share a past but have since diverged into separate people.[630] This is far from simplifying what we observe; it is a speculative reconstruction that alters the very nature of observation itself.

After leading his audience through this narrative, he leaves them with the idea that we must accept the multiverse because science has no alternative. He claims that contemplating the multiverse can provide psychological insight, allowing people to reflect on choices they could have made differently, but then he closes with a deterministic statement that we must focus only on the present because we cannot undo the past. The irony here is that he began by presenting a theory in which infinite versions of ourselves exist elsewhere, yet he ends by insisting that our choices are fundamentally locked into a single timeline. This contradiction exposes the true nature of asserting many worlds.

Despite the countless chicken-and-egg systems in the universe, where it is inherently impossible to determine which precedes the other, many scientists continue to hope that all of existence can one day be encapsulated within a singular deterministic framework. This presumptuous goal called the Theory of Everything carries an unshakeable belief that every facet of reality, encompassing all phenomena, is fully quantifiable and conceptually explainable through matter alone.[631] For them, existence is not merely to be observed or marveled at but decoded, as if reality itself were a system meant to be fully computed and mapped.

However, critics of this view, such as James A. Hughes, a philosopher and critic of *scientism*, highlight the limitations of reducing all aspects of reality to scientific testing. In his essay, "The Folly of Scientism," Hughes critiques the overextension of scientific reasoning into domains like consciousness and meaning, where empirical methods often fail to capture subjective dimensions of existence.[632] How can the mind entirely perceive reality when that reality is shaped by the mind's own perceptions, which it uses to interpret itself? In this way, the mind is both the tool and the subject of this inquiry, creating a logical gap.

In physics and cosmology, the role of scientific laws and mathematics in explaining the universe's origins often sparks debates centered on the interpretation and application of these natural laws. A notable instance in this discussion is Stephen Hawking's statement in *The Grand Design*, where he suggests that, given the existence of gravity, "the universe can and will create itself from nothing."[633] To examine Hawking's assertion, it's necessary to understand the nature of scientific laws.[634] Physical laws—such as gravity—are not agents; they are descriptive models, like adjectives, not entities. These laws summarize consistent patterns observed in nature, offering predictive precision, but they do not possess causal power. They cannot act, create, or initiate events; they describe what happens, not why it happens.

As the Austrian philosopher Ludwig Wittgenstein astutely noted:

> The great delusion of modernity is that the laws of nature explain the universe for us. The laws of nature describe the universe, they describe the regularities. But they explain nothing.[635]

This sentiment directly pushes back on the widespread belief that scientific laws explain the universe—they describe its patterns, but offer no account of why those patterns exist at all.

Think about the simple scenario of a ball rolling across the floor. In this case, the rolling ball is a reaction to an action–perhaps someone pushing it. This illustrates Newton's third law of motion: for every action, there is an equal and opposite reaction.[636] The law itself, however, only describes what consistently happens under these circumstances. It tells us that if there is an action (the push), there will be a reaction (the ball rolling). But this law does not cause the ball to roll. The actual event of the ball rolling is contingent on a real action taking place. It would be illogical to state that 'because we have a law such as Newton's third law of motion, the ball can and will create itself from nothing.' The ball's material composition transcends the laws that govern its interactions.

Expanding on our conversation, the conceptual hurdles faced by materialists become evident. Materialist explanations for the origins of the universe, such as nebular formations, stars, moons, water, DNA, and life, often assume certain conditions already exist without addressing how these conditions themselves came into being.[637][638] By their explanations, one must presuppose the existence of the material conditions that Materialism seeks to explain, leading to circular reasoning. This inherent circularity shows a significant snag within their worldview.

Another example of this circular reasoning is evident in a recent presentation Sean Carroll gave at Ohio State University as part of the 15th annual Biard Lecture of the Center for Cosmology and AstroParticle Physics (CCAPP), where he explained the universe's origins. He begins by describing the early universe as existing in a quantum mechanical state but then casually asserts that it later "collapsed" into a definite state. Carroll claims that the universe "measures itself," triggering this collapse—despite his strong advocacy for the Many-Worlds framework, which explicitly denies that collapse ever truly occurs.[639]

But, overlooking the claim that the universe began as an indefinite quantum system, how could the universe, while still in an indeterminate quantum state, act as its own measurement device—collapsing its own wavefunction without any external observer or defined mechanism? Would this not presuppose a definite state before one could even be defined?

He then claims that we no longer need to worry about quantum mechanics because gravity (of course) "takes over." This assumes a seamless transition from quantum uncertainty to classical determinacy without providing an explanatory mechanism. Carroll presupposes the same process that needs explaining—how quantum indeterminacy gives way to classical determinacy—resulting in a circular assumption rather than an actual solution. His shifting narrative reflects not simply confusion but the underlying tension of defending a model whose assumptions no longer align with observable structure. It reveals a deeper impulse—one that resists the possibility of embedded purpose.

Carroll's words in context:

> So, the universe, according to our favorite theories right now, at very early times, was essentially featureless. There was a quantum mechanical state the universe was in where everything was literally exactly the same from point to point. But what happens is, and again, this is complicated quantum theology involved here, but essentially, the universe measures itself. It collapses from a superposition of many different things that add up to smoothness to a very particular specificity.[640]

As our understanding broadens, these narratives on cosmic, chemical, and biological evolution are increasingly recognized as so inconsistent that it seems inevitable this school of thought will collapse within the next few decades. This will be an intellectual climate change that these scientists did not anticipate, one that will make it clear the evidence was right in front of them all along, yet overlooked due to their entrenched dogmatic views.

When that time comes, adherents of radical Materialism may double down, gravitating toward new models of a self-creating, self-actualizing universe. Others may defer the question further, speculating about meta-universal alien intelligence or proposing that we exist within a simulation—one intentionally designed by beings like us, who themselves exist within an infinite series of simulations. Yet no matter how far they push the problem, the conclusion of intentional design will become unavoidable.

Henry Stapp, a prominent physicist, challenges these interpretations. He argues that consciousness has a fundamental role in quantum processes. According to Stapp, the observer effect is not simply a problem to be explained away but a phenomenon that points toward *free will* and top-down causation, where consciousness influences physical states. He notes:

> Orthodox quantum mechanics...allows the mentally described aspects of the agent to be cleanly separated from his physically described body and brain...Each mental element is now conceived to be created, not by a brain process, but rather by a mental process.[641]

Stapp emphasizes that quantum mechanics' formalism actually requires this active role of consciousness. This interpretation, grounded in experimental evidence, confronts the bias against incorporating non-physical elements into physics. From this perspective, the "measurement problem" is not a problem at all. Consciousness is an essential component of reality's structure, woven into the quantum process itself.

## Quantum Technology

History shows that the most profound breakthroughs arose from pushing the limits of observable reality with both sharp analysis and deep intuition. Celebrated for his substantial contributions to quantum mechanics, Richard Feynman emphasized the unconventional pathways leading to scientific breakthroughs.[642] He offers an important perspective on the relationship between intuition and scientific discovery.

**Figure 20.1** Richard Feynman at Paine Mansion Woods, 1984. **Image source:** Tamiko Thiel, Wikimedia Commons.

While I give credit to Behe and Meyer for introducing me to the idea of design in nature, it was Richard Feynman, a physicist and materialist, who awakened my conviction that the world could be understood by simply imagining how things work. That's the subtle but critical difference I draw between materialism and radical materialism. Feynman never closed the door to wonder. His attitude invited curiosity rather than shutting it down behind jargon or prestige. He didn't hide behind complexity; he used simplicity to illuminate it. In my opinion, Feynman is the most fascinating scientist of the past century. That's what sets him apart from many of today's pop-culture scientists, who act as if their credentials give them license to speak with authority on subjects they either distort or fundamentally misunderstand. These figures often use esoteric abstractions to shield poorly reasoned assumptions—often under the guise of "settled science."

A prime example is Lawrence Krauss, who wrote a critique titled "Ben Carson's Scientific Ignorance." In it, Krauss accuses Carson of misunderstanding the second law of thermodynamics and the nature of the Big Bang. Carson had expressed skepticism about the idea that an "explosion" (the Big Bang) could lead to the highly ordered structure we see in the universe today. Krauss responds with an explanation about local order arising from environmental entropy, citing snowflakes and stars as examples of complexity produced by simple natural processes. But in doing so, he uses examples of what Carson assumes to be designed systems to argue that complexity can arise "naturally."

What he never addresses is the origin of that natural capacity itself: Where did the ability of nature to produce such order come from? Krauss reframes the question, offering mechanisms rather than origins, and relies on effects to prove causes. In essence, he's pointing to complexity and saying, "See? It happens," without acknowledging that Carson's point is precisely that these phenomena appear too organized to be the result of unguided processes alone. Krauss uses selective reasoning throughout the article to portray Carson as illogical for believing something that, in truth, no one has proven false.

He treats his specialized knowledge as if it grants him general infallibility. This is the same Lawrence Krauss who argued in his own book that "nothing is really something"—a philosophical. Unlike Carson, who performed the world's first successful separation of craniopagus twins in 1987—a feat of immense biological precision—Krauss has contributed no comparable achievement in the experimental sciences.

Feynman's genius was that he never pretended to know more than nature would allow. He said:

> In this age of specialization men who thoroughly know one field are often incompetent to discuss another. The great problems of the relations between one and another aspect of human activity have for this reason been discussed less and less in public. When we look at the past great debates on these subjects we feel jealous of those times, for we should have liked the excitement of such argument. The old problems, such as the relation of science and religion, are still with us, and I believe present as difficult dilemmas as ever, but they are not often publicly discussed because of the limitations of specialization. ... To make progress in understanding we must remain modest and allow that we do not know. Nothing is certain or proved beyond all doubt. You investigate for curiosity, because it is unknown, not because you know the answer. And as you develop more information in the sciences, it is not that you are finding out the truth, but that you are finding out that this or that is more or less likely.[643]

That humility, grounded in intellectual honesty, is sorely lacking in the kind of argument Krauss makes—where the credentials do the heavy lifting and disagreement is treated as heresy. In the end, this isn't about whether Carson is right or wrong on every point. The issue is how scientific authority is wielded: as a searchlight, or as a weapon. Feynman used it to light the way. Krauss uses it to guard the gate.

Feynman once remarked about the origins of certain scientific equations, saying, "Where did we get that from? It's not possible to derive it from anything you know. It came out of the mind of Schrödinger."[644] This statement sheds light on Erwin Schrödinger's development of his fundamental equation in quantum mechanics.[645] [646] Schrödinger's equation, essential for understanding particle behavior at the quantum level, wasn't an outcome of linear progression from existing scientific knowledge. Instead, as Feynman pointed out, Schrödinger's equation was not just a stroke of luck or a conclusion that could be reached through a linear sequence of steps, but the result of an active engagement with the universe's underlying principles.[647] This is where the pursuit of answers draws forth insights beyond what was initially sought.

Nevertheless, one issue I have with Feynman's perspective is that, despite his brilliance, he often presented mathematics as a privileged gateway to reality—as if those who "speak math" possess an elite epistemic access to what is. This reflects a major misconception of materialism

If history is any guide, dismissing the observer's role may hinder breakthroughs that require confronting, rather than circumventing, the unresolved mysteries at the heart of quantum mechanics. Our growing understanding of the cosmos and our technological advancements consistently reveal a deeper truth: the truth is that we share a profound relationship with the cosmic technology that surrounds us and permeates our lives.

Even before modern computers and quantum discoveries, our technology has long reflected hidden structures in reality. The telegraph allowed for instantaneous connections across vast distances. The telephone transmitted voices in real time, echoing the universe's interconnectedness. Even a photograph—freezing a continuous, uncertain reality into a single, definite state—can be seen as a form of collapse, selecting one possibility from countless potentials. But these discoveries are not strange at all—they are signposts, nudging us toward a greater realization of the fundamental, interconnected framework that has always existed.

Now, let's reconsider science fiction technology from shows like the Jetsons and their face-to-face video communication across vast distances. This concept, once a fantasy, now parallels our reality with technologies like FaceTime. This speculative bridge between fiction and reality, in many ways, extends to computing itself. The rise of cloud technology reflects a deeper truth about the universe—one that quantum mechanics is only beginning to unveil. Cloud-based automation redefines traditional notions of separation and distance, much like quantum entanglement. In advanced cloud systems, processes synchronize across multiple nodes, ensuring that an action in one location is instantly mirrored elsewhere.[648]

While quantum computers are functioning prototypes, they are far from fully operational due to issues like decoherence, noise, and scalability challenges. Despite these hurdles, they harness the principles of superposition and entanglement, offering glimpses of exponential computational potential compared to classical computing systems.[649] It's quite a feat to achieve within a century of the first PC. Our technological and conceptual advancements represent a convergence with the cosmic blueprint, as if reliving an ancient memory—reaching the apex of the Tower of Babel or the Tree of Knowledge—where our reach and understanding could become boundless.[650]

By unraveling the principles of quantum mechanics, we are developing technologies that can revolutionize various fields. Quantum cryptography leverages entanglement to create theoretically unbreakable encryption, ensuring secure data transmission resistant to eavesdropping or tampering.[651] This reinforces the fundamental role of information in quantum mechanics. Likewise, quantum computers offer unprecedented computational power, tackling problems that are currently intractable for classical computers, such as drug discovery, optimization, and complex system modeling.[652]

Additionally, quantum sensors exploit quantum states to achieve unprecedented levels of precision in measurements.[653] These sensors can measure physical quantities with extreme accuracy.[654] In 2023, physicists achieved the most precise measurement of the electron's *magnetic moment* to date. This property, which reflects the electron's intrinsic magnetic strength, was determined with an accuracy of 0.13 parts per trillion.[655] Such precision allows for stringent tests of the standard model of particle physics, as any deviation between the measured value and theoretical predictions could indicate new physics beyond the current understanding.[656] As quantum experiments refine our understanding of fundamental particles and interactions, they force us to reconsider the nature of reality itself.

The ability to measure physical properties with extreme precision is not just transforming our understanding of fundamental particles—it is also shedding light on the intricate processes within living systems. Inside each cell, an array of sophisticated mechanisms and processes operate continuously, facilitating life's essential functions like energy production, replication, and repair. Factoring in the potential influence of quantum phenomena makes the picture even more fascinating.

As quantum computing advances, this developing technology could help decode and illuminate mysteries surrounding DNA. With its ability to process vast amounts of information simultaneously through qubits and quantum entanglement, it could revolutionize our understanding of genetic information and its regulation. By simulating the quantum mechanical behaviors of molecules, quantum computers may help us understand DNA replication, repair, and gene expression in unprecedented detail.

A recent breakthrough has been proposed, where DNA itself operates as a perfect quantum computer, as detailed in the article "Scientists Propose DNA as Perfect Quantum Computer, Paving Way for Advances in Medicine."[657] The research suggests that DNA's aromaticity and the oscillatory resonant quantum state of correlated electron and hole pairs allow it to function similarly to a Josephson Effect between two superconductors, demonstrating the intrinsic quantum nature of genetic material.[658] This discovery unlocks new possibilities for rapid and reliable genome analysis and could significantly advance personalized medicine.[659]

Consider the dense network of biochemical reactions that sustain cellular life: enzymes catalyze reactions at remarkable speeds, DNA strands unwind for replication, and cellular signals cascade to trigger various responses.[660] Classical biology teaches us these processes follow precise biochemical pathways. Still, quantum mechanics introduces a new dimension of variability and unpredictability.[661] [662] For instance, quantum tunneling could affect electron behavior in enzymatic reactions, making these processes more efficient or occurring in ways classical physics cannot fully predict.[663] Cells coordinate myriad functions and respond to environmental changes with precision and adaptability.

If cells do utilize quantum computation, it would disrupt the materialist framework that underpins much of scientific discourse.[664] Nothing could be more convincing than the complex information-processing capabilities inherent in biological systems which show that life operates on principles beyond chemical reactions and deterministic interactions. The notion that quantum computation occurs within biological systems remains a cutting-edge research area, with much still to be understood and proven.[665] A major challenge in *quantum biology* is how quantum states—typically prone to rapid decoherence—can persist in the warm, wet, and noisy environment of living cells. While some life forms exhibit quantum effects without possessing consciousness, it is possible that the mind, in more complex organisms, helps stabilize or direct these processes.[666]

## Cognitive Information Threshold

As for consciousness, contrary to the conventional view of the brain as the creator and sole facilitator of consciousness, the human brain, in reality, acts as a limiter, restricting the self's ability to perceive reality as it truly is. Just as quantum measurement collapses a range of possibilities into a definite outcome, our brain functions as a biological filter or interface, reducing a broader consciousness into a structured, veiled experience of reality. Like a computer, it translates data into specific formats, offering a limited, simplified view of existence. This controlled output is a necessary limitation, focusing our awareness on aspects of reality that are immediately relevant to human experience.

In other words, this confinement allows us to interact meaningfully with the system without being overwhelmed by its complexity. Without this filtering, our brains—acting as fuses—would simply blow. Although human-made technology benefits us in many ways, the hubris born from these achievements distances us even further from perceiving true reality.

Some experiences—like meditation, psychedelics, or other altered states—can offer glimpses beyond the usual "translation" that our brain provides. In these states, it's as if the brain temporarily loosens its controls, allowing access to a broader flow of information or voltage. People often report a direct connection to a larger, unified reality, a different experience of time, or complex, patterned structures within ordinary objects. This demonstrates an ability to see past the usual confines of consciousness into the veiled, engineered layers of reality, if only temporarily. Although meditation and natural practices may safely enhance perception, the use of psychedelics carries notable risks. These substances have the potential to damage the critical interface for perceiving and interpreting reality. Unregulated use amplifies these risks, particularly for individuals with a predisposition to mental health vulnerabilities.[667]

Figures like Newton or Einstein might have had an unusual capacity to access concealed layers of information, as if their fuse or filter was set differently from the norm. Their brains allowed more of the fundamental information, with higher voltage or bandwidth, to pass through, granting insights that most of us typically cannot access. In this way, their minds were tuned to process information that exceeded ordinary constraints. Where most people have filters that limit perception to immediate, practical realities, Newton and Einstein were wired to receive more of the complex, underlying structure—patterns, mathematical insights, and abstract connections—directly.

This extra voltage or bandwidth, which produces an intelligence many would envy, might come with a cost. For some individuals, it could lead to social isolation or behaviors that others interpret as obsessive-compulsive. However, the person may not be obsessive; rather, they are overwhelmed by the sheer volume of insights and patterns flooding their mind, consumed by this constant influx, struggling to filter and process what they're perceiving. Everyday interactions and routines might feel trivial or distracting, leaving them isolated or misunderstood, as they aren't processing information in the same way as those with a more typical brain fuse. This information overload could drive them inward and, in a more colloquial sense, "crazy," driving them to explore these insights even as they struggle with the mental toll.

Even what we see as mystical, like alchemy, may be attempts to interpret and work with this deeper flow of information. Alchemy, often dismissed as pseudoscience or primitive chemistry, could represent a compelling exploration of the universe's hidden layers, an effort to tap into forces and connections beyond ordinary perception. If true, this realization underscores another materialist problem, which seeks to wholly dismiss or diminish Newton's spiritual pursuits. It's worth considering that even a subtle form of telepathic communication is possible, with the act of measurement itself introducing inconsistencies that seem to refute it.

By rigidly insisting that reality is confined to the material, materialists— believing themselves to be exploring reality—are, in fact, creating and fortifying barriers to greater insights and reality. So, what we consider genius may not be just about intelligence or creativity in the conventional sense, but rather a different wiring and an ability to receive and interpret information from deeper layers of reality, which for most of us are filtered out or obscured.

We can apply this idea to many of history's great composers, artists, poets, and visionaries. Just like scientific geniuses, these individuals might have had an unusual receptivity to the layers of reality most people cannot access. Their minds may have been tuned to channel cached structures of beauty, emotion, and universal patterns, expressed through music, art, and literature, in a way that naturally resonates with others. Figures like Beethoven, Mozart, and Bach come to mind. Their compositions often feel timeless, as though they're tapping into something fundamental, beyond the human organism. It's as if they had a direct connection to the patterns and harmonies that underlie reality itself, a language of sound that speaks to emotions and concepts too abstract for words. For instance, Beethoven continued to compose even after he lost his hearing, suggesting that his connection to music came from a stream that perceived something more than the physical sense of sound.

Artists like Leonardo da Vinci, Michelangelo, and Van Gogh seem to have had an unusual openness to the deeper structures of form, light, and movement. Da Vinci's works are artistic and scientifically insightful, as if he saw the beauty and the mechanics of life through the same lens. Van Gogh's paintings, with their vivid colors and swirling forms, seem to capture an emotional and energetic depth that resonates with people on an almost spiritual level. It's as if these artists were able to see something beneath the surface of reality that they could then express visually.

If we think of reality as layered, with the material world being one accessible aspect and a "realer" or deeper reality beneath it, then the actions of Jesus—the most impactful figure in the history of the world—could be seen as expressions of a direct connection to that foundational layer. This would indicate that his consciousness wasn't limited by the usual "interface" or filter that constrains most human perception.

What we might perceive as physical limitations, like gravity, the forces behind storms, or even the boundaries of matter, might not apply in the same way to such one connected to this cached reality. Walking on water, calming storms, or even transforming matter could represent acts that, rather than defying natural law, resonate with the most fundamental layer of existence, where the limitations we experience no longer apply. This casts Him as a transcendent observer, demonstrating a different type of observer effect, operating on a higher dimension, enabling Him to manipulate the behavior of quanta. Perhaps this is an aspect of the fundamental reality Paul alluded to when he wrote, "what is seen was not made out of what was visible." In this context, Jesus was attuned directly to the underlying reality that structures and governs our more superficial reality that we call the physical world, able to interact with it in a way that bypasses the usual physical constraints.

## The Transcendent Bridge

In a technological system, certain components are designed to be secure and self-contained, protected from disturbances. Earth is one of these secure nodes within the universe's immense structure, a place shielded from extreme cosmic forces where life and consciousness can thrive. The stability of gravitationally bound systems like the solar system, largely unaffected by cosmic expansion, points to this layered design. Earth's distance from the chaotic galactic core, its unique position relative to the sun, and the stability of the solar system are all specifications within this larger framework.

Earth acts as an interface for consciousness, providing the ideal conditions for awareness to explore and reflect on the greater system. It is the locus and focal point of perception, interpretation, and understanding, revealing the splendor of the universe and life. This gives Earth a role beyond sustaining life: it is a protector and a revealer of insight. Human perception is a lens for existence itself, making consciousness central to the universe's architecture. The vast structure of the cosmos surrounds and shelters consciousness, with Earth as the hidden place where this awareness resides.

Because of this purpose and integral relationship to existence, the search for consciousness elsewhere in the universe is like trying to locate consciousness within the human body or brain. Earth is the bridge between the physical and the transcendent, the point where experiences are realized. Our trials, joys, and quest for understanding are woven into this framework. Our search for meaning, awareness of beauty, and drive to explore the cosmos are expressions of the universe's own impulse to be understood and reflected upon by conscious beings. Though elusive, consciousness is more vital to the universe than even space, time, or matter. Without consciousness, space, time, and matter have no existence.

# CHAPTER 22

# Multiplex Complex

*The universe that we observe has precisely the properties we should expect if there is, at bottom, no design, no purpose, no evil, no good, nothing but blind pitiless indifference.*
—Richard Dawkins

To further elucidate biological technology, imagine a colossal, living structure akin to a high-tech, responsive building. Each room within this vast structure represents a cell, contributing to the body's overall function and vitality. Unlike a static building, this structure is dynamic, constantly adapting and responding to its environment. Every room works in harmony, functioning similar to how the body's network of cells and systems communicate, adjust, and maintain the intricate balance required for life.

Imagine walking through this structure when suddenly, you hear something strike a window. You look up just in time to see the glass fracturing before your eyes. Jagged shards hang precariously, revealing the damage. But before the tension fully takes hold, the broken pieces dissolve, swept away by an unseen force. In their place, new shards materialize, seamlessly knitting together until the window is whole again. You didn't just observe a repair; you witnessed a regeneration. It's as if the window itself understood and corrected the flaw. This self-repairing ability, seen in various biological forms such as muscle regeneration, bone healing, and skin recovery, speaks to the human body's remarkable regenerative abilities. Just like cells in the body collaborating to repair a wound, the rooms in this metaphorical building work together to mend the crack.

As we continue our walk through this living, metaphorical structure, imagine stepping into a room where the air becomes stiflingly warm. The moment discomfort starts to settle in, vents open, and a rush of cool air restores balance. This is no random adjustment but a precise response, much like the body's thermoregulation, where sensors detect temperature changes and prompt the brain to signal sweat glands and blood vessels to recalibrate the body's temperature. These mechanisms work in harmony, demonstrating the body's ability to adapt to internal and external shifts.

This interplay of systems is a prime example of homeostasis, the body's ability to maintain internal stability despite external fluctuations. Homeostasis is not limited to managing variables like temperature or pH; it involves coordinating the recalibration of multiple physiological processes that maintain balance throughout the body.

One critical aspect of this balance is fluid regulation. Every cell requires a specific equilibrium of water and electrolytes to function properly. The body is constantly adjusting, balancing fluid intake and loss through processes like sweating, urination, and even breathing. Without careful regulation, cells would either become overfilled and burst or shrink and malfunction.

The kidneys take on a vital role in maintaining this balance, guided by hormones such as ADH and aldosterone.[668] They filter waste while preserving essential nutrients and fluids, ensuring the body stays hydrated and its internal environment remains stable. Cellular mechanisms like osmosis and active transport further support this process, ensuring substances stay within optimal ranges, while excess waste is removed efficiently. Just like in a home, waste elimination and fluid balance are essential to each cell's health. Cellular by-products, created through metabolic activity, must be quickly removed to sustain functionality. Lysosomes act as the body's internal waste disposal, breaking down unwanted materials, which are then expelled via exocytosis, similar to an automatic trash system keeping the body clean and operational.[669]

The Golgi apparatus is an essential player in cellular homeostasis, with a role that defies simplistic classification. First observed in 1898 by Camillo Golgi, who referred to it as the "internal reticular apparatus," it has since been recognized for its multifunctional sophistication.[670] Not just a simple tool or an apparatus, the Golgi apparatus is central to vital cellular processes.

The processes within the Golgi are fascinating. Proteins, for instance, undergo glycosylation: a sophisticated modification that can be compared to a master chef preparing dishes tailored to specific tastes.[671] Just like a chef meticulously organizing ingredients and sides, the Golgi adds sugar molecules to proteins, fine-tuning them for their distinct roles within the body.[672] Each protein is custom prepared with precision, like a chef crafting each dish to match the preferences of the diner.

As proteins and lipids move through the Golgi, they continue their transformation. Like raw ingredients arriving at a busy kitchen, biomolecules enter the Golgi from the endoplasmic reticulum, where they are carefully processed, sorted, and dispatched, similar to how a chef prepares complex meals with various sauces and sides. But the Golgi's function outshines simple culinary preparation. It ensures that each molecular 'dish' is sent to its precise destination, whether within the cell or throughout the body to other cells or tissues. The Golgi is an efficient logistics hub. This processing center coordinates dispatches as well. This is comparable to a chef who prepares the meals and ensures that each plate reaches the correct table in the restaurant but also arranges delivery to the correct addresses throughout the city.

But here's where the Golgi apparatus transcends even the most advanced culinary analogies. It goes where no human technology has gone before. Unlike any kitchen or system we know, the Golgi has the extraordinary ability to completely disassemble and reassemble itself. Imagine a restaurant that doesn't just serve meals but can dismantle its entire structure—breaking down its walls, tables, and kitchen equipment— dispersing them across a city block, only to reassemble itself on demand.

Like a scene from Star Trek, it breaks down into smaller molecular components and seamlessly reconstructs itself. Recognizing the Golgi apparatus as a sophisticated system of NatTech could lead to transformative advances in medicine and technology. It's clearly not just an "apparatus" but a sophisticated technological processing system. Applying our understanding of how the Golgi sorts, modifies, and dispatches molecules could advance targeted drug delivery, enabling treatments that reach specific cells with pinpoint accuracy and minimal side effects. Its self-reassembling nature could inspire breakthroughs in tissue regeneration, nanotechnology, and material science, leading to engineered systems that can adapt, repair, and reassemble themselves—just as the Golgi does in cells.

# CHAPTER 23

# Crystals and Evolving Junk

*The greater part (95 per cent in the case of humans) of the genome might as well not be there, for all the difference it makes.*[673] —Richard Dawkins

It is remarkable how it is still asserted with certitude by many that the cell originated through self-organization—a claim that would never be made for any other complex system executing such specific and intricate functions. In fact, no human-made system matches the sophistication and robustness of life. The more we uncover about the technological operations within living systems, the clearer it becomes that Darwin's theory would hardly gain the same acceptance if it were first introduced today.

The true depth of life's design cannot be fully appreciated through complexity alone. Even though complexity might explain a system composed of numerous patterns or parts, the functional sophistication we see in organisms suggests a higher level of refinement and development.

In both nature and engineering, beauty is not entirely ornamental; it often enhances function by promoting structural integrity, efficiency, and adaptability. The elegance of a bird's wing, the symmetry of a flower, or the aerodynamic design of an engineered vehicle all demonstrate how aesthetics and purpose are intertwined. This deeper level of sophistication in life suggests an intentional framework where function and beauty are not separate but rather complementary aspects of design. For an analogy, abstract art might be considered complex, but Michelangelo's The *Last*

*Judgment* surpasses complexity; it embodies sophistication with detail and meaning.

However, acknowledging that comparisons between human creations and living organisms are false analogies is warranted, as such comparisons overlook the superior functionality, recalibration, and self-repair inherent in living systems, which actively strive to maintain function. Understanding this fundamental yet often downplayed difference between complexity and elaborate sophistication is necessary to fully appreciate the scope of life's design.

## What About Crystals?

Materialist explanations often misrepresent the functional complexity required for the origin of life by drawing false parallels to non-living processes. A common example is the tendency to retroactively rationalize the origin of the cell by likening it to the simple, law-governed formation of structures like crystals. This would be like attributing the development of a steam engine to the existence of deep-sea vents. The repetitive, non-functional patterns seen in crystals arise entirely from well-understood physical and chemical laws.[674]

Thus, using the formation of crystals as an analogy for the complexity necessary to initiate life is a misleading proposition. Living organisms, in contrast, are dynamic systems that actively *fight* to maintain functionality: they are robust preservation systems entirely absent in non-living matter such as crystals. Explaining the formation of crystals does not begin to describe how life could arise through unguided, natural processes. Materialists, however, often control the terms of debate by restricting discussions of design to only the most complex and sophisticated aspects of nature, such as DNA or molecular machines. This framing raises the bar for what qualifies as evidence of design while simultaneously dismissing simpler patterns—like crystals or snowflakes—as irrelevant.

It's ironic, however, that they use the existence of orderly, design-like structures as evidence against design, arguing that patterns—such as crystals—arise naturally through physical laws, without considering the origin of those laws themselves. This strategy allows them to exclude discussions of intentionality in simpler structures and reinforces the illusion that only the most advanced systems are worth debating. In doing so, they avoid engaging with the broader implications of design arguments and overlook how even the simplest features of nature exhibit function, relational and relative purpose, and intentionality within a larger framework encompassing natural existence.

Take, for example, the snowflake, a natural ice crystal. Although snowflakes form through well-understood physical processes, they play functional roles within Earth's broader ecological systems. Snow, made up of countless snowflakes, acts as a temporary reservoir for water, gradually releasing it during seasonal melting, which supports ecosystems, agriculture, and groundwater replenishment.[675] Snow also serves as an insulator, protecting plants and animals from harsh winter temperatures while reflecting solar radiation to help regulate Earth's climate.[676] These roles make clear snow's purpose-driven contribution to sustaining life on the planet.

The wintry technology of snowflakes not only serves functional roles but also embodies inherent beauty. In both living and non-living aspects of nature, beauty is an embedded feature, allowing for subjective experience, individual preference, and emotional engagement. Similarly, other crystalline structures—such as diamonds, rubies, and emeralds—contribute to a world that is both functional and visually captivating. This exemplifies *relative purpose*, a concept often overlooked. But as stated, Materialists, who vehemently oppose design, tend to control what is deemed acceptable for debate—much like a biased bouncer who arbitrarily enforces a dress code and decides who gains entry into a club. While materialists attempt to limit the conversation, true design is not just about function but also very much about engagement.

Authors, for instance, always consider their audience when constructing a narrative. Likewise, in human technology and product development, audience is a key consideration in design. Every aspect of a product—its function, aesthetics, and user experience—is carefully refined through a process of iteration and optimization. This is evident in companies like Apple, Tesla, and Dyson, where *industrial design* teams ensure that aesthetics, form, and ergonomics are not merely byproducts of function but integral to the user's experience and engagement. As Apple itself states on its page titled, "Make the result beautiful.

And the effort invisible:"

> At Apple, the Industrial Design team plays a pivotal role at every stage of the product development process, from the preliminary concept to the production of the painstakingly crafted final product. This team is world-renowned for their meticulous attention to detail and the high-quality standards they use to select materials, manufacturing processes, and final colors and finishes. Their projects extend further than product categories to also support the design of accessories, packaging, and the Apple retail experience.[677]

This mirrors what we see in nature: a balance of utility and beauty, appealing to a diverse range of observers, with each element refined and interconnected like an elaborately crafted final product. This suggests that what we call industrial-design thinking in human innovation is a reflection of a deeper principle embedded in the fabric of reality itself.

Relative purpose of natural products is evident in our cultural expressions. What would the concept of a king be like without precious metals or stones? Or consider how diamonds are presented in commercials for mothers or brides during spring weddings. Their associations with love and renewal show their relative role in our cultural expressions. While they are not presented to fathers or grooms in the same way. Diamonds, with their sparkle, rubies with their deep red, and emeralds with their vivid green—none of these qualities are necessary for the physical functioning of the universe. However, their aesthetic appeal makes the world an interesting place.

Consider luxury watches made by companies like Rolex or Omega. Though their primary function is to tell time—a role fulfilled by far simpler and cheaper devices—their craftsmanship, materials, and aesthetic design elevate them into symbols of status, style, and artistry. Their purpose appeals to audiences who value beauty, heritage, and exclusivity. These qualities may not be desired by all, essential for survival or mechanical function, but they enhance the human experience and reflect purposeful intent within the broader order of nature—similar to how Spring's blue skies and green grass, and Fall's colorful leaves, each are personal favorites for many.

Crystals, nevertheless, form under precise conditions understood by the fine-tuned constants of nature, such as temperature, pressure, and atomic bonding properties. Their predictable structures contribute to geological stability, chemical balance, and even technological applications, such as the piezoelectric properties of quartz used to convert mechanical energy into electrical energy.[678] Thus, the painting on the wall or the sprinkler that waters a campus is no less designed than the high-tech building it serves.

Proponents of ID should challenge the materialist framework more, rather than continuing to concede to its theoretical rules and regulations. By broadening the argument, they would show how design permeates every level of nature—not just in advanced biological systems but also in the natural laws and processes that produce simpler patterns. Recognizing the functionality, interconnectedness, and beauty of these elements strengthens the case for intentionality in the natural world. Given the growing body of evidence there is less justification for walking on eggshells when it comes to addressing these broader implications of design.

## Darwin's Finches

Beyond analogies like crystal formation, certain adaptive examples have been used for years to support evolutionary theory.[679] One of the most iconic examples include Darwin's finches, which were once a key symbol of evolution and natural selection. These birds are notable for their diverse beak shapes and sizes, which were noted to adapt to different environmental changes.[680] These adaptations were extrapolated to support his broader hypothesis of common descent, suggesting that all life forms might have evolved from a shared ancestor, diversifying into the myriad forms we see today. However, as new insights continue to unfold, the initial excitement surrounding beak variations in Darwin's finches and their role in evolutionary theory appears to have stemmed from a flawed extrapolation.[681]

The changes in finches are better explained as an inherent adaptive response to environmental changes, rather than as a step in a broader evolutionary process of speciation. Recent discoveries, such as the roles of genes like ALX1 and HMGA2, suggest that these variations arise from a reservoir of existing variability, enabling rapid adaptation within species.[682] This pre-existing genetic information allows Darwin's finches to exhibit shifts in specific traits, such as beak shape, in response to environmental pressures. However, these shifts do not indicate a progression toward evolving into fundamentally different kinds of organisms. They reflect evidence of pre-existing biological technology rather than the appearance of entirely new traits.

Major structural transformations cannot occur through the same kind of genetic shifts seen in finch beaks. They fall short of accounting for the large, coordinated shifts in developmental pathways required for one type of organism to evolve into another. Any changes during early embryonic development, when fundamental body structures are being established, are likely to be lethal or highly disruptive due to the precise regulatory mechanisms orchestrating this process.[683] Any major restructuring—such as altering foundational body plans during early embryonic development— would be like attempting to reconfigure a skyscraper's foundation after construction has already begun, a process likely to result in catastrophic failure rather than a successful transformation.

Despite Darwin's finches traditionally being presented as a clear example of evolution in action, they actually highlight the limitations of natural selection. These findings reinforce the notion that natural selection can act on existing traits but lacks the creative power to generate entirely new forms of life. Despite the dead end in explaining how the extraordinary functional systems of the cell or different body plans could arise by chance on a prebiotic Earth, Darwin's 19th-century folktale still dominates scientific discussions on the origin of life.

## Accumulated Junk

This same framework has also led to misplaced assumptions in other areas of biology, such as the long-standing dismissal of vast portions of the genome as functionless remnants of evolution. The term 'junk DNA' arose from the expectation that large portions of the genome were mere leftovers of evolutionary processes.[684] However, this so-called junk DNA, as we now understand, is anything but useless.[685]

Non-coding DNA does not encode proteins, but it became clear that many of these DNA sequences play critical roles in gene regulation and other essential cellular processes.[686] They determine how genes are turned on or off in different cells and at various times.[687] They contribute to maintaining genome stability and are integrally involved in the dynamic choreography of cellular development and differentiation.[688] Ongoing research continues to reveal their importance in maintaining cellular function. The revelation of these functions illustrates how previous scientific assumptions, rooted in a materialistic worldview, have often underestimated the natural technology inherent in biological systems.

The "C-value *enigma*," formerly known as the "C-value *paradox*," has further highlighted inconsistencies in materialist assumptions.[689] This enigma refers to the discovery that the amount of non-coding DNA varies widely among different eukaryotic organisms. Contrary to earlier predictions that more complex organisms should have proportionately larger genomes filled with this left-over "junk," we now know that some simpler organisms not only have larger genomes but also possess significantly more non-coding DNA than more complex ones.[690] Rather than being an enigma, this exposes a fundamental flaw in the assumption that randomness can give rise to the highly ordered and functional complexity observed in living organisms. It directly contradicts the expectation that increased randomness should lead to greater biological sophistication.[691]

Additionally, the "G-value *paradox*," which deals with the number of protein-coding genes and its lack of correlation with organismal complexity, further complicates this understanding.[692] For example, humans do not have significantly more protein-coding genes than rice, which is a much simpler organism.[693] These unexpected discoveries about genome size and complexity disrupted Darwinian assumptions about the relationship between an organism's DNA content and its phenotypic complexity.[694] Despite the common appeal to science's self-correcting nature, the classification of portions of the genome as 'junk' serves as yet another example where accumulating evidence goes against the notion that life's complexity arose through undirected processes.

One of the key figures contributing to a shift is biochemist John S. Mattick, who has advanced understanding of non-coding DNA.[695] His research calls into question the long-held belief that genes primarily encode proteins. He emphasizes the role of regulatory RNAs in gene expression.[696] Mattick's discoveries of epigenetic control reveal complexities in hereditary processes that extend beyond DNA sequences.[697]

Similar to software in technology, this non-coding DNA forms a complex and dynamic regulatory network.[698] It finely tunes gene expression with precision, adapting to environmental changes and internal cues, like advanced sensor and processor systems.[699] While these DNA segments play critical roles in error correction and redundancy, they also ensure the integrity and stability of genetic information.[700]

In engineered systems, redundancy strategies mitigate potential failures by duplicating vital components or data.[701] This approach, seen in hardware configurations, data replication, and network redundancy protocols, preserves essential information—just as the replication of genetic sequences safeguards biological function.[702] Whether in cloud computing, network redundancy, or fault-tolerant hardware, engineered systems prioritize stability, mirroring the way genetic mechanisms ensure resilience against mutations and disruptions.

The adaptability of repetitive DNA sequences suggests a sophisticated genomic mechanism for rapid environmental recalibration. This mirrors the capabilities of artificial intelligence systems, which continuously refine their algorithms in response to new data. Like AI adapting to its environment, these genomic mechanisms exhibit *biological intelligence*, dynamically adjusting to ensure survival under changing conditions.[703]

Building upon this, systematic patterning—common in software engineering—finds a compelling parallel in genetics.[704] Just like software code dictates an application's functions, genetic sequences operate like programming logic, determining when, where, and how specific genes activate.[705] This level of control and precision reflects a sophistication beyond structural composition.

Beyond their regulatory roles, DNA segments also contribute to structural stability, much like engineered patterns that reinforce bridges and skyscrapers.[706] These sequences likely maintain genome integrity, ensuring proper organization and function. Additionally, their role as engineered buffer zones or spacers suggests a deeper purpose.[707] Just as mechanical designs incorporate non-functional spaces to enhance overall efficiency, these genetic elements may facilitate spatial organization within the genome, optimizing both packaging and accessibility of genetic material.[708]

Darwin did not know about DNA—the molecular basis of heredity—the intrinsic nature of adaptability, the intricacies of cellular machinery, or the complexities of ecological systems that we understand today. As considerate as Darwin was of his theory's potential flaws, it's possible that with this knowledge, Darwin might have strayed from his broader conclusion. He once stated that if any complex organ existed that could not have formed through numerous, successive, slight modifications, his theory would be invalidated.[709]

By that standard, the theory should have been abandoned decades ago. But rather than letting go, Darwin's adherents have kept it on life support, refusing to acknowledge the growing body of evidence has concluded it is long past resuscitation.

As scientific understanding advances, the trend becomes clear: no biological component remains without purpose. Every element of life, once dismissed as superfluous, is gradually being recognized for its role within a broader, highly integrated system. Just as assembling a model electric car requires every piece in the kit, with none left over, biological systems also appear to be fully accounted for, with no truly "useless" parts. The deeper we investigate, the more evident it becomes that nothing is truly without purpose—only awaiting discovery.

# CHAPTER 24

# How Life Works

*We do not yet have a genuine theory of biology. The Theory of Evolution is not a theory in the sense in which I am using the term. It is more an historical account, itself standing in need of explanation.*[710] —Denis Nobel

Imagine an observer from the 1960s who is suddenly transported to our modern era. Because he is only familiar with medical technology from the '60s or earlier, he would find the robotic surgeon both captivating and puzzling. It would be like encountering a form of 'mechanical life,' autonomous and responsive in ways he never thought possible.

In this hypothetical scenario, the robotic surgeon is heavily reliant on a complex network of external servers for its decision-making and for its precision control. Now, suppose a cyberattack on the network's infrastructure were to cause the robotic surgeon to lose its vital connection. In that case, the system would enter a 'safe mode.'[711] This mode might involve more than simply pausing its procedure; it could also mean shutting down critical operational functions to prevent errors or unsafe movements, effectively rendering the robotic surgeon non-operational until the connection is restored or the system is manually overridden and checked by specialists.[712]

For all intents and purposes, the robotic surgeon is no longer functioning as it was engineered to. It's no longer has the 'life' it had when fully connected to its guiding systems. It is as if the machine experiences a form of 'death,' transitioning from a dynamic, interactive entity to an inactive assembly of hardware.

Our observer from the 1960s might examine the robotic surgeon, study its components, and even attempt basic troubleshooting. However, believing its animation was the result of an "emergent" property, this observer would never get the robot to function in the extraordinary way it did before losing its 'signal.'

Similarly, when life departs from a biological organism, the body remains—with all its complex components intact—yet it no longer functions as a living being. It is as if the 'signal'—that mysterious, animating force—has simply vanished. What remains is a biological 'safe mode,' a form we commonly refer to as death. We can study the body, understand its mechanics, and even identify the chemicals that once contributed to its life processes. Still, the essential 'signal' is gone forever.

This leads us to a fundamental question: If every physical component of life remains after death, then what was the animating force that departed? The transition from a functional, dynamic system to an inert collection of molecules suggests that life is more than just the sum of its parts. However, most scientists attempt to answer this question while only considering the biological, chemical, and physical processes that sustain life. Philip Ball explores this notion in *How Life Works: A User's Guide to the New Biology*.[713] He argues that life is not simply the product of genetic instructions or molecular mechanisms but instead emerges from a highly coordinated, multi-layered system of interactions.

Ball does not assert outright that life emerged from thermodynamic processes, but he reports the claims of researchers who argue that life is likely to arise under the right physical conditions. He presents their view as part of a broader reconsideration of life's origins in terms of physics rather than biology alone:

> Far from evading entropy's demands, then, life might be especially adept at granting them. Biophysicists Eric Smith and Harold Morowitz have argued that for this reason life is highly likely to arise, purely on thermodynamic grounds, in any environment that has the necessary chemical ingredients (whatever they are!) along with concentrated reservoirs of energy.[714]

This quote illustrates Ball's interest in the idea that life may emerge not as a miraculous exception to physical law, but as a thermodynamically favored outcome—one among many ordered structures that form to dissipate energy in nonequilibrium environments.

Yet, while the title of Ball's book suggests a deep understanding of life, Ball himself admits a limitation, stating, "Happily, I'm not obliged here to try to offer a definition of life."[715] This deliberate avoidance highlights a fundamental issue: understanding how life operates does not resolve the deeper question of what life is. As Ball acknowledges, "The problem of defining life has bedeviled biology throughout its history, and still there is no consensus."[716] Ball's book, then, is not just an exploration of biological mechanisms; his book unintentionally demonstrates how elusive life's essence remains.

This "new" biology that Ball describes does more than reveal complexity; it implicitly challenges the old biology's foundational assumptions. If life is not just a set of molecular mechanisms but a self-organizing system governed by coherence and internal logic, then the cornerstone of biology—that life can be fully explained through gradual evolution—is fatally flawed. Evolutionary biologists now discuss emergence because they must. However, they have not developed a testable, predictive framework that explains how emergent properties arise and operate in living systems. To draw an analogy, calling the Cambrian Explosion emergent does not explain why or how it happened—it merely acknowledges that something unexpected and complex arose.

What's most striking is that even the best explanations of emergence— like Ball's—rely exclusively on living systems and bioproducts of agency to describe how emergence functions. Whether it's gene networks, metabolic cooperation, or organism-level coordination, the only examples that materialists can point to are systems already governed by embedded logic.

Even beyond biology, so-called emergent systems—like bee swarms, ant colonies, traffic patterns, or flocking birds—do not support the materialist claim that order can arise from chaos. These are not raw, unguided physical systems. They all involve agents—biological or human— with built-in behaviors, rule sets, memory, communication, and purpose. The circular reasoning of emergence is always illustrated using systems that already behave like they were designed—because they are. The theory denies agency, yet borrows its outcomes to prove its point. (As a follower of *Closer to Truth*, hosted by Robert Lawrence Kuhn, I find the dramatic irony almost unbearable. In episode after episode, systems rich in coordination, constraint, and embedded logic are presented as proof of emergence.)[717]

What does "emergence" even mean under materialism? The word itself implies *a coming forth* or *into view*, but from what—and by what mechanism? Are we supposed to accept that the universe, DNA, and consciousness can all be explained by the "coming forth" theory? If materialists are so devoid of answers they resort to rebranding "It Happened" as a theory, then they aren't explaining anything at all.

Ball's description of life as a "qualitatively unusual" nonequilibrium system makes life appear otherworldly, as he states:

> We are a qualitatively unusual sort of nonequilibrium system, and until we have a theory that accounts for that difference, we will not have a proper understanding of how life works.[718]

This statement strongly implies that evolution alone cannot be the cornerstone of biology because it does not account for life's nonequilibrium nature.

Ball suggests that when an organism dies, it is not a single force that vanishes, but rather the structured, self-sustaining processes that allowed it to resist entropy. When this process ceases, the body's matter, once so precisely and actively arranged, resigns itself to disorder and decay. As Ball puts it:

> Living things seem to ignore this entropic imperative. Their animation relies on the creation and maintenance of order ... whereas once it ceases—when we die—the same matter resigns itself to a steady slide into disorder and decay.[719]

While Ball does not argue for a non-material explanation, his own premise underscores how inadequate purely physical definitions of life are. If life were simply the sum of its molecules, its cessation should be no more mysterious than shutting off a machine. Yet, what disappears at death is not just molecular activity but an entire goal-directed, adaptive system—a level of complexity that current scientific models cannot capture. This distinction is not about complexity: life operates according to principles that are foreign to the standard laws governing non-living matter.

His use of the word emergent acknowledges this sophistication but does not ultimately define what the term means. In fact, even he recognizes the limits of the term, stating:

Some researchers, though, are suspicious of this notion of emergence. It seems to them an appeal to quasi-mysticism, as though a magic ingredient has been added to the system beyond the forces and exchanges of matter and energy between its constituents.[720]

While Ball explains life as an emergent phenomenon—arising from the intricate cooperation of numerous biological systems—his guide to *how life works* inadvertently refutes *abiogenesis*. If life truly depends on the cooperation of complex, pre-existing systems, then how could the first life have arisen? The *law of biogenesis* states that all living organisms come from other living organisms. Where did the first organism *alone* acquire these layered systems with their entropy-defying, otherworldly quality, while also establishing the law of biogenesis? And if the first attempt at life failed to survive, how could life emerge again?

This presents an inescapable contradiction: abiogenesis requires life to emerge from non-living matter in the absence of the intricate systems that Ball himself acknowledges are necessary for life to function. If Ball is correct, then abiogenesis is not simply unlikely—it is conceptually incoherent. In an attempt to explain how life works, Ball has instead made it clear that explaining how life began and evolved is a damning indictment of *methodological naturalism*.

## The Selfish Gene

Ball also calls out the once-prominent metaphor of the "selfish gene," famously introduced by Richard Dawkins. Ball makes fun of the contradiction in Dawkins' language, noting that describing genes as "selfish" inherently implies agency. "Oh, but look: the language of agency has crept in again," Ball mocks, pointing out Dawkins' contradiction. Ball continues sarcastically, mimicking Dawkins' likely response: "No, no, genes don't actually 'care' about anything; this is just a manner of speaking."[721]

While Dawkins insists that genes are "selfish," implying intention, Ball suggests that Dawkins' metaphor could be flipped to describe genes as "cooperative," considering their actions contribute to the survival and reproduction of the organism.[722] This exposes the contradiction of using language that suggests agency in a framework that vehemently denies it. According to Ball, Dawkins' "selfish" genes resemble a broader idea discussed in economics, where individuals acting in their own self-interest can unintentionally benefit society as a whole. This concept is similar to what economist Adam Smith called the "invisible hand" of the market.[723] Ball uses this analogy to point out that even Dawkins himself admits his book's title, *The Selfish Gene*, could have alternatively been *The Cooperative Gene*.[724]

However, despite the strengths of his work, Ball's frequent use of engineering metaphors and design-oriented language reveals his own underlying inconsistency. Interestingly, as he critiques outdated and oversimplified biological metaphors, Ball believes that his perspective is superior to those who see life as rudimentary machines because he views life through a more systems-oriented, technological lens.[725] He suggests that traditional metaphors do not fully capture the masterful engineering inherent in biological systems, asserting:

> There is so far *no technological artifact* that provides a good analogy for living systems. These are a different kind of entity, with *their own logic*, and they have to be their own metaphor.[726]

This adds a layer of intrigue to his critique, which is quite the balancing act, critiquing the simplification in biological discourse while inadvertently embracing design-oriented language himself. This is a stance of hypocrisy.

Even the title of his book, *How Life Works: A User's Guide to the New Biology*, suggests a technological perspective of life.

Ball says that a cell *"thrives* on noise and diversity, on chance accidents and fluctuations."[727] At first glance, this captures something remarkable: the cell operates without centralized control, yet with an efficiency that surpasses any engineered system. It's like imagining a city where every car moves unpredictably, with no traffic lights or central authority—yet there are no traffic jams. Everything flows rapidly, and even the crashes contribute to progress, helping the whole system adapt and recalibrate.

But this is where Ball's interpretation falters. He calls it noise, yet elsewhere admits that living systems operate with "their own logic," and that no current machine provides a suitable analogy. If that's the case, he should consider that what he labels as randomness may only appear random because we haven't yet grasped the underlying organizing principles. The cell isn't surviving despite disorder—it's operating through a structured, embedded intelligence, or as he puts it, logic. What Ball presents as chance is better understood as adaptive specificity: variability harnessed by a deeper system logic. His description of this natural technology is quite remarkable, but his continued allegiance to evolutionary randomness prevents him from fully recognizing what he's describing. He even concedes that each level of biological organization functions with its own distinct logic:

> Life is a hierarchical process, and each level has its own rules and principles: there are those that apply to genes, and to proteins, to cells and tissues and body modules such as the immune system and the nervous system. All are essential; none can claim primacy.[728]

As François Jacob observed, "There is not one single organization of the living, but a series of organizations fitted into one another like nests of boxes or Russian dolls. Within each, another is hidden."[729]

Furthermore, Ball critiques the resistance within scientific circles to any notion of purpose or intentionality in nature. He emphasizes the importance of not shying away from these concepts when discussing biological systems. He argues that being afraid of the words "agency" and "purpose" *has held science back* and that acknowledging purpose does not necessarily imply a retreat into mysticism or religion but rather a recognition of the intricate and goal-oriented nature of life processes.[730] In other words, using metaphors such as agency and purpose could be seen as just stating what is observed despite the undesired implications. Effectively, Ball points out that the supposed conflict between religion and science has been a projection. Ball asserts that life forms can be seen as "generators of meaning," systems that create and interpret significance in ways that transcend mechanistic explanations.[731] However, if life forms are truly generators of meaning in a way that transcends mechanistic explanations, then this is no longer just a biological or scientific claim; it's a metaphysical one.

When describing evolutionary processes. Ball writes:

> Evolution works with what is already there, even if this means redirecting it to new ends. We might (with great caution!) compare it to an electronic engineer who uses pre-existing circuit components like diodes and resistors, and standard circuit elements such as oscillators and memory units, to create new devices. Thus, life possesses a modular structure.[732]

Ball inadvertently mirrors the critique he levels against Dawkins, as the language of design and engineering subtly creeps into his description. Though Ball likely does not hold a deistic view, his description of evolution working with "what is already there"—rather than blindly assembling from scratch—suggests an implicit metaphysical perspective that he does not explicitly state.

Richard Dawkins, Philip Ball, and all who attempt to explain life's origin inevitably—though reluctantly, for some—resort to analogies involving agency or design. By cautioning readers not to take his analogy literally, Ball avoids committing to the design-like implications of his metaphor. This is ironic because while mocking Dawkins for using contradictory language, Ball does the same by implying evolutionary processes work like an engineer *repurposing parts*, but then he qualifies it with "great caution!" to avoid the same criticism.

Observing Ball's tendency to point fingers, it becomes apparent that his own elaborate descriptions deserve similar criticisms. While observing Ball's use of language, one might, as he did Dawkins, note: "Oh, but look: the language of design has crept in again." In response, Ball might clarify his position, emphasizing a metaphorical intent of his language: "No, no, of course natural processes don't 'purpose' or 'design' anything—this is just a manner of speaking..."

This inconsistency is further revealed in Ball's discussion on *synthetic morphology*. Synthetic morphology is the study of how living shapes and structures develop based on inherent rules regulating their growth and interaction, viewing them as engineering templates that can be understood, replicated, or modified.[733] It reveals a deeper logic within biological systems. But as is customary, materialist interpretations attempt to frame synthetic morphology as merely an extension of evolutionary theory, redefining the design logic observed in nature as nothing more than "emergent" properties of blind processes.

Ball argues that synthetic morphology "demands a new view of engineering, in which we assemble objects from their basic components not in a simple assembly-line manner according to a blueprint but by exploiting rules of interaction to enable a desired structure to emerge—as if, you might say, by cellular consent."[734] Ball describes traditional engineering as crude, inflexible, and ill-suited to the complexity of living systems. But where do these self-organizing, adaptive, and entropy-defying properties—and the laws that govern them—actually come from?

He presents this shift as necessary because, as Marta Shahbazi puts it, "the process of building a structure changes the very nature of the building blocks." Development, in this framing, is not top-down but dynamic and distributed—what René Doursat calls "morphological engineering," where one should "shape [the] building blocks in such a way that they do it for you."[735]

But in calling for engineers to shape systems so that a desired structure *emerges—to guide outcomes* that are "versatile, adaptive, and robust," and functional—Ball quietly reintroduces the qualities of coordination and purpose he claims are absent in nature.[736] The language of emergence becomes a stand-in for otherworldly engineering by way of "It Happened"— allowing intentionality to slip back in through the side door. He rejects design in theory, yet his entire model of engineering relies on the assumption that we can abstract and apply nature's design-like processes—despite not knowing what those principles are, how they function, or why they reliably produce complex outcomes at all. His language of "emergence," "rules," and "consent" effectively reintroduces design under another name, treating nature as if it has discovered solutions without needing a designer—while allowing us to treat those solutions as if they were designed all along.

Ball quotes biologist Michael Levin and computer scientist Josh Bongard, who argue that "we might turn this fact on its head and reconsider our notion of what a machine is: to regard living things as 'machines as they could be.'"[737] Levin expands on this idea, stating, "We view life as an especially interesting class of machines that is making us expand the limiting old ideas of what machines are and how to make them."[738] Once again, by endorsing the view of living organisms as advanced machines, Ball implicitly acknowledges that biological systems exhibit engineering qualities far beyond our current technological capabilities.

Imagine you're Philip Ball at a symphony, but there's a twist: the musicians are entirely invisible. However, every note, every harmony fills the hall, perfectly on cue, revealing the skill of the performers and the guiding hand of the conductor. Still, you cannot see any of them. The music is real; you can hear it; you can feel it. Yet, the lack of visible players makes you question how it's all happening. Like the observer at the concert, Ball firmly acknowledges and highlights the intricate coordination of life's processes; like the music from an orchestra, he hears the harmony and observes the apparent guidance in biological systems. Yet, because he cannot detect where this orchestrating signal originates, he stops short of declaring it all deliberately orchestrated.

Like Ball, grappling with the beautiful music of life, many who study biological systems may tacitly struggle to accept that such elaborate processes are adequately recognized or that they are fully described within our current understanding. For example, Denis Noble points out that genes do not function in isolation but operate within a dynamic and interactive network, with regulatory feedback loops exhibiting a level of sophistication that surpasses advanced technological systems.[739]

Both, Ball and Noble observe that these regulatory networks and system-level responses are not adequately explained by traditional views. They point to more coordinated, system-level recalibrations that enable organisms to respond to environmental challenges in real time. However, the dramatic irony is that just when Ball's and Noble's arguments seem indistinguishable from those of intelligent design proponents they reaffirm their belief in Darwinian evolution, insisting that 'we know' these advanced technological systems some how arose through blind Darwinian processes.

## A Manner of Speaking

While critiquing modern biology, Ball argues that it has worked hard to distance itself from any notion of purpose, goals, or design in life to avoid associations with intelligent design or religious thinking. He admits:

> Biology keeps that danger at bay with 'as ifness': we can speak 'as if' purpose, intention and goals exist ... but we must maintain that this is just a manner of speaking.[740]

Ball's statement suggests a form of willful ignorance—or at the very least, a conscious refusal to accept the implications of the evidence. By admitting that biologists must speak as if purpose, intention, and goals exist—while simultaneously insisting that this is merely a manner of speaking—he acknowledges the necessity of teleological language yet refuses to accept its implications, precisely because doing so would destabilize the strict naturalistic framework of modern biology.

Even as Ball insists that "evolution really doesn't have goals," he concedes that "agency is arguably the fundamental feature [of life]"—a claim that does not quite follow.[741] He acknowledges that organisms "use their genetic resources to achieve goals, but not to be defined by those resources," a statement that directly undermines the idea that life is purely mechanistic.[742] If agency is inherent to living things, then life is not just a collection of molecules obeying blind physical laws—it is something more. How could agency evolve or emerge from non-living matter?

Saying that Darwinian evolution has "nothing to do" with life's origins is misleading because evolution requires a transition from non-life to life for it to have any meaning at all. Ball admits, "One philosophical difficulty with the Darwinian view of evolution is that there's no way of defining a well-adapted organism except in retrospect." If fitness can only be judged after the fact, then evolutionary theory cannot explain how life began—it only describes what survived. Again, this means that calling anything about life an "emergent property" is just restating that biological phenomena appeared, without explaining how.

Ultimately, Ball concedes more than he may realize. He concludes, "The real reason biology frets so much about irruptions of teleological thinking is that again these expose the fundamental lacuna: it can't deal systematically with agency."[743] This is the inescapable problem. Life is coordinated, adaptive, and goal-directed. If biology cannot account for this without borrowing the language of purpose, then perhaps purpose is not an illusion, but a reality.

Though possibly wary of exposing their doubts about Darwinism, it's hard to imagine Ball and Noble never quietly questioning its legitimacy. In fact, it's plausible that many seemingly unlikely evolutionary leaders harbor doubts. Indeed, when contemplating these complexities and implications, a cold shudder might often run through some. But they rather choose silence over acknowledging skepticism, fearing exclusion from the community. A love for the praise of others may outweigh the desire for acknowledging truth.

CHAPTER 25

# ID by Another Name

*It invented the molecules that made life possible. It invented the schemes of evolution. It invented all sorts of amazing things.* —Michael Levitt on biological intelligence (BI)

The concept of flying was once relegated to the domain of birds and myth until the Wright brothers turned it into reality.[744] Similarly, landing on the moon was only a fantasy until the Apollo missions made it a historical fact. In each of these instances, humanity's achievements were initially inconceivable, only to become tangible through ingenuity and a shift in perspective. The development of computers and advanced AI followed a similar trajectory. A century ago, the idea that an electronic machine could compute complex calculations and mimic human thought processes was the stuff of science fiction. Yet today, we have created computers and advanced AI systems inspired by the natural world, particularly the human brain.

Despite the consistent progression toward seemingly far-fetched achievements, skepticism has always accompanied such bold endeavors. In 1901, George W. Melville, former Engineer-in-Chief of the U.S. Navy, wrote a scathing article stating, "There probably can be found no better example of the speculative tendency carrying man to the verge of the chimerical than in his attempts to imitate the birds...."[745] Similarly, the New York Times predicted in 1903 that achieving manned flight would take up to 10 million years, dismissing the idea as futile.[746] Despite this dismissive skepticism, just nine weeks later, the Wright Brothers proved the skeptics wrong by achieving the first successful manned flight on December 17, 1903.[747]

**Figure 26.1** This historic image captures the Wright brothers' first successful flight on December 17, 1903, at Kitty Hawk, North Carolina. **Image source:** The National Park Service (NPS), 'Historic image colorized for the National Park Service.'

The Wright brothers' success was a landmark in aviation history, but they weren't the first to dream of flying. Long before their triumph, many pioneers had attempted to conquer the skies. One of the earliest recorded attempts at flight was made by Abbas Ibn Firnas in 852 AD. Manifesting humanity's long-standing desire to soar, he crafted wings and tried to emulate bird flight.[748] This fascination with flight persisted through the centuries, from Roger Bacon's (c. 1250) speculation on flying machines to Leonardo da Vinci's (c. 1505) detailed designs inspired by the movements of birds.[749] By 1648, John Wilkins, a prominent English clergyman, natural philosopher, and founding member of the Royal Society, was also studying bird flight patterns to explore the potential for human aviation.[750] [751]

The inventor Leonardo da Vinci wrote in his *Codex on the Flight of Birds*: "A bird is an instrument working according to mathematical law, which instrument it is within the capacity of man to reproduce with all its movements."[752] These early pioneers all shared the inspiration of soaring like birds. These first forms of biological mimicry reflect an innate fascination with achieving the engineering feats inherent in nature's blueprint.

This fascination with NatTech has now culminated in a broader trend. Inspired by the regenerative abilities of biological tissues, scientists are creating self-healing materials for construction, automotive, and aerospace industries, cutting down on maintenance and extending lifespans.[753] Architects are mimicking termite mounds to design buildings with natural cooling and ventilation, slashing energy use while keeping interiors comfy.[754] Engineers are crafting robots that move like animals, from the speed of cheetahs to the flexibility of octopuses, with exciting uses in search and rescue, medical procedures, and environmental monitoring.[755] [756]

As scientific understanding grows, more researchers are recognizing the remarkable optimization, design, and purpose found in nature. They are beginning to see that technology's foundations are already laid out in nature. We just need the insight to unlock these endless possibilities and harness the genius that exists in the natural world.

Researchers in the mid-20th century began drawing more direct parallels between living systems and machines. The rise of cybernetics, pioneered by Norbert Wiener in the 1940s, laid the foundation for comparisons between biological and computing systems.[757] The shift became particularly evident with the discovery of DNA in the 1950s and 60s, which led scientists to describe life in engineering terms, such as "code" and "blueprint."[758] Although the term 'biotechnology' was first coined by Károly Ereky in 1919, it gained renewed significance in the 1970s, when the integration of genetic engineering and molecular biology marked a significant turning point, subtly shifting scientific language and further reinforcing the connection between biology and technology.[759]

This is another reflection of our natural, machine-like learning, which has driven the advancement of our technology. This confluence sets the stage for today's scientific narrative, which is quietly being rewritten, reshaping our understanding and interaction with the natural world. Just as a child grows according to an intrinsic blueprint, everything needed for the growth and development of our technology has always been present and waiting in the natural world. This connection transcends our creativity. Our advancements are less a product of human ingenuity and more a fulfillment of a providential design, guiding us toward discoveries that reflect the fundamental order of the natural world.

We have played the role of recognizing and unlocking these possibilities, allowing them to take shape in ways that embody an inherent purpose and human will. This parallels our instinctual drive to transfer technology—like the innate need to extract nutrients from food for sustenance. While we may enjoy it, it is also essential. We have invented nothing entirely on our own; we learn to follow nature's recipes—some we see, others we discover. A hidden framework has guided our progress, like how a seed naturally grows into a tree. When it came to the creation of the first aircraft or machine, we did not invent them from scratch; rather, we reimagined and harnessed natural technologies to suit our needs and desires.

One of the most striking examples of nature's embedded logic is found in Physarum polycephalum, a slime mold that exhibits a remarkable capacity for decentralized problem-solving.[760] Though brainless and composed of a single cell, this organism can navigate mazes, optimize networks, and adapt its structure in real time to find the most efficient paths between food sources. Scientists have mimicked its behavior to design transportation systems and computational algorithms, noting that it often replicates or even improves upon human-engineered solutions.

In one famous experiment, Physarum recreated the layout of the Tokyo rail system with striking accuracy, achieving a balance between efficiency and redundancy—something central to resilient design.[761] This behavior is not the product of randomness or trial-and-error, but of a built-in capacity to process environmental feedback through physical structure. In this way, Physarum functions as a living demonstration of the principles behind civil engineering technology—where computation, adaptation, and intelligence are distributed across form and function rather than imposed by external control.[762]

Understanding nature as a source of inherent potential has led to a deeper appreciation of its engineering marvels, paving the way for the growing field of biomimicry—a scientific discipline that is essentially *intelligent design by another name.*

Janine Benyus, in her seminal 1997 work *Biomimicry: Innovation Inspired by Nature*, brought the field of biomimicry to the forefront, demonstrating the potential of nature-inspired design to solve human problems and enhance technology.[763] Benyus argues that nature's solutions offer sustainable and efficient models for human innovation.[764] On her website, biomimicry.net, Benyus defines biomimicry as "learning from and then emulating nature's forms, processes, and ecosystems to create more sustainable designs."[765]

Her website includes the following quote:

Biomimicry ushers in an era based not on what we can extract from nature, but on what we can learn from her. This shift from learning about nature to learning from nature requires a new method of inquiry.[766]

Similar to Ball's arguments on morphology, biomimicry calls for a new approach, as reconciling undirected causes with purposeful design marks a major shift for Materialism—from learning about nature to becoming its pupil. Like Ball, Benyus contrasts top-down engineering models with nature's design processes, which she argues surpass human-made systems. Effectively, Benyus takes a *natural technology* approach. However, she explicitly names nature as *the creator*, suggesting that through evolutionary processes, nature has engineered solutions with sophisticated and intentional design.[767] She also positions nature as an influential "mentor" to us, from whom our technology can significantly benefit.[768]

In a podcast interview hosted by Appalachian State University, Benyus discusses the concept of biomimicry, which she champions as a method for solving some of the most significant scientific and social challenges of our time.[769] During the conversation, she expounds on how nature is not just a repository of solutions but an active participant in problem-solving, often demonstrating an innate capacity for innovation and design. She states:

> Some of the people who make our world—the chemists, the architects, the engineers, and physicists—are turning to the natural world as a whole new library of innovations that are already sustainable.

Continuing her thoughts, Benyus extends this perspective:

> You asked earlier about social innovation and is there anything we are learning from the natural world on that front. It's really interesting. Every few years, especially in the last 20 years, a new group of people who are trying to redesign something come to us all of a sudden—all of the engineers came, and then all of the architects came. Now, social innovation people are coming, and they are saying, 'What can we learn about the natural world about how we organize ourselves?'

Benyus' portrayal of nature as an engineer and an architect reflects her belief that there is purposeful, intentional design within its workings, an idea she interprets as evidence of nature's inherent role as a designer. However, her interpretation has a fundamental flaw. She suggests a circular logic, implying that nature's complexity both creates and arises from itself, assuming that complexity and creativity are self-originating.

By naming nature as its creator, Benyus offers a narrative that is spiritually akin to, yet semantically contrasts with, religious texts such as the Biblical Psalmist's declaration: "...the LORD he is God: it is he that hath made us, and not we ourselves..." (Psalm 100:3, KJV).[770] [771] In Benyus' view, however, nature acts as a type of Presiding Mind, assuming a godlike role in shaping and sustaining its own systems. In this case, it is "nature that hath created nature, and nature that hath made itself.

Although her perspective resonates with themes of intelligent design, she inadvertently articulates a pantheistic faith, expressing reverence for the natural world as both the source and architect of life. This cosmogony, which posits that nature itself possesses creative power, also aligns with the perspectives of Stephen Hawking and Max Tegmark. As Hawking asserts, "the universe can and will create itself from nothing," while Tegmark envisions the universe as a mathematical structure that gives rise to all existence.[772]

These spiritual-like perspectives within Materialism resemble a phenomenon often unnoticed in politics, where leaders take strong, uncompromising positions to rally their constituents: the people they depend on to extend their contract of employment. These politicians press the right buttons, vowing to obstruct their opponents at all costs, assuring their supporters that blocking the other party is itself a victory. Yet, at the same time, they court the groups those policies claim to help—arguing that real progress can only come under their leadership rather than the opposition, who they claim have failed to deliver on their promises.

This is sometimes true because they had previously worked to block everything proposed by the opposing politicians. But once in power, these same leaders often support or enact the policies they claim to reject, benefiting the opposition's constituents more than their own. Or even worse, they implement policies that directly harm their own voters—yet their supporters overlook, justify, or even defend these actions, as long as their 'team' remains in power. This reflects how objections often have less to do with the initiatives themselves and more with which team presents them. Similarly, materialist perspectives often incorporate elements that resonate with spiritual or religious beliefs, and no one seems to object or even notice.

## BIO-X

Biomimicry represents just one facet of a broader movement within the scientific community. Another example of this trend is BIO-X at Stanford.[773] This program is a leading example of how modern science is adapting to address the complexities of biological systems through interdisciplinary collaboration. Founded in 1988, it aims to merge engineering, computer science, physics, chemistry, and other fields to tackle some of the most complex problems in the biological sciences.[774] While biomimicry explicitly acknowledges the design inherent in nature, Stanford's BIO-X program operates within a more traditional framework.

Although BIO-X's approach is undeniably innovative, it remains bound to a predominantly materialistic framework. This adherence to Materialism limits the initiative's potential to fully explore or acknowledge purposeful design in the biological systems it studies. In their view, signatures of design are byproducts of biological systems rather than intrinsic to them.

Michael Levitt, a Nobel Laureate in Chemistry and a key figure within BIO-X, embodies this begrudging acknowledgment.[775] [776] His work in computational biology has advanced understanding of molecular dynamics and protein folding, both of which are foundational to bioengineering and medicine. His discussions on "biological intelligence" (BI) and the self-assembling nature of proteins, though framed within evolutionary theory, often use language that echoes metaphysical concepts of design and purpose.

Consider for a moment the implications of the following statements by Levitt:

> *It invented* the molecules that made life possible... *It invented* the schemes of evolution... *It invented* all sorts of amazing things... And when you see these things, it's all about self-assembly.
>
> So basically, a *protein is a machine*... And this machine *can do really important things* in a [tiny] size like that, but most importantly, the machine *assembles itself*. And all of biology, if you think about it, an egg and a sperm... from there it's all done by itself. There's no one outside saying, ... 'Gee, your arm is on wrong, we're going to put it right.'"
>
> So, we still have a future ahead of us of self-assembly... I'm sure we'll get there. Chemists are getting more and more informed by biology.
>
> *We did not create biology*... And therefore, it's an alien form of intelligence from which we can learn a great deal...
>
> Bacteria started evolving three billion years ago, quite a soon time... Then about a billion years ago, *we invented eukaryote cells*.... [777]

This language strongly suggests a conscious, purposeful agent behind nature's complexity. Levitt credits evolutionary mechanisms with this creative power, yet in doing so, he inadvertently grants them a godlike ability to invent, design, and assemble "all sorts of amazing things." Levitt understands the implications of his own words so well that he describes biology as an "alien form" of intelligence—something beyond our world.

He even states, "this could be created by God in four days or in a longer period, depending on your religion." Yet, he insists this isn't the point—even though his own words suggest otherwise. If biology operates with an intelligence so foreign that even he calls it an "alien form," and if he can casually invoke God—whether in four days or whatever one believes—then perhaps the real point is one he'd rather not admit. This is not unique to Levitt. Notably, every biological system is described in research as technology, and every biological system we emulate is stamped as technology. The difference is that we know how ours was made.

Thinkers, such as Ball, Noble, Benyus, and Levitt bring to mind the ancient parable of the blind men and the elephant, where each man touches a different part of the elephant and comes to a different conclusion about what it is. Ball, Noble, Benyus, and Levitt touch on different aspects of the "elephant," which is life's elaborate complexity. Though they may describe these aspects in terms of natural processes and evolution, their language and conclusions hint at a larger, cohesive whole.

In this way, modern science, driven by its discoveries and the need to explain the sophistication of life, is being subtly steered towards an intelligent design approach. The frameworks and terminologies initially developed to support a materialistic understanding of the world are now employed in ways that increasingly contradict Materialism. Whether these scientists realize it or not, their work contributes to a quiet revolution in our understanding of life, gradually bringing us closer to distinct yet undeniable forms of creationism.

# CHAPTER 26

# Achronological Record

*Those who reject evolution live in a parallel universe where the normal rules of science do not apply... Their surety is rooted not in evidence or experiment but in a conviction that their views are right, regardless of contradictory facts.*[778] —Alan D. Attie, Elliot Sober, Ronald L. Numbers, Richard M. Amasino, Beth Cox, Terese Berceau, and Thomas Powell, from "Defending Science Education"

## The Cambrian Explosion

The *Cambrian Explosion* is a term used to describe a pivotal event in Earth's history.[779] Scientists estimate that it occurred around 541 million years ago, when there was a sudden and remarkable proliferation of complex multicellular animals, particularly in the oceans. It is widely accepted that for billions of years before the Cambrian Explosion, life was limited to bacteria and other simple organisms.[780] However, the claim that bacterial life existed before this time is based primarily on indirect chemical signatures and the interpretation of ambiguous microfossils, both of which remain subject to debate.[781]

The Cambrian Explosion is widely regarded as the period during which the first major animal groups appeared, including many that resemble modern species. This event is significant because it represents a short geological period when phyla (major animal body plans) appeared seemingly out of nowhere. These include groups such as arthropods (which include insects and crustaceans), mollusks (such as snails and clams), and chordates (which include vertebrates).[782]

At once, animals with jointed legs, segmented bodies, eyes, gills, and hard exoskeletons burst onto the scene.[783] These organisms displayed advanced body plans with specialized tissues, organs, and nervous systems. The first predators appeared alongside burrowers, swimmers, and filter feeders, rapidly transforming ecosystems. Suddenly, animals were crawling, swimming, and burrowing across the earth.

**Figure 27.1,** A fossil Drotops armatus on display in the Sant Hall of Oceans in the Smithsonian Museum of Natural History. Drotops armatus is a species of trilobite. Trilobites are arthropods, a kind of animal that includes shrimp, lobsters, scorpions, crabs, and crayfish. Trilobites are thought to have first appeared about 542 million years ago (the early Cambrian period). **Image source:** Tim Evanson.

**Figure 27.2,** A fossil Cheirurus ingricus on display in the Sant Hall of Oceans in the Smithsonian Museum of Natural History. **Image source**: Tim Evanson.

Imagine delving into Earth's past—layer by layer—like a paleontologist uncovering ancient civilizations. According to evolution, Earth's geological layers are supposed to resemble the pages of a family photo album, with each layer adding snapshots of life's progression over time.[784] But upon opening the first pages of this geological album, something is missing—there are no baby pictures, no awkward teenage transitional years, no gradual transitions into adulthood. Instead, fully grown "family members" appear all at once, who are said to share a common ancestry, yet without the expected developmental history. This describes how the fossil record opens, with an abrupt appearance of fully formed animals and no clear evolutionary precursors leading up to them.

In any other context, we would immediately recognize such an abrupt appearance as unnatural. Imagine if the same pattern occurred with artifacts of human history. No one would seriously expect to find modern technology buried among primitive artifacts without any trace of gradual development leading up to it. Compare the Cambrian period to the Neolithic era, when human tools and technologies first surfaced.[785] Similar to how the Cambrian saw the sudden rise of new body plans, imagine archaeologists sifting through layers of stone tools and primitive artifacts, only to suddenly uncover elaborate devices like watches, smartphones, and radios. What would be even more confounding is if this technological leap were not confined to a single site but occurred simultaneously across multiple, unrelated regions—including those with no prior evidence of such advancements. Likewise, the Cambrian Explosion was not a localized phenomenon but a widespread event, appearing across multiple fossil sites worldwide.

**Figure 27.3** This image features basic types of Neolithic or Copper Age stone tools: knives, axes, a hammer-axe, and a hammerstone from The City of Prague Museum, creatively remixed with modern objects such as an iPhone and a watch to simulate an anachronistic discovery. **Image source:** Zde Wikimedia Commons, CC BY-SA 4.0.

Figure 27.3 illustrates how finding advanced technology—such as a modern smartphone and watch—among primitive Neolithic tools would contradict conventional timelines of technological development. Likewise, the Cambrian Explosion defies evolutionary expectations of gradual change. In fossil beds across the globe—from Canada's Burgess Shale and China's Chengjiang Biota, to Australia's Emu Bay Shale and even Greenland's Sirius Passet—the records reveal a sudden and simultaneous appearance of complex life forms, with no gradual transitions leading up to them.[786]

This is common knowledge among the scientific community, yet, to much of the public, it remains like a well-kept secret within the evolutionary narrative. Much of the public is led to believe that abiogenesis is a fact, but the closest evidence we have for abiogenesis is the spontaneous appearance of fully formed animals around the world.

The most widely cited explanation for such phenomena is a rise in oxygen levels, based on the so-called Neoproterozoic Oxygenation Event (NOE, ~800–540 million years ago), which is claimed to have suddenly enabled complex life to flourish. However, the evidence used to support this event is more plausibly explained by massive sediment mixing, volcanic activity, and hydrothermal events triggered by a catastrophe. Such shifts could occur in a matter of days or months rather than millions of years. [787] [788]

But this NOE explanation presents a contradiction: Oxygen-producing cyanobacteria are said to have existed for billions of years and are believed to have triggered the Great Oxidation Event (~2.4 Ga), long before the NOE. [789] If a precise oxygen threshold was required for complex life, then what mechanism finely regulated its levels at just the right time? If the NOE was truly responsible for stabilizing Earth's atmosphere, then what maintained atmospheric stability for billions of years before it? This narrative consistently shifts the problem rather than solving it.

Biochemist Nick Lane describes the Great Oxidation Event as an "apocalyptic extinction"—wiping out most early life due to its toxicity—yet insists that life rapidly adapted and became dependent on this once-deadly gas. [790] If oxygen was originally lethal, why didn't life simply remain anaerobic? What selective pressure forced life to evolve in a way that depended on what had just nearly destroyed it?

Evolution is supposed to be a gradual process, yet Lane's explanation requires life to make an extraordinary leap—one that just so happens to conveniently align with the oxygenation story. Instead of a natural, stepwise process, this is yet another retrofitted explanation that turns contradiction into progress. These explanations resemble those of a suspect whose story keeps changing, creating a tangled web of inconsistencies. Each attempt to explain one issue away only contradicts another.

Moreover, the Cambrian Explosion is not an isolated deviation—it is part of a recurring pattern that defines the fossil record. Rather than showing slow, stepwise evolution, the record is plagued by sudden bursts, where fully formed organisms appear abruptly with no definitive precursors. Yet, discussions of this broader pattern are largely avoided in scientific literature, while focus remains on isolated events rather than on the overarching trend. From the Avalon Explosion to the Mesozoic Marine Revolution, life erupts in rapid bursts. The Great Ordovician Biodiversification Event (~485–443 Ma), the Carboniferous Terrestrial Expansion (~345–300 Ma), and the Cenozoic Mammalian Radiation (~66 Ma–Present) all follow this same pattern.[791] [792] The expected transitions are missing, replaced by sudden rollouts of new life forms, ecosystems, and biological innovations. If this pattern has been upheld for over 150 years despite concealed contradictions and its conflict with gradualist predictions, how can we be certain that the dating of these events is reliable?

To avoid acknowledging the reality observed in the fossil record, evolutionists apply the euphemistic term "radiation" to describe these bursts. While "radiation" is used in other biological contexts, it is often misused to downplay the abrupt fossil appearances of new life forms. This term serves two purposes: it hides the fact that the fossil record contradicts gradualism and makes the sudden burst of new body plans sound like a natural process. It also allows both gradualism and abrupt appearances to shift the discussion toward spectacular diversification rather than addressing the absence of transitional forms.[793]

Stephen Jay Gould, a well-known paleontologist, argued that the fossil record overwhelmingly supports leaps in biological complexity. He acknowledged that the first major organisms appeared abruptly in the fossil record, followed by these patterns of sudden bursts. Although these bursts are thought to be separated by up to hundreds of millions of years, he did not consider them to represent entirely new forms of organisms but instead hypothesized that they were evidence of evolutionary jumps. He called these long periods of no change "stasis." However, Gould's so-called jumps are not evolutionary change at all, as they have no documented ancestral forms leading up to them.

While accepting the bursts observed in the fossil record, Gould and Niles Eldredge replaced Darwin's gradualism with a modified framework, introducing their theory, which is so contrary to Darwin's that the term *evolution* lost its original meaning.[794] [795] They conceded that Darwin's fossil-based predictions—rooted in gradual, small steps—failed and had to be reinterpreted to maintain the evolutionary framework, effectively smuggling in special creations under a different name: *punctuated equilibrium.* [796]

In his book, *The Structure of Evolutionary Theory*, Gould critiques this central assumption of Darwinian evolution—that small, gradual changes (microevolution) can be extrapolated over vast geological time scales to explain the full diversity of life.[797] He argues that Darwin relied on Lyellian uniformitarianism, which assumes that evolutionary change occurs at a slow and steady rate, rejecting the idea of sudden, catastrophic shifts. Gould points out that the fossil record does not support this assumption, stating:

> particularly in Darwin's embrace of Lyellian uniformity, and his denial of catastrophism (through arguments about the imperfection of the fossil record to allay the literal appearance of such rapidity in geological data),

for even a fully consistent, intellectually sound, and operationally potent theory will not regulate actual events if surrounding conditions debar its operation.[798]

Gould's verbose prose avoids clearly stating the obvious contradiction. That is, Darwin was so committed to gradualism that he blamed the Earth for not being able to preserve transitional forms. However, Gould acknowledges that this does not line up with the reality observed in the fossil record.

In keeping with Darwin, modern Darwinians continue to explain abrupt appearances as mere artifacts of an incomplete record. However, as Gould acknowledged, no matter how well-reasoned a theory may be, it cannot override the reality that life's history is marked by sudden, large-scale changes. In other words, Darwin's assumption cannot be considered valid simply because it's ideologically appealing—it must be demonstrated in nature. Even the prominent paleontologist Eugenie C. Scott has argued:

> Science requires the testing of explanations of the natural world against nature itself, and discarding those explanations that do not work. What distinguishes science from other ways of knowing is its reliance upon the natural world itself as the arbiter of truth.[799]

Further, Gould and Niles Eldredge argue that phyletic gradualism was never derived from fossil evidence but was instead an a priori assertion shaped by 19th-century liberal ideology. They state:

> Phyletic gradualism was an a priori assertion from the start—it was never 'seen' in the rocks.[800]

They criticize how most paleontologists precluded a fair assessment of evolutionary "tempos and modes" by dismissing contradictory fossil evidence as an imperfection in the record rather than recognizing it as valid data:

> It could not be refuted by empirical catalogues constructed in its light because it excluded contrary information as the artificial result of an imperfect fossil record.[801]

They further argue that no gradual transitions have been detected within any hominid taxon, writing:

> The record of human evolution seems to provide a particularly good example: no gradualism has been detected within any hominid taxon, and many are long-ranging.[802]

Instead of continuous transformation, they believe large traits such as brain size appear to arise through species selection rather than slow, incremental mutations. Their conclusion is clear: "We think that it [gradualism] has now become an empirical *fallacy*."[803]

For much of modern history, the "ape-to-man" progression has been central to evolutionary theory, widely presented in textbooks, museums, and media as a straightforward lineage. However, as the lack of definitive transitional fossils has become increasingly evident, many evolutionists have reframed the narrative. Rather than acknowledging the failure to find a direct human-ape ancestor, they now claim that those who still argue against humans evolving from apes simply misunderstand evolution—while simultaneously insisting that chimpanzees are our closest relatives. But this revision assumes that most evolutionists of the 20th century also misunderstood evolution, given that they explicitly framed human evolution in precisely these terms.

However, Gould himself is a key example of a revisionist who once tolerated the popular "March of Progress" depiction of evolution but later sought to distance himself from it. He admitted this contradiction, pointing to his own book jacket that featured the image, and he emphasized that it had become the defining icon of evolution:

A personally embarrassing illustration of our allegiance to the iconography of the march of progress. My books are dedicated to debunking this picture of evolution, but I have no control over jacket designs for foreign translations. Four translations of my books have used the 'march of human progress' as a jacket illustration. ... The march of progress is the canonical representation of evolution—the one picture immediately grasped and viscerally understood by all. ... Hence we continually make errors inspired by unconscious allegiance to the ladder of progress, even when we explicitly deny such a superannuated view of life. ... [W]e are virtually compelled to the stunning mistake of citing unsuccessful lineages as classic 'textbook cases' of evolution.[804]

Given that Gould has stated that there have been virtually no successful "examples brought forward to refute" punctuated equilibrium, and that the absence of evidence for gradualism in the fossil record is the "trade secret of paleontology," it is peculiar that so many materialists still insist the fossil record provides overwhelming evidence for evolution.[805] That assertion clearly arises from dogmatism. The sudden appearances of new life forms force radical materialists to invoke explanations that sound quite similar to inflationary cosmology.[806] [807]

While disconnected leaps in technology are plausible, the sudden appearance of isolated body plans defies natural explanation. However, these bursts seen in the fossil record parallel the pattern of human-made technological advancements. Indeed, technological progress has rarely been a slow, gradual climb—it comes in sudden leaps, much like what Gould described as punctuated equilibrium in evolution. Throughout history, civilizations relied on stone tools for millennia, then suddenly, metallurgy arose, sparking the Bronze and Iron Ages. Writing appeared independently in Mesopotamia, China, and Mesoamerica, not as a slow spread but in distinct jumps. The same pattern holds for agriculture, pyramids, the bow and arrow, and even mathematical concepts like zero—each appearing in rapid bursts across unconnected societies.[808] These technological 'explosions' show that innovation does not always follow a smooth, linear path but often stagnates for long periods before accelerating in sudden, transformative shifts.

The Cambrian Explosion and subsequent extinctions can be seen as systematic updates in Earth's natural 'hardware.' It resembles a massive deployment of new specifications and components, similar to a technological rollout, resulting in the sudden introduction of diverse and complex life forms. The initial hardware served its purpose for a period, supporting the ecosystem's needs. Species extinctions are like overhauls, phasing out old modules in a vast system, representing the natural process of introducing and updating the hardware components of the planet's ecosystem.

## Mutations

Since evolution cannot rely on a gradual accumulation of transitional forms in the fossil record, proponents now largely argue that genetics provides the answer. Mutations are claimed to serve as the engine of evolutionary progress, supposedly driving the development of new traits and body plans. Yet, just as the fossil record fails to demonstrate slow, stepwise development, so too does genetics fail to provide a clear, workable mechanism for these large-scale transformations.

Mutations can be neutral, harmful, or, in rare cases, beneficial. Beneficial mutations, which are exceedingly rare, typically result in minor alterations rather than significant changes.[809] These rare beneficial mutations are insufficient to account for the full range of biological systems and the vast array of distinct structures in nature.[810]

Harmful mutations can be likened to glitches in a biological software system—where errors with the code can disrupt the organism's complex processes.[811] This genetic entropy, where the accumulation of detrimental mutations occurs over generations, does not drive innovation; it drives the gradual erosion of the genetic code's benevolent precision. Rather than paving the way for progress, these mutations lead to degeneration, manifesting in aging, disease, and ultimately death. Cancer and other genetic diseases arise from corrupted genetic instructions, where the cellular 'programming' malfunctions, leading to uncontrolled growth or other harmful effects.[812]

This brings us to another contradiction within evolutionary theory: If evolution is purely unguided and every trait is just a neutral or adaptive variation, then the concept of 'deformity' contradicts this framework. Why do we instinctively recognize certain mutations as errors rather than just differences? Even materialists unwittingly assume there is an inherent design or ideal biological form rather than a purely random process.

Under evolutionary theory, all traits result from genetic variation, yet society selectively labels most deviations from the "norm" as "deformities" while insisting others are just stamps of individuality. By Darwinian assumptions, deformity as a concept should have no meaning. And yet, genetic disorders, congenital anomalies, and structural malformations are treated as deviations from a biological standard: conditions to be corrected whenever possible. Medicine operates on the implicit assumption that there is a baseline for human biology—a design that, when disrupted, must be restored.

And yet, when it comes to sex and gender, this principle is abandoned. Here, the standard shifts—not based on biology, but on ideology. In all other cases, medical intervention seeks to align or restore the body to its *natural* function, but in this one instance, where chromosomal sex and physical sex are clear, it seeks to override them. When a person's mental state directly conflicts with their chromosomes and physical body, it is the chromosomes that are deemed wrong, and the body is subjected to deformation to fit the perception. However, Materialism holds that physical reality is all that exists, governed by natural laws alone. By these standards, this perception that conflicts with biology is, by definition, metaphysical. It is rooted in a subjective, internal experience.

## Missing Links

The debate over transitional organisms in evolution often focuses on the fossil record. But we should also consider the absence of such forms among currently living species. If the gradual evolution by trillions of modifications were accurate, we should expect to observe a gradient of transitional forms in today's biodiversity. However, we don't see these living transitions between animal groups in the living or the dead.

In her 2004 book *Evolution vs. Creationism,* the prominent evolutionist Eugenie Scott avoids the question of direct evidence of ancestry by instead focusing on broad claims of common ancestry. Explaining why humans are not descended from monkeys, Scott emphasizes the inference of shared ancestry, stating:

> Did man evolve from monkeys? No. The concept of biological evolution, that living things shared common ancestry, implies that human beings did not descend from monkeys, but shared a common ancestor with them, and shared a common ancestor farther back in time with other mammals,

and farther back in time with reptiles, and farther back in time with fish, and farther back in time with worms, and farther back in time with petunias. We are not descended from petunias, worms, fishes, or monkeys, but we shared common ancestors with all of these creatures, some more recently than others. The inference of common ancestry helps us make sense of biological variation. Humans are more similar to monkeys than we are to dogs because we shared a common ancestor with monkeys more recently.[813]

Scott attempts to clarify a common "misunderstanding" about evolution by asserting that humans did not evolve directly from monkeys but instead share a common ancestor with them. However, this distinction is largely semantic, as it still assumes humans and monkeys descended from a shared, less developed primate—functionally making little difference to her initial question.

More problematic is the passage's digressive link into the past, where it continuously asserts that humans share ancestors with reptiles, fish, worms, and even petunias, yet fails to provide a single concrete reference point for these supposed transitions. The argument amounts to an endless chain of "we share, they share, all life shares," but never substantiates these claims with specific fossil evidence, genetic markers, or transitional species that would serve as actual reference points. Instead, she relies on sheer assertion, expecting the reader to accept this unbroken lineage without any supporting details.

Furthermore, this reasoning is circular, assuming common ancestry to explain biological variation while presenting that variation as proof of common ancestry. The claim that humans are more similar to monkeys than to dogs because of a closer common ancestor also lacks an independent basis, treating morphological similarity as definitive proof of relatedness rather than considering alternative explanations, such as common design. Ultimately, this passage presents an illusion of continuity, when in reality, it offers no direct evidence to bridge the evolutionary gaps—relying instead on an unchallenged narrative that demands acceptance without demonstration.

If up to a trillion species have existed throughout Earth's history, then, to account for the gradual changes proposed by Darwin, an estimated quintillion to sextillion individual organisms must have lived and died over the past 3.5 billion years, each supposedly contributing incrementally to the evolution of these species.[814] This presents an insurmountable problem that one would expect Darwinists to address.

Within evolutionary discourse, it is widely accepted that 99% of all species that ever existed are now extinct.[815] This staggering assertion originates from Darwin's identification of missing transitions among living organisms and within the fossil record. He presumptuously concluded that most transitional species would inevitably become extinct—a necessary claim—over time as they were replaced by more adapted or better-suited forms.[816] Although Darwin did not specify an extinction rate, his discussions about missing intermediate forms set the stage for later estimates.

The claim that 99% of all species that ever existed are now extinct is a theoretical reconstruction grounded in the presupposition that Darwinian evolution is the correct explanation for the diversity of life and its historical progress. Without this assumption of Darwinian evolution as the only valid framework, these estimates would lack any basis. It's like claiming an empty jar once held spectacular things, but now it's empty—a game of 'make-believe,' or in this case, 'let's believe.' As Louis Bounoure, professor of biology at the University of Strasbourg, sternly stated, "Evolution is a fairy tale for grown-ups. This theory has helped nothing in the progress of science. It is useless."[817]

Paleontologist David M. Raup asserts in his work, *The Role of Extinction in Evolution*, that the fossil record captures less than 1% of all species that have ever lived. This claim stretches the idea of an incomplete fossil record to a fantastical assumption. Such staggering claims of widespread extinction demand compelling evidence.[818]

Claims such as Raup's show the common issue of circular reasoning in evolutionary biology: the assertion that a vast majority of species are extinct is frequently justified by the gaps in the fossil record; these gaps are, in turn, explained by the theory of evolution, which predicts such extinctions. In other words, the lack of evidence is used to support the theory, and the theory is used to explain the lack of evidence. As another example, this is also similar to the conceptual problem-solving seen in inflationary theory. However, in paleontology, these explanations function as a fossil "Deflationary Theory."

Raup's claim that the fossil record represents only a tiny fraction of all species that have ever lived reveals a bias in relying on less than 1% of preserved fossils to make broad assertions about species extinction.[819] But I guess natural selection, after all, isn't just a creator of extraordinary biodiversity but, apparently, a master magician, performing the greatest vanishing act right before our eyes, sweeping up the supplanted like a great flood.

Similarly, the interpretation of genetic data, often used to fill the gaps in the fossil record, relies on a comparable set of assumptions. Constructed pathways, grounded in materialist assumptions, extrapolate backward to bridge significant gaps with theoretical links.[820] Scientists estimate these missing links by assuming a constant rate of species turnover over time.[821]

However, this approach also misses the mark by ignoring the episodic nature of mass extinctions that have periodically reshaped life on Earth.[822] The geological record shows that species extinction does not follow a smooth, gradual path but is marked by sudden, catastrophic episodes. It's an intriguing idea that life is resilient enough to progress through trillions of transitions, yet none of those transitions ever seemed to stick around long enough to leave clear evidence of their existence. What's more interesting is that the absence of transitional fossils is often framed as an expected feature—as if a law—of evolution rather than a problem for the theory.

## Standards

Although similar anatomical structures or genetic sequences across different species are often interpreted as evidence of common ancestry, modern discoveries suggest that these similarities are better explained as functional standardization within biological systems—akin to recurring design principles in engineering.[823] Engineers reuse effective designs to solve similar problems across diverse systems. Whether in computer, electrical, or mechanical engineering, they don't throw out standard schemas just for the sake of reinventing the wheel.[824]

According to the webpage "Engineering 101" on drawer.caddi.com, design standardization is a critical practice for manufacturing companies due to its numerous benefits.[825] Standardization improves quality control by minimizing errors and defects, as consistent standards reduce inconsistencies during production.[826] It increases manufacturing efficiency and productivity by optimizing equipment, tools, and processes, and workers become more adept with familiar standards, leading to faster and more accurate work.[827]

Standardization simplifies logistics and inventory management by reducing the need for varied parts and materials.[828] Maintenance and servicing are also streamlined, as standardized components can be easily interchanged between products, making repairs quicker and more cost-effective.[829] Ultimately, standardization is essential for maximizing quality, efficiency, and performance.[830]

More fundamentally, organisms themselves belong to identifiable categories that share common frameworks. If life did not adhere to underlying biological standards, the concept of biological diversity would lose meaning—there would be no recognizable distinctions between species, no identifiable traits that define different forms of life. Instead, the presence of shared biological features across species reflects a system of structured organization, not the haphazard results of an unguided process.

Consider J.K. Rowling's *Harry Potter* series. Each book contains unique plot points and character developments but shares a consistent writing style, recurring characters, and thematic elements.[831] If pages from these different books were mixed up on the floor, someone unfamiliar with the series might struggle to determine which pages belong to which volumes. They might even assume all the pages are from a single volume due to the recurring elements and similar style across the books.

But there's something more intriguing about the Darwinian narrative. When it comes to the future of species, particularly human beings, the focus tends to shift to potential threats and existential risks rather than the potential for new species to emerge through evolutionary processes. There is a noticeable lack of confidence in the predictive power of Darwinian mechanisms to chart a progressive future for life. This may be because their supporting evidence exists only in retroactive justifications of past events. The same fervent believers in life's supposed evolution from nothing— through catastrophic beginnings of inanimate matter into today's biological complexity—are paradoxically skeptical of the resilience of life, despite its supposed journey from nothing to its proven ability to thrive in extreme environmental conditions.[832]

If Darwinians can conclude past events such as bird feathers evolving from reptilian scales, it should logically follow that the same esoteric reasoning could predict future evolutions of current species. If evolutionary processes truly drove life from nothing to the vast array of species we observe today, undoubtedly, anticipation of future diversity would be common within scientific literature. This includes the development of fundamentally new species and substantial speculation on how humans might continue to evolve. Yet, there is a deafening silence on this front. I guess it is far easier to rely on confirmation-biased assumptions than to make predictions. Or, maybe, there is no true scientific explanation of how this works.

We've mapped the entire human genome, yet our understanding of how Darwinian mechanisms drive the complexity and diversity of life is just as uncertain today as it was in 1859?[833] In other scientific fields, such as astronomy, our understanding of past celestial movements allows for precise predictions of future events like solar eclipses and planetary orbits, showing how concrete knowledge of historical patterns can reliably forecast future occurrences. If evolutionary theory cannot begin to predict future species' diversity or the development of new traits, then maybe these stories are just stories.

Though these *evolutionary biologists* accuse thinkers like Behe and Meyer of pseudoscience, the accusation is the epitome of projection. Behe and Meyer make precise, empirical claims—that systems like the bacterial flagellum exhibit irreducible complexity, and that the specified information encoded in DNA cannot be explained by any known undirected or even determined material process. These are not theological arguments masquerading as science; they are scientific challenges that expose the limits of materialist explanations. The inference to a mind is not their starting point—it is the conclusion drawn from the consistent failure of all known physical causes to account for the data. Behe did right by changing his title from evolutionary biologist to (R)evolutionary biologist.[834]

Evolutionary biology, by contrast, is not science in any meaningful or traditional sense. Not a single evolutionary biologist can offer a step-by-step, mechanistic explanation for the origin of the systems they claim expertise over. The field points to broad similarities, invokes long ages, and assumes blind processes did the work—but it cannot demonstrate how. It explains nothing beyond what its framework already demands must be true. In that sense, it is no more a science than astrology—both assign causes to events they cannot observe, test, replicate, or even logically explain, and both survive by cloaking speculation in the sleek language of expertise.

# CHAPTER 27

# March of Technological Progress

*"On the great principle of hereditariness, of which he himself was the prophet and expounder, Mr. Darwin could not help being a remarkable man."* — G. W. Bacon, The *Life and Labours of Charles Darwin* (1882) page 7.

*"The laws governing inheritance are quite unknown; no one can say why the same peculiarity in different individuals of the same species, or in different species, is sometimes inherited and sometimes not."* — Charles Darwin, On the Origin of Species *(1859), Chapter 1.*

C + B. The finest gradation, B and D rather greater distinction.[835] —Charles Darwin

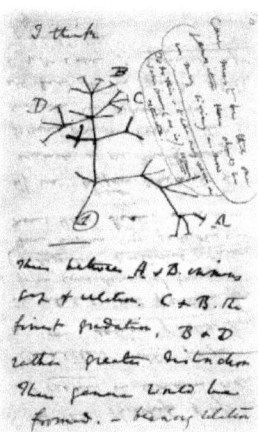

**Figure 28.1** This is the—*prophet and expounder*—Charles Darwin's famous sketch from, often referred to as his *Tree of Life* diagram, drawn in one of his notebooks as he began developing his theory of evolution by natural selection.

Darwin's sketch contains an evident error when analyzed. The letters B, C, and D are positioned on separate branches—each branching off from three main upper divisions of the tree. Though these branches differ in the number of splits (e.g., the middle branch splits into 4, while the left and right split into 3), all branches ultimately trace back to the same "trunk" through the same number of generations or "joints."

Darwin scribbles that B and D display "greater distinction" compared to B and C, yet based on the tree's structure, all letters (B, C, and D) share an equal hereditary distance from their common ancestor. This means their genealogical distinction cannot differ. The added splits on the middle branch (housing B) do not alter its generational distance to the trunk. It is noteworthy that the NCSE, an organization that promotes itself as a

327

defender of evolution, as many other organizations, uses Darwin's Tree of Life sketch in its logo. By relying on a romanticized iconography, similar to the *March of Progress*, these organizations show their faithfulness to promoting an ideological narrative rather than science literacy.

Gould was also critical of Darwin's Tree of Life, specifically as a branching model of evolution. He saw it as another romanticized depiction lacking empirical support. Darwin's early sketch reveals that his theory relied more on inference than direct evidence, as seen in Figure 28.1. Gould saw this as a reflection of the scientific ignorance of the time, where sweeping conclusions were drawn without sufficient fossil evidence to support them.

As Gould also noted, the grandeur of Darwin's theory has been reduced to scientifically misleading and inaccurate imagery, such as the famous march of human progress, which falsely depicts a linear progression from a hunched-over ape to an upright human.[836] It is presented through a series of familiar silhouettes that have subtly influenced our perceptions since childhood.[837] This linear depiction has become so ingrained in our minds that many subconsciously think of it as the definitive representation of evolution, creating a nostalgic place in some worldviews. To much of the uninitiated, it still appears logically sound. The *March of Progress* serves as a visually compelling "proof" of a concept that goes unexplained.

**Figure 28.2** Evolutionary Progression from Ape to Human. This illustration is inspired by the famous "March of Progress" from *Early Man* (1965) by Rudolph Zallinger. It visualizes the conceptual sequence of evolutionary stages from ape and ape-like ancestors to modern humans, reflecting popular representations of human evolution in scientific media. **Image source:** S. A. Cooper.

It's a potent example of how suggestion can shape our perspective of human history.[838] This smooth and uninterrupted narrative from monkeys to humans did not arise from insights provided by molecular biology or the fossil record, but from a distorted notion of comparative anatomy.[839]

Genetics, however, arose during the same era as Darwin but was largely overlooked. It provided an empirical foundation for heredity—unlike the speculative assumptions of natural selection. Instead of forcing a reassessment of Darwin's theory, genetics was adapted to cover natural selection's growing contradictions.[840] However, bringing genetics into Darwin's theory didn't resolve its problems—it created new ones.

To help explain the lack of transitional forms, theorists concocted hypothetical *neutral mutations*, which, while attempting to fit the genetic data into the existing framework, only deepened the theoretical inconsistencies.[841] To account for the origin of complex traits, theorists increasingly posited gene duplication and regulatory DNA—not because these mechanisms were shown to produce new, functional biological systems, but because the existing framework demanded some explanation.[842] These were not natural extensions of Darwin's ideas but reactive adjustments to keep the theory intact. Though once dismissed as junk, this regulatory DNA has now been appropriated by the same theorists to support evolutionary claims— rebranded as a primary driver of complexity—yet another example of turning contradiction into progress.

As is customary, rather than addressing conflicts directly, each contradiction is treated as an opportunity for theoretical expansion rather than a challenge to its validity. These conflicts include, but are by no means limited to, the complexity of the cell, the distinct roles of coding and noncoding DNA, the Waiting Time Problem, and Haldane's Dilemma—all of which pose significant challenges to the gradualist framework of evolution.

The Waiting Time Problem refers to the fact that the time required for specific mutations to arise and spread through a population is far longer than the available window in the fossil record.[843] To paper over the insurmountable genetic gaps, theorists have increasingly relied on speculative mechanisms, such as adaptive radiation and accelerated mutation rates, to create the illusion of plausible transitions.

Imagine a vast library, supposedly filled with billions of books, each containing precise, meaningful content. However, instead of being written by authors, these books supposedly formed through a process where letters were randomly placed on blank pages over time. But there's a catch: each letter-change happens only once every few generations, and for the book to reach a meaningful final form, every change must be functional and contribute to the intended message. If a random letter disrupts meaning, the sentence or passage is discarded, making the process even slower and less likely to produce coherent new content.

Now consider this: some books contain paragraphs of highly specialized information, which, according to the process, could only arise by accumulating multiple coordinated changes. But the real problem is that the time needed for even a single coherent sentence to form is astronomically longer than the supposed age of the library itself—let alone the time required for billions of fully written and published books to appear. Even if we assume that a single functional phrase miraculously appeared, the probability of the next necessary phrase appearing before the first one is lost or replaced is even lower. By the time a full paragraph forms, the timeframe required already surpasses the age of the universe.

This is the crux of the Waiting Time Problem in evolutionary genetics: the time needed for multiple coordinated mutations to arise and fix within a population far exceeds the available evolutionary timeframe. Scaling this to biological evolution, the problem is even worse—because life does not require just one meaningful sentence, but millions—if not billions—of interdependent genetic instructions, all precisely arranged to form functional biological systems.

"Haldane's Dilemma" is a variation of the Waiting Time Problem, but instead of focusing solely on how long it takes for beneficial mutations to appear, it addresses the cost of natural selection.[844] Haldane argued that natural selection comes with an inherent cost because the spread of a new advantageous gene requires the removal of organisms that lack it.

This process cannot occur too rapidly without significantly reducing overall reproductive success. If too many genetic changes are selected at once, the resulting population loss would be unsustainable, potentially leading to extinction. He calculated that in a typical vertebrate population, the rate at which beneficial mutations can be incorporated is highly constrained. His estimates suggested that under realistic conditions, a population could only fix a limited number of advantageous mutations per generation without an excessive burden of deaths. This limitation places a severe restriction on the speed at which evolutionary change can occur, contradicting the assumption that major transformations could happen within the available timescales.

Imagine a city mandated to upgrade its taxi fleet to newer, more efficient models or face shutdown. However, instead of replacing entire cars at once, the city must swap out one part at a time—first the alternator, then the braking system, then the fuel injectors—while keeping every taxi in service. Since city regulations require upgrades to occur at random intervals, older taxis without new parts must be removed as newer models take over. If a new upgrade begins while some taxis are still catching up with the previous one, they fall out of sync, becoming obsolete and requiring earlier removal. If upgrades happen too quickly, too many taxis break down before the process is complete, leaving the city with too few vehicles to operate. Even if the process is slow, outdated taxis still become incompatible with the evolving system and must be phased out. If replacements of taxis don't arrive fast enough, the system collapses.

At first glance, these regulations might seem counterintuitive—why must the upgrades happen this way? Why couldn't the system allow for a smoother transition, ensuring that no taxis are lost too quickly or left incompatible with the rest? The reason this process feels unnatural is because it is. No functioning system, whether mechanical or biological, operates under such rigid and arbitrary constraints.

Yet, this is precisely the kind of process required under the assumption that life diversified as Darwin proposed—from simple microbial life to complex organisms, including humans. The logic of step-by-step evolutionary change demands that beneficial mutations arise randomly, spread gradually, and replace previous traits, all while the organism remains fully functional at every stage. But this creates an unavoidable contradiction: if evolution is too slow, necessary changes never accumulate fast enough to produce new forms of life; if it is too fast, transitional populations are constantly eliminated, leaving no room for the gradual progression that evolution depends on.

But even so, in this scenario, these taxis have intentional mechanics actively working to troubleshoot and integrate each upgrade, ensuring that no vehicle is permanently taken out of service due to incompatibility. Meanwhile, the city also needs engineers to force parts from new models that don't even fit in the older taxis in the first place. By contrast, in evolution, no such oversight exists—mutations occur randomly, without a built-in system to coordinate changes or ensure they integrate properly. The issue isn't just that the process is inefficient; it's that the way it is framed ensures that it can never truly work as a creative force.

Haldane's Dilemma is rarely discussed today, not just because his numbers disproved Darwin's timeline, but because it reveals a far deeper issue—his calculations were based on a process that was never real to begin with. He assumed beneficial mutations must be appearing and replacing old traits, then calculated whether this could happen at the necessary speed. When the math didn't work, it should have been a red flag about the process itself. Instead, it became an isolated "problem" to be worked around, without questioning whether the underlying assumptions had any foundation.

The real issue is not just that his findings contradicted evolutionary expectations—it's that his entire dilemma was built on fictional premises. Beneficial mutations, as needed for Darwinian evolution, have never been observed to create the kind of large-scale transformation the theory demands. While some mutations may offer minor optimizations or adaptations within an existing framework, this is far from the step-by-step development of new structures that Darwin's model requires. Haldane accepted evolution as fact and sought only to explain how it worked. But if beneficial mutations aren't occurring in the way required, then debating their spread isn't just a dilemma—it's meaningless.

This is why modern evolutionary thought has quietly moved away from Haldane's conclusions. Instead of directly addressing the contradictions his work exposed, the focus has shifted toward alternative explanations like neutral evolution or punctuated equilibrium—attempts to avoid the problem rather than confront it. But none of these models answer the real question: If the mechanism of evolutionary change does not exist, what is left of the theory? It's like calculating how long it would take for pigs to fly across the Atlantic when the real question is: Can pigs even fly in the first place? Some evolutionists are still trying to solve a "dilemma" that never existed, all while downplaying Haldane's construct—one that paradoxically supports their case by assuming such beneficial genes exist in reality.

Therefore, not only does the available time fail to account for sudden bursts like the Cambrian Explosion, but even if such events were ignored, there still wouldn't be enough time, plausible probabilities, or a viable mechanism to explain the gradual evolution of life as we see it today. As a result, the term "evolutionary genetics" is a misnomer, serving only to reinforce a predetermined conclusion that is not grounded in reality. What is called "evolutionary genetics" could more accurately be described as *biological eisegesis*, where genetic data is interpreted to fit preexisting evolutionary narratives, rather than serving as an independent and predictive framework.

## The Mechanical Evolutionist

Consider the perspective of an observer from a distant, unknown land, untouched by modern technology but well-versed in Darwinian thought. Imagine this observer, venturing out of his homeland for the first time after a mass extinction, with no way to verify the origin of technology. Embracing Materialism, this visitor takes it upon himself to explain the "evolution" of automobiles—from the simple Model T Ford to today's sophisticated Tesla.[845] [846]

As the observer picks his way through rows of rusted and abandoned cars, he marvels at what seems like a clear lineage. "Of course," he mutters to himself, "it all makes sense." These machines, from their wheels to their engines, must have evolved over time. The old Model T, with its basic four-cylinder engine, seems almost primitive compared to the sleek designs he now stands before. He pauses at a manual stick shift. "Ah," he thinks, "a relic of the past, clearly selected out by the more efficient automatic transmission." He's confident in this theory—natural selection at work.

Moving further, he finds a six-cylinder engine and chuckles knowingly. "Here it is," he says softly, "another transition." To him, it's obvious—this engine is proof of the gradual evolution from four to eight cylinders. He's convinced he's uncovering the secrets of mechanical evolution, the same way he believes species adapt in nature. The progression seems undeniable, each innovation simply a step in the unrelenting technological march toward the Tesla in the distance, the pinnacle of automotive sophistication. The shift from steel bodies to fiberglass seems like a perfect example of evolutionary refinement—lighter, faster, better.

His perspective blinds him. He is so sure of his materialist framework, he fails to see the complexity before him. The purpose and design behind each feature elude him. "The evidence is overwhelming," he assures himself, glancing at the gaps between two-wheeled motorcycles and massive sixteen-wheeled trucks. The missing links don't bother him—he sees only the similarities, convinced that he's piecing together a grand evolutionary narrative. He misses the car show, distracted by the parts, just as others once did when claiming natural selection could explain entirely new body plans.

What escapes him, though, is that the similarities he points to aren't the product of gradual, mindless processes. Each car, though sharing a common structure, is the result of intentional design. The uniformity in engine placement, the number of doors, even the aerodynamics of the windshield—all carefully crafted for function, efficiency, and safety. Yet, to him, it's all further evidence of a slow, evolutionary march forward. He does not see that these shared features persist because of practical standards, not evolutionary inheritance.

He walks past the Tesla, seeing it as another notch on the evolutionary ladder. "From a Model T to this," he marvels. But the Tesla isn't simply an improved Model T—it operates on an entirely different system. The transition from gasoline engines to electric motors isn't just a swap of parts; it requires a whole new set of blueprints, factories, tools, and skills. The observer never realizes that each step in the progression he imagines demands intentional planning, innovation, and human ingenuity. Yet, as he continues his journey, he remains blissfully confident in his theory of mechanical evolution.

However, this car analogy, while helpful in illustrating technological development, is ultimately too simplistic. Cars do not replicate themselves; they rely on human designers, engineers, and factories to produce each new model. Now imagine if cars could replicate themselves—not just by making replacement parts, but by creating entirely new cars. This would require a completely different set of mechanisms, far beyond what we see in human engineering. Yet, in the biological world, self-replication is not only possible but a standard feature, controlled by highly complex biological systems that function with intricate precision.

## Darwin's Tree of Life

In 2007, W. Ford Doolittle and Eric Bapteste shook up the world of evolutionary biology with their paper, "Pattern pluralism and the Tree of Life hypothesis."[847] Their paper takes a hard look at Darwin's Tree of Life (TOL) hypothesis.[848] Described in their paper, Horizontal Gene Transfer (HGT) is the process by which genetic information is transferred between unrelated organisms, bypassing traditional inheritance from parent to offspring. This process allows organisms to rapidly acquire new traits and adapt to their environment, creating complex genetic relationships that challenge the linear, branching process suggested by the Tree of Life model.

The crux of their critique lies in the fundamental issue they raise: "…there is no strong expectation that a universal hierarchy that embraces all life should be produced with molecular markers." In other words, the claim implies that the molecular data do not necessarily support a clear, hierarchical pattern of descent from a common ancestor. If evolution were strictly tree-like, we should see a consistent pattern across different genes and proteins, but instead, molecular phylogenies sometimes produce conflicting trees.

This means that the Tree of Life model is inadequate for representing the evolutionary relationships of prokaryotes due to the complexities of horizontal gene transfer (HGT).[849] Doolittle and Bapteste also note that "Microbial phylogeneticists have not in general taken it to be their duty to confirm the existence of a natural inclusive hierarchy or tested the TOL hypothesis that this hierarchy is to be explained by an historical branching process." This statement highlights that scientists who study microbial evolution have not actively tested or questioned the assumption that all life forms, including microbes, fit into this branching hierarchy.

In the early 2000s, Doolittle was already critiquing the tree-like structure of evolution due to HGT, but the 2007 paper was particularly groundbreaking because it brought together several ideas into a more coherent challenge to the Tree of Life model. While other scientists contributed to the broader field of HGT, Doolittle was the key figure in applying it specifically to the Tree of Life model prior to 2007. Unlike the typical parent-to-offspring gene transfer, HGT is like a horizontal swap of genetic information between different organisms.[850]

This is like entering a public place, such as a classroom, and picking up new traits from fellow students instead of inheriting them only from parents. Imagine your genetic makeup changing after leaving this space. This type of issue is particularly significant in microbiology, where prokaryotes dominate life on Earth. It necessitates a more interconnected, web-like model of evolution to provide a viable explanation.[851]

Darwin's framework assumes that genetic similarity reflects evolutionary relationships, with closely related species sharing closely related genomes. However, the widespread occurrence of HGT undermines this assumption. Genes can cross boundaries between unrelated species, creating inconsistencies in genetic distances. These discrepancies caused by HGT show that genetic similarity does not validate the assumption of evolutionary proximity.

To reconcile these conflicts, evolutionary biology often resorts to mechanisms like *convergent evolution* or *incomplete lineage sorting* (ILS), which are used on a case-by-case basis rather than stemming from a cohesive, predictive model. But if such significant genetic exchange occurs outside of Darwin's proposed mechanisms, how can one still be certain that the Darwinian version is the correct explanation for the diversity of life? It's worth noting that Doolittle and Bapteste have expressed regret over the implications their findings may have cast on Darwinism. To be clear, the views expressed in this book are solely a reflection of those outlined by their paper—not necessarily their own opinions, just their findings.

Instead of saying a contradiction breaks the model, scientists often describe it as evidence that the system is "more complex than we thought." However, one can't just point to newer interpretations that function as an ever-moving target while maintaining that the original theory is still intact. Doolittle's and Bapteste's critique cannot be framed as part of an iterative process of refinement. It is not building upon or adjusting existing ideas other than maintaining all life has a common ancestor. It's not just a critique of a vague interpretation of the theory; it's a fundamental break from how the evolutionary process has been understood in Darwinian terms for 150 years, despite later variations like the "bushes" model. The Tree of Life, rooted in Darwin's early sketch, still remains the ultimate icon for evolutionary theory.

One recurring issue with the Darwinian framework is its reliance on ad hoc explanations to account for phenomena. When Doolittle and Bapteste revealed that genetic material could be transferred laterally, disrupting the linear Tree of Life model, the response was not to rethink the core assumptions of evolutionary theory but to expand and adapt the existing framework. As Doolittle and Bapteste note:

> We have come to appreciate the plurality of evolutionary processes of lineage diversification. But most of us hold on to the first two tenets, that there is a real and universal natural hierarchy, and that descent with modification explains it, in much the same way as Darwin did. We may be process pluralists, but we remain pattern monists.

This quote underscores how the scientific community has acknowledged a plurality of evolutionary processes, such as HGT, yet remains committed to a single hierarchical pattern of relationships, "much the same way as Darwin did" with the original Tree of Life model. This unwavering commitment to the Tree of Life reveals a cult-like adherence to Darwinian assumptions. For example, Doolittle and Bapteste further explain:

> The tendency to treat phylogeny as a simple tree arises from the basic assumption that the distribution of life's features reflects inheritance, and this assumption is so entrenched that it is often taken for granted.

In a way, Doolittle's and Bapteste's own words indict the evolutionary framework they are defending. This entrenched assumption about inheritance helps explain why the Tree of Life model persists in evolutionary biology, even when evidence such as HGT points to a more complex, interconnected evolutionary process.

Their findings suggest that the Tree of Life model is not just in need of adjustment, but is fundamentally flawed for certain organisms, particularly prokaryotes.[852] They overlooked, however, a more profound concept. The processes observed in these organisms, particularly HGT, point to natural technology. Prokaryotes are not the products of evolution at all. Instead, they operate on a completely different system, akin to a distributed network or collaborative technology. This is not a deviation from evolutionary theory; it reveals the true nature of life's design—a complex, interconnected system of genetic information exchange that functions as adaptive technology, not a branching tree.

In this context, HGT is an intentional mechanism that enhances the adaptability and resilience of organisms in an intelligently engineered biosphere. It is a kind of *genetic engineering* built into the fabric of life, allowing organisms to exchange genetic material rapidly and efficiently to respond to environmental changes. This mechanism could be seen as a method of maintaining balance within ecological systems and ensuring the survival of diverse life forms, not resulting from millions of years of slow, gradual evolutionary changes as posited by Darwin's theory.

Therefore, HGT in NatTech can be seen as a more interconnected and flexible network of life, where organisms are equipped with the ability to adapt quickly through horizontal gene transfer. HGT is analogous to a software update in a complex system, where organisms can 'download' beneficial traits to improve their functionality and chances of survival.

For example, bacteria can rapidly develop antibiotic resistance not by evolving it slowly over generations, but by directly acquiring resistance genes from other bacteria through plasmids. Similarly, thermophilic microorganisms in extreme environments can gain heat-resistance genes from unrelated species, allowing them to survive in conditions that would otherwise be lethal. In another case, pea aphids have obtained carotenoid-producing genes from fungi, granting them a capability that is otherwise absent in animals. These examples illustrate how HGT functions as a precise and efficient transfer of information, rather than a blind, trial-and-error evolutionary process. These engineered features showcase the foresight and ingenuity embedded within natural systems, affirming that life's complexity is not a byproduct of randomness but the outcome of a purposeful design.

Similarly, Bluetooth mesh networks enable devices to update each other when in proximity, even without direct internet or server connections.[853] This allows for firmware updates and data exchanges between different devices, a process deliberately designed for interoperability.[854] Likewise, HGT operates as a biological mechanism that enables genetic information to be shared between organisms, facilitated by vectors such as viruses.[855] [856] In the Internet of Things, smart home devices like thermostats, light bulbs, and security cameras can communicate and update each other through proximity-based connections, despite their different primary functions.

## Beyond the Ape to Man

Despite what Gould called the visceral effect of the iconic ape-to-man image, it's important to remember that Darwin's theory goes far beyond just proposing a pathway for how humans evolved from apes. According to this assumption, we—along with our consciousness and all living organisms—are said to have evolved from the same microbial ancestor, where HGT is prevalent. This means that everything—without exception—from humans to broccoli, grass, mangoes, snakes, oak trees, elephants, slugs, butterflies, ginger root, frogs, peacocks, jellyfish, kangaroos, whales, cows, rabbits, spiders, mushrooms, bees, starfish, lobsters, penguins, cacti, eagles, walruses, bats, chameleons, turtles, sunflowers, ants, octopuses, and even apple pies— along with the extrapolated 1 trillion species believed to have ever existed— are all claimed to trace their origins to a single common ancestor.

I have mentioned this more than once, but it's worth emphasizing: the materialist evolutionary process inherently includes the creation of human ingenuity, and by its own logic, *our machines and technologies are extensions of the same blind progression.* No matter how counterintuitive it may seem, there is no defining line in this string of dominoes. Yet, our own technology provides the answer to a riddle that has puzzled thinkers for centuries. While the shared characteristics of our innovations do not arise from one another as in biological evolution, their common denominator is a mind. And in this, the riddle begins to unravel—materialist logic fails to follow through, but design logic remains coherent, whether forward or in reverse.

Imagine someone saying, "the house did not evolve from the tent but rather they share an ancestor." Or consider if every piece of machinery or technology ever created by humans—all with their specialized functions, sizes, and complexities—were thought to originate from a single invention. Imagine assuming that the Boeing 747, John Deere tractors, the steamboat, the submarine, the typewriter, the microwave oven, the bicycle, the combine harvester, the MRI machine, the clock, the telescope, the air conditioner, the electric guitar, the roller coaster, the dishwasher, the excavator, the wind turbine, the sewing machine, the cotton gin, the kiss cam, the chainsaw, the satellite, the bulldozer, the elevator, the escalator—and all technology supposedly emerged from one rudimentary machine.

Why does the prevailing biological model focus so singularly on a common ancestral source, yet remain unable to definitively and empirically pinpoint an unbroken lineage in the present—despite evidence pointing toward independent sequences, repurposed structures, and standardized components? Just one single outlier or anomaly—a confirmed, uninterrupted lineage—would help solidify their model. But the fact that not even one exists says enough.

# CHAPTER 28

# Ecological Schematics

*I look at the term species, as one arbitrarily given for the sake of convenience to a set of individuals closely resembling each other, and that it does not essentially differ from the term variety, which is given to less distinct and more fluctuating forms.*[857] —Charles Darwin

The conventional method of taxonomy was established by Carl Linnaeus: the 18th-century Swedish botanist often called the Father of Taxonomy.[858] He believed that the *Systema Naturae* (The System of Nature) was a reflection of the working plan underlying Creation.[859] Linnaeus saw the classification of plants, animals, and minerals as a means of uncovering God's intentional design in nature. He saw his scientific work as guided by divine providence and considered his discoveries a form of understanding the Creator's wisdom.[860]

A central tenet of Linnaeus' worldview was that species were fixed and unchanging, having been created directly by God. He explicitly rejected the notion that new species could arise over time, stating that no new species are produced, and all existing life forms owe their origin to "some Omnipotent and Omniscient Being, namely God." He wrote:

> As there are no new species; as like always gives birth to like; as one in each species was at the beginning of the progeny, it is necessary to attribute this progenitorial unity to some Omnipotent and Omniscient Being, namely God, whose work is called Creation. This is confirmed by the mechanism, the laws, principles, constitutions, and sensations in every living individual.[861]

His perspective aligned with the idea that nature was divinely arranged, with each species possessing an essential, immutable identity.

Linnaeus also regarded the study of nature as a form of worship, believing that observing the intricate details of the natural world allowed humans to admire and praise the Creator. He saw the process of plant reproduction as particularly significant, referring to it as revealing the "very footprints of the Creator."[862] This conviction led him to develop his system of classification. His system became the foundation of modern classification, assigning each species a two-part Latin name: *genus and species*, a structure still in use today.[863]

Despite Linnaeus' clear belief in a Creator, materialist historians often commit what could be called the 'Peter Denied Jesus' fallacy, in which an isolated statement or minor adjustment in scientific understanding is exaggerated to suggest that a scientist never truly believed or ultimately abandoned their faith. Linnaeus' recognition of variation within species has sometimes been reframed as a rejection of divine creation, despite his lifelong belief that nature was the work of an Omnipotent Creator.

The Linnaean view of classification, which emphasized fixed and divinely created species, was upended in the 19th century by Charles Darwin's theory of evolution by natural selection, which redefined biological classification in terms of common ancestry and gradual change. Traits were now examined through the framework of evolutionary relationships.

However, if natural selection were the sole driver of such similarities, it would require random mutations to produce identical solutions multiple times in separate lineages—a statistical improbability. When similar traits appeared in species classified as unrelated, evolutionary theory explained them retroactively as cases of *convergent evolution*, despite the lack of a clear mechanism on how this occurs. Theorists of convergent evolution proposed that environmental pressures could independently produce similar complex biological structures—such as the eye—in different lineages. According to the theory, eyes did not arise once from a single common ancestor but instead evolved independently in multiple species.[864] This makes it especially daunting, however, when evolutionists downplay how improbable it is for the eye to have arisen even once through natural selection, while often avoiding the fact that their own model requires it to have happened multiple times.

Because the repeated appearance of complex traits like eyes relies on convergent evolution, it implies that such features are, in some sense, inherent—or even 'predestined'—within the evolutionary framework—an assumption that contradicts the idea of a blind, undirected process. This is even implied by conserved genetic regulatory mechanisms, such as the Pax6 gene, which suggests a common underlying framework rather than truly separate evolutionary events.[865]

The limitations of morphology-based taxonomy led scientists to seek a more precise method, shifting classification to molecular data through systems biology. However, molecular classification presents contradictions, as it often assumes genetic similarity indicates shared ancestry. DNA sequencing has revealed unexpected genetic similarities that contradict traditional groupings, raising questions about the assumption that similar traits evolved independently.

Horizontal Gene Transfer dismantles the assumption that genetic similarities always arise from independent adaptation to external conditions, instead demonstrating that genes can be directly exchanged between species. This weakens claims of convergent evolution, showing that supposedly "independent" traits can result from shared genetic material. Aphids, for example, have acquired carotenoid-producing genes from fungi, and certain beetles have gained bacterial genes to help digest plant material.[866] [867]

Although conventional taxonomy helps categorize biological diversity and structural relationships, it often overemphasizes physical similarities at the expense of deeper functional roles. It emphasises the 'what' and the 'who'—that is, what organisms are made of and who they're related to in an evolutionary sense. It organizes life into a structured hierarchy ranging from Kingdoms to Species. This system helps scientists classify species by common traits but misses the bigger picture of how they relate.

Imagine walking into a vast library and deciding to categorize the books based on their most obvious traits. Hardcovers over here, paperbacks over there, and magazines in a separate section. You start to notice patterns in the typesetting—grouping together books that share a similar font or spacing, as though these choices reveal something profound about their content or purpose.

Next, size catches your attention. You begin sorting the books by dimensions, forming neat transitions from small pocket-sized editions (4.25" x 7") to midsize volumes (5.5" x 8.5" or 6" x 9"), and finally to the larger ones (8.5" x 11"). The gradual shift in height and width seems meaningful, as though physical size somehow mirrors the importance of what's inside.

You refine the process further. Maybe word count becomes a criterion, or the presence of glossaries or indexes. Even cover art, divided between cartoonish illustrations and realistic depictions, starts to form neat groups. Without ever turning a page, the collection now seems perfectly ordered by these physical attributes. Yet, as your knowledge of the words within grows, you begin to recognize deeper patterns—grouping books not just by their appearance but by the morphology and syntax of their contents, revealing unexpected connections.

Similarly, in biological classification, organisms are grouped based on their observable traits. For instance, the animal kingdom is divided into various phyla according to body plans and structures.[868] Some, like those in the phylum Chordata, have backbones, distinguishing them significantly from phyla without vertebral structures, such as Arthropoda, which includes insects and crustaceans with exoskeletons.[869] This would be like classifying books in a library by whether they're paperback or hardcover. Within the Chordata, specific features like the structure of ear bones and the pelvis are used for further differentiation.[870] Mammals, for example, are characterized by their three unique middle ear bones—the malleus, incus, and stapes— that aid in hearing, a key trait in the Darwinian narrative of evolution.[871]

Similarly, the pelvis structure varies across species, supporting different modes of movement; the pelvis in bipedal organisms supports an upright posture, essential in human evolutionary biology, whereas in quadrupeds, it supports a more horizontal body alignment.[872] These anatomical details are seen as essential in understanding evolutionary lineage and functional adaptations within the Darwinian framework. Additionally, some phyla are distinguished by their developmental patterns, such as Protostomes, where the mouth forms before the anus during embryonic development, and Deuterostomes, where this process is reversed.[873]

Though these traits are essential for classification, this focus can downplay the more significant characteristics of the roles and functions that organisms play within ecosystems. Recognizing these roles offers insights into the delicate balances and intricate relationships that characterize natural systems. The concept of roles and functions is touched upon in general biology classes, but it often doesn't receive the attention it deserves beyond the classroom, typically being limited to ecological studies where plants and animals are observed in food chains.

Likewise, in our library analogy, we first concentrated on the 'what' and 'who' aspects. Books were sorted by categories that seemed meaningful on the surface but lacked depth. But now, let's shift to the 'why'—viewing this library from a different perspective, one where we no longer presuppose that there's no broader meaning behind these traits.

Imagine discovering that Mass Market Paperbacks (4.25" x 7") often contain fast-paced, widely accessible fiction. Trade Paperbacks (5.5" x 8.5" or 6" x 9"), on the other hand, tend to feature more in-depth works, whether literary novels or nonfiction. Hardcovers (6" x 9" or 8.5" x 11") may be reserved for substantial, detailed books—often with richer content, indexes, or glossaries for deeper exploration. Here, the typeface isn't just an arrangement of letters but a carefully structured vehicle for ideas, guiding readers through paragraphs, chapters, and complex narratives.

The physical features—once considered arbitrary—now reveal a clear purpose. The cover, often a reader's first impression, is deliberately designed to attract a specific audience. A shirtless man on a horse, for instance, unmistakably signals romance, setting the stage for a love story. On the other hand, a cartoonish image of a character in a futuristic landscape might hint at a science fiction adventure, sparking anticipation of exploration and discovery. The cover serves a crucial role in shaping expectations and guiding readers toward the narrative before they even turn the first page.

The content within the pages gives context to the formats. Expository nonfiction, for instance, unfolds across chapters, offering informative content with real facts and details, whereas fiction uses its structure to weave narratives that reflect on human experiences. The setting of the book, or its 'environment'—in literary terms and, by extension, in biological terms—serves a crucial purpose in shaping its meaning. This understanding shifts the focus from external traits to the real story unfolding inside, revealing connections and purpose that cannot be seen by examining its physical attributes.

Likewise, purpose is quintessential to understanding these organisms and their roles within the ecological systems they inhabit. Why do these organisms exist in their specific niches, and how do they perform their roles? Exploring the 'why' reveals the fascinating and deeper significance of ecotechnology (EcoTech). Shifting focus on organisms' classifications based on their functional roles within ecosystems is more pertinent to understanding the system-level dynamics of this EcoTech.

If taxonomy is meant to reflect biological reality, then function-based classification offers a more accurate framework than rigid phylogenetic trees. The natural world operates as an interconnected system, where organisms interact through roles rather than fixed hierarchies. Whether as producers, consumers, or decomposers, life forms are defined not just by inherited traits but by their contributions to the larger ecological network. A NatTech perspective acknowledges this systemic organization, classifying organisms by their functional roles rather than forcing them into artificial ancestral groupings.[874] This functional perspective offers a clearer, system-based approach than traditional taxonomic classification.

- **Producers:** The foundation of all life, producers, primarily found in the Plant Kingdom, and certain bacteria, such as cyanobacteria, convert inorganic substances into organic materials through photosynthesis.[875] This group illustrates how taxonomic characteristics, like chloroplasts in plants and light-absorbing pigments in cyanobacteria, are critical for their roles in energy capture and conversion.

- **Consumers:** These are defined within various taxonomic groups based on their dietary needs, representing their roles in energy movement through the food web. They are the energy translators. For instance, herbivores like deer (Family: Cervidae) and carnivores like lions (Family: Felidae) are categorized not only by their physical and genetic attributes but also by their interactions within the food chain.

- **Decomposers:** Including fungi and certain bacteria, decomposers break down dead matter and recycle nutrients back into the ecosystem. Functionally diverse, these organisms share essential traits that enable them to decompose complex biological materials, thus sustaining nutrient cycles vital for ecosystem health. They ensure that nothing goes to waste.[876]

## Keystone and Pollinators

Some roles in nature are so important that they transcend simple taxonomic boundaries. Two prime examples are keystone species and pollinators.

**Keystone Species:** Imagine an arch made of stones. There's one stone at the top, called the keystone, that holds everything together. If you remove it, the whole arch collapses. In nature, keystone species play a similar role. They have a huge impact on their environment, keeping everything in balance.[877] [878]

- Wolves were reintroduced to Yellowstone National Park, they controlled the deer and elk populations, allowing plants to grow back. This new plant growth created habitats for other animals, like birds and beavers. Beavers then built dams, which created ponds, supporting even more life. This ripple effect, started by the wolves, shows how one species can influence the entire ecosystem.[879] Similarly, a software update in a complex system can recalibrate functions, patch vulnerabilities, and optimize performance, ensuring stability and efficiency across interconnected processes.[880]

- Starfish prey on sea urchins and mussels. Without starfish, these prey species can dominate the seafloor and reduce biodiversity. Starfish help maintain a diverse and balanced ecosystem by keeping these populations in check.[881]

**Pollinators:** Animals like bees, butterflies, and hummingbirds are vital for the reproduction of flowering plants.[882] Although these species are spread across various orders and families, they share the common function of facilitating pollination.

- Bees are the most well-known pollinators. They transfer pollen as they collect nectar, essential for plant reproduction.

- Hummingbirds also play a significant role in pollination, especially for flowers that bees cannot access. Their rapid wing flapping allows them to hover and feed on nectar, transferring pollen in the process.

## Eco Engineering

The sophistication in the engineering behind ecosystems is evident in the diverse interactions among species, each playing multiple roles reinforcing the system's resilience. Redundancy in ecological roles, such as multiple species performing similar functions (e.g., several pollinator species), provides a buffer against environmental fluctuations and perturbations, ensuring ecosystem stability even when individual species populations fluctuate. Ecosystems demonstrate remarkable adaptability, adjusting to changes through dynamic feedback mechanisms. For example, predator-prey relationships often exhibit oscillatory dynamics that prevent either population from overwhelming the other, maintaining a balance that supports the health of the broader community.[883]

Moreover, several case studies highlight how natural technologies manifest in different ecosystems. Coral reefs, often referred to as the rainforests of the sea due to their complex structures, provide habitats for a myriad of species.[884] The architectural design of coral reefs facilitates nutrient cycling, water filtration, and species diversity, showcasing an intricate balance of structural and functional engineering.[885]

Wetlands function as natural water purification systems, with plants and microbial communities working together to filter pollutants and store carbon.[886] Their design mitigates flooding, processes nutrients, and supports a diverse biological community.

Forest ecosystems are exemplary in demonstrating ecological engineering. Trees provide oxygen, store carbon, moderate temperatures, manage water cycles, and create habitats for countless other species. The layered structure of forests—from the canopy to the understory to the forest floor—exemplifies a multi-tiered design optimized for light distribution, habitat diversity, and energy flow.[887]

Understanding ecosystems as engineered systems shifts how we approach conservation. Maintaining the health of an ecosystem is akin to keeping a complex machine running smoothly. A study by Qing Yang, Grace L. Salter, and Chloë Schmidt, titled "Marine Biodiversity and Ecosystem Functioning," examines how species richness boosts its productivity and stability.[888]

Imagine an ecosystem as a complex technological system where each species is an integral component with a specific function. These interactions function like natural algorithms, adjusting dynamics based on resource availability and environmental conditions. Yet, the more diverse the components, the more efficiently and resiliently the system operates. The research shows that having a diverse mix of producers, consumers, and decomposers helps ecosystems thrive.[889] Just as removing one cog from a machine can cause it to malfunction, losing species can disrupt the entire system. This study suggests that to keep ecosystems healthy, we need to focus on saving individual species and preserving the whole network of life that supports them.[890]

Moreover, efficient resource distribution and nutrient cycling are hallmarks of well-engineered systems. Ecosystems demonstrate this through nitrogen, carbon, and water cycles, ensuring that essential elements are reused and redistributed to sustain life across various forms of existence.[891] These cycles are like circuit designs in technology, where each component provides the continuous flow of energy and materials. Biodiversity enhances ecosystem resilience, enabling systems to adapt to changes and environmental pressures. The variety of genetic information available within an ecosystem's species pool functions like multiple backup systems or alternative pathways in an engineered design, ensuring stability under diverse conditions.

Several ecosystems exemplify dynamic stability and engineered resilience. The Amazon rainforest, often described as the Earth's lungs, plays a vital role in global carbon storage and oxygen production.[892] The interdependence of species within the rainforest creates a robust system capable of self-regulation and resilience, showcasing an ecosystem designed for large-scale environmental stabilization. Similarly, the ecological engineering seen in the Great Barrier Reef precisely balances species interactions and water chemistry to support biodiversity.[893] The reef's structure provides defense against storm damage to coastal regions. It supports marine life and helps stabilize aquatic food webs.[894]

The Serengeti Plains offer another example.[895] The migratory patterns seen in the Serengeti result from precise ecological calibrations that allow massive herds of herbivores to graze across the plains, supported by predators that help maintain the health of the herd populations. This migration ensures nutrient redistribution across the ecosystem, exemplifying dynamic stability in these natural ecological technologies.

For too long, conventional science has judged the book of life only by its cover. Looking beyond physical characteristics to consider the purpose and function of each species allows for a more comprehensive perspective—one that reveals the systemic engineering underlying life and its biodiversity.

CHAPTER 29

# Relational Reality

*It from bit symbolizes the idea that every item of the physical world has at bottom — at a very deep bottom, in most instances — an immaterial source and explanation; that what we call reality arises in the last analysis from the posing of yes-no questions and the registering of equipment-evoked responses; in short, that all things physical are information-theoretic in origin and this is a participatory universe.*[896] —John Archibald Wheeler

Language is more than communication. It's a cognitive interface that reflects the deeper structure of the universe. It doesn't just express thought; it also encodes relation, hierarchy, memory, and recalibration. This is a structured system of relational alignment that sustains symbolic meaning. We even see this embedded in biology. More accurately stated, DNA does not hold structure as code, but as memory embedded in orientation. Proteins behave the same way: their function does not come from sequence and shape alone, but from how that sequence locks into shape—and stays put. Each of these systems coheres to a deeper syntax: a structural logic that preserves form through alignment. Neural patterns sustain memory through coherence. Homeostasis, cognition, and regeneration all depend on systems that remember how to hold together. Life doesn't just run on memory—it is memory, structured into motion.

Seen this way, language functions as an operational mirror. Our ability to use symbols, construct abstract meaning, and translate logic into form reflects that we are embedded in a system that encodes, remembers, and reorients.

Mathematics and physical law are expressions of this symbolic layer—not detached formulas, but structured protocols for coherence preservation. In the same way sentences are ordered to carry meaning, physical systems are ordered by orientation to sustain structure.

The precision of mathematics confounds materialists because it describes the most exquisite structure, which they presume was built by chaos. They see the immaculate syntax but deny an encoder. To them, the system's elegance becomes its own illusion of authorship. As discussed earlier in relation to Chomsky, syntax is not learned but surfaced—a structural inheritance. Orientation functions the same way: not imposed, but revealed through coherence.

The patterns we interpret as laws are the visible expressions of deeper coherence systems already at work. We do not decode a symphony—we participate in its feedback loop. Even our most innovative technologies are grounded in a universal syntax. These are expressions of orientation translated into function. Language is the heart of reality—but it's not mathematical elegance or string vibrations. It is ordered structure.

Once fully acknowledged, it is here where the materialist will instinctively begin grasping at straws. They will abandon the premise of blindness, but only to relocate agency. Rather than admit that syntax originates from a Supreme Intellect, they will attribute intelligence to beings like us—or to systems modeled in our image.[897] But coherence doesn't imitate our cognition—it precedes it. The syntax isn't modeled after us. We are modeled after it.

## Real Time Recalibration

Materialists seek a Theory of Everything that not only unifies the fundamental forces—gravity, electromagnetism, and quantum mechanics—into a single equation but also attempts to reduce all aspects of reality, including consciousness and existence itself, to purely material processes defined by matter and energy. However, nature operate through interconnected, self-regulating processes that exhibit intentional design.

To ground this idea, we need to explore the quantum world and its fundamental role in shaping reality. To illustrate how this NatTech integrates with what we call the physical, consider a GPS device: it continuously recalibrates its position using signals from multiple satellites, refining its location dynamically through constant interaction.

Without these relationships, the GPS's position would be meaningless. Similarly, in the quantum world, properties like a particle's position or spin don't exist independently. According to Carlo Rovelli's Relational Quantum Mechanics (RQM), these properties are defined only through relationships, comparable to how the GPS's location depends on its exchanges with satellites.[898] Each quantum interaction updates the system, capturing a snapshot of reality specific to that moment and context.[899]

As Michio Kaku explains,

> The GPS system itself is a marvel of modern technology. Each of the thirty-two GPS satellites orbiting the earth emits a specific radio wave, which is then picked up by the GPS receivers in my car. The signal from each satellite is slightly distorted because they are traveling in slightly different orbits. This distortion is called the Doppler shift. (Radio waves, for example, are compressed if the satellite is moving toward you, and are stretched if it moves away from you.) By analyzing the slight distortion of frequencies from three or four satellites, the car's computer could determine my position accurately.

As Michio Kaku explains, the GPS system determines location not by referencing a fixed point, but by analyzing signals from multiple satellites, each contributing to the final calculation. Likewise, in quantum mechanics, properties like position or spin do not exist in isolation—they are defined through interactions, existing only in relation to the system that measures them. This use of RQM reflects its observational strength, not its philosophical commitments; the relational behaviors it describes are treated here as expressions of coherent structure, not indeterminacy.

This concept extends beyond quantum mechanics. The entire structure of reality—from what are interpreted as fundamental particles to large-scale cosmic phenomena—operates through interactions rather than isolated absolutes. Analogous to how GPS adjusts for obstacles like tall buildings or weak signals, quantum properties calibrate based on conditions during each interaction. There is no absolute "truth" about where a particle—or even a car—is; there's only the truth of its relationship at the moment it is measured. [900]

RQM offers a coherent framework for describing the world we live in: a dynamic web of interactions, from what are interpreted as expanding spacetime to shifting relational structures—continually reshaped by the relationships between their components. A static view of reality fails to account for essential processes like cosmic and biological homeostasis, DNA replication, ecosystems, and the interplay of forces over time. While RQM offers a powerful reframe of quantum behavior, it remains confined within a naturalist worldview. NatTech adopts RQM's insight on relational properties—but sees in that structure something more: not just interdependence, but intention woven into the framework.

Consider how quantum interactions create a sequence of updates that define the state of reality, moment by moment. In this sense, time unfolds as a relational property. Just as a GPS dynamically recalibrates its position through exchanges with satellites, the flow of time is marked by the continual recalibration of quantum properties through interactions. What we perceive as aging could be understood as an ongoing process of calibration.

Only without interactions does the concept of "before" and "after" dissolve, leaving time as a feature of the dynamic, irreducibly complex web of relationships within the universe. This dynamic nature of time is mirrored in the essential processes that define life and reality. Each process reflects the interplay and relationships that orchestrate reality—and with it, time itself—in a constant state of becoming.

Even consciousness operates relationally. We don't perceive the world in isolation; our thoughts, awareness, and experiences are shaped by interactions with our surroundings and the connections between our ideas. In this sense, consciousness reflects the relational nature of the quantum world. It actively participates in defining reality, much like a quantum observer determines the state of a particle.

What about free will? In a deterministic universe, every event is only the inevitable result of prior causes. But the relational framework of RQM departs from this view, showing that outcomes are probabilistic and shaped dynamically through interactions. In this sense, free will isn't an illusion— it's an intrinsic feature of the relational universe. Analogous to how a quantum particle exists in a superposition of states until observed, our minds often hold multiple possibilities—emotions, decisions, or outcomes— before we collapse them into a definitive choice. This dynamic interplay reflects the relational nature of quantum systems and human experience, suggesting that even states like joy and love are not fixed but arise through ongoing interactions, shaped by relationships, perception, and the observer's focus.

When faced with a setback, we find ourselves in a state of emotional superposition, from sentiments of despair to resilience. The "collapse" into one state depends on our observation: a conscious focus on growth transforms failure into opportunity, while dwelling on negativity reinforces defeat. This ability to choose, to recalibrate our emotional and cognitive states in real-time, positions free will as a participatory force within a relational universe. We are active agents shaping the fabric of reality.

As explored in Chapter 14, these same principles governing quantum interactions may apply to larger systems.[901] Quasi-integrals of motion are self-regulating mechanisms that stabilize chaotic systems like the solar system. These quasi-integrals, though not perfectly conserved, subtly adjust planetary trajectories to maintain balance over billions of years. This parallels the relational dynamics of quantum systems, as described in RQM, where interactions define properties and ensure coherence despite underlying uncertainty. Both frameworks suggest a universal principle of real-time stabilization, a cosmic homeostasis, where the interplay of relational adjustments bridges quantum randomness and classical determinism.

If quasi-integrals of motion are described as emergent, that demands a clear answer: emergent from what? And if that can't be clearly specified, how can we justify calling them approximations rather than ad hoc descriptions? The Hamiltonian function provides the connective thread between quantum and classical systems. It expresses the internal coherence structure that regulates system behavior—deterministic when coherence is fully resolved, and transitional when alignment is incomplete. What appears probabilistic in quantum systems reflects underlying coherence constraints, not randomness. This consistency across vastly different domains is not incidental. It reveals an underlying principle: precision and adaptability, pointing to a coherent architecture woven into the fabric of reality.

This returns us to the concept of Dark Gyroscopes, previously introduced as a reinterpretation of gravity—not as a pull from mass, but as the visible effect of an unseen orientation system embedded in space. When exploring the relational nature of particles, quantum recalibration, and coherence from subatomic spin to galactic structure, this idea finds its place.

Materialist science assumes gravity is caused by mass because the two correlate. But correlation is not causation. Mass accompanies gravitational effects, but that doesn't prove it generates them. It is more coherent to see both as effects of a deeper stabilizing dynamic—an embedded orientation system that governs spin, structure, and systemic coherence. The belief that mass bends space has led to endless theoretical patches—dark matter, inflation, quantum gravity—none of which unify the system. This is not failure from complexity. It is a category error: mistaking matter as the origin of structure, when there is no structure without relational logic.

The NatTech perspective reveals that spin is not a byproduct of matter—it is a behavior woven into the operating system of reality. From electrons to black holes: everything spins, everything stabilizes, everything orients. These are not coincidences; they are signatures of system-level rules. Gravity is not a force acting on mass—it is a structured expression of embedded orientation, a gyroscopic stability function that manifests as curved paths, orbital locks, and galactic cohesion.

# CHAPTER 30

# Spatial Intelligence

*My views are considerably more radical than those of either of my predecessors, however, because they have been sharpened by recent events. I am increasingly persuaded that all physical law we know about has collective origins, not just some of it. In other words, the distinction between fundamental laws and the laws descending from them is a myth, as is the idea of mastery of the universe through mathematics alone. Physical law cannot generally be anticipated by pure thought, but must be discovered experimentally, because control of nature is achieved only when nature allows this through a principle of organization.* —Robert B. Laughlin[902]

For centuries, gravity has been defined through the frameworks of Newtonian force and Einsteinian curvature. Newton saw gravity as an invisible force acting at a distance—a pull between two masses proportional to their product and inversely related to the square of their distance. He wrote, "A centripetal force is that by which bodies are drawn or impelled... towards a point as to a centre."[903] But he never claimed to know why this happened. "Hypotheses non fingo," he said—I frame no hypothesis.[904] His predictions held, but the cause remained a mystery—an honest admission. The system behaved as if pulled inward, but Newton couldn't explain the cause.

As Enlightenment mechanism and empiricism took hold, science drifted from Newton's caution. Correlation hardened into assumption: mass was no longer just associated with gravity— it was presumed to cause it. The search for explanation gave way to the ease of mathematical convenience. Thus, a *pattern was mistaken for cause*—and that mistake became dogma.

Dennis Sciama, a physicist who explicitly addressed the failure of general relativity to explain inertia, puts it plainly: general relativity assumes the coupling between mass-energy and curvature—it doesn't explain it. As he wrote:

> Einstein showed that his field equations imply that a test-particle in an otherwise empty universe has inertial properties. [905]

In other words, general relativity assigns structure to space without requiring any relational input—it builds inertia into the formalism, rather than deriving it from interaction with the rest of the universe. Sciama makes this relational point more explicit: "Inertia is not an intrinsic property of matter, but arises as a result of the interaction of matter with the rest of the matter in the universe." [906]

Philosopher of science Jonathan Fay, expanding on Sciama's critique, writes: "In general relativity, the source of inertia is not explained... We are told that objects follow geodesics, but why those geodesics exist, or what determines inertial resistance, is not derived from deeper principles." [907] In another work, Fay extends this further—arguing that spatial geometry itself may not be fixed by convention or measurement, but arises from relational magnitude and internal coherence. [908] He sees this as a direct challenge to Newtonian mechanics, which makes no reference to the distribution or amount of matter in the universe.

In 1973, Brill and Deser demonstrated that closed spaces in general relativity suffer from instability—even infinitesimal perturbations can violate the constraint equations, meaning the theory cannot guarantee coherent evolution for these geometries. [909] Their work reveals that Einstein's model, elegant as it is, lacks an internal mechanism to ensure stability. But this wasn't a mathematical flaw; it was a conceptual limitation of that era.

Relational coherence is assumed—but not explained. What holds the system together? If curvature (of space, of motion, of trajectory) is used to describe the system's structure—then what regulates that curvature to prevent it from becoming infinite, unstable, or inconsistent with structural coherence? General relativity answers with boundary conditions or symmetry assumptions, but not with underlying logic. As Einstein acknowledged, "The concepts of classical mechanics afford no way of expressing this," referring to the field-like properties of spacetime needed to explain inertia.[910] He admitted that "classical mechanics offers no explanation for this equality" of inertial and gravitational mass, despite its experimental confirmation.[911] This is where Coherence Theory takes over: the structure is not imposed by curvature; it's organized by alignment. The stability is encoded. General relativity was not the final word.

From my coherence perspective, the missing word is orientation. The equivalence principle, rather than pointing to a geometric artifact, may be revealing a deeper structural rule: that mass itself is an expression of how tightly a system is aligned within a coherent lattice. Inertia and gravity are not two phenomena—they are two aspects of the same orientational logic.

This reverses the causal arrow: mass is not the cause of gravity. What we call mass is a node of rotational inertia—an anchored condition within a coherent spatial locus. Free fall is not a fall—it is alignment. In the NatTech model, motion arises from internal balance within the orientation lattice. No attraction is required. No spacetime must bend. As Einstein himself put it, spacetime "acts upon" matter but "cannot be acted upon" in return—meaning spacetime is a dynamic background with no reciprocal interaction, a one-way causal structure that lacks explanatory closure.[912] Objects follow stable paths not because of pull, but because their position sustains alignment with the system's internal logic.

If spacetime isn't curved—and gravity isn't a force—but instead a function of coherence within an orientation lattice, then time dilation isn't caused by warping. An atomic clock near Earth's surface ticks more slowly than one in orbit—not because time is bending, but because the denser rotational coherence near Earth requires more resistance to maintain alignment, slowing the rate at which coherence can be recalibrated.[913]

As shown in GPS systems, what's called time dilation reflects this same relational effect: systems deeper in the orientation field require more effort to update their state. Gift (2021) demonstrates that GPS clock corrections can be derived without invoking relativistic time dilation at all—supporting this coherence-based perspective of temporal recalibration.[914] This sheds light on what Newton and Einstein could not: the origin of inertia, the conservation of angular momentum, and the coherent spin of galaxies.

Newton, as a creationist, was pioneering the early foundations of physics, taking inertia as a given, but his framework didn't yet explore its deeper cause. Einstein, building on those ideas, tied inertia to spacetime curvature, yet still missed the underlying relational dynamics. Neither they nor the materialist paradigm that has dominated scientific thought has had the conceptual tools to explore how relational coherence could explain these phenomena.

Einstein admitted a hidden assumption within general relativity:

> We can think of no cause for this preference for definite states of motion to all others... it must be regarded as an independent property of the space-time continuum.[915]

In my Coherence Theory, this isn't an unexplained preference—it's a structural outcome. The universe doesn't arbitrarily prefer certain motions. Definite states are sustained by a coherence lattice that enforces orientation memory and stabilizes motion. What Einstein treated as a given, this framework sees as the output of an active spatial intelligence.

## The Solar *System*

Our solar system operates as a coherent orientation system structured by rotational balance, inertial memory, and phase coordination.[916] Planets and moons align through embedded logic, the same principles used in our most stable technologies because they reflect universal behaviors.[917] The fact that every planet spins is not simply a byproduct of accretion or random collisions of planetesimals—it is a fundamental requirement for maintaining spatial coherence, a necessary condition for the system's stability and order.

Researchers like Ali Seraj and M. Oblak have observed "orientation memory" in gyroscopes during what they interpret as gravitational wave events. But this interpretation, too, assumes a spacetime-curvature model I reject. In my view, these gyroscopes aren't being acted on; they are recalibrating in real time to preserve orientation. It is not gravity as force— but gravity as restored coherence: *spatial coherence*. This reframes the search for quantum gravity: if gravity is not a force or a field but a behavior of coherence, there's nothing to quantize. No gravitons. No mediation. No force to unify.

In Principia (Book II, Prop. LII), Newton described a sphere rotating in a uniform fluid.[918] He found that the orbital periods of surrounding fluid layers increase with the square of their distance—an elegant geometric law.[919] He saw the layering, the symmetry, the rotational coherence. But he never asked why that symmetry exists in the first place. He measured the stability, but not the stabilizer. To Newton, the behavior was induced by the body. In the NatTech model, it's not the body that causes coherence—it's the body locking into an already coherent framework.

What Newton described mathematically, I recognized structurally: spin is the visible resolution of orientation logic. A gyroscopic lattice does not emerge from matter—it organizes it. Many later mistook effect for cause, reverse-engineering motion into imaginary mass—while coherence lay in plain sight. Spin is the mechanism by which a body remains synchronized within a larger orientation lattice. Without spin, coherence breaks down.

Axial tilt is also functional. Earth's 23.5° tilt produces stable seasonal patterns and climate rhythm. In engineering, tilt is adjusted to control scanning fields, power distribution, or signal targeting.[920] It's an orientation setting—an embedded directive, not a leftover from chaotic collisions.

Earth's Moon rotates once per orbit, always showing the same face. This is a phase-lock condition rather than a gravitational effect. In electronics, phase-locked loops stabilize frequency and prevent drift.[921] The Earth–Moon system performs the same function.

Richard Feynman, confronting the infinities of classical electrodynamics, proposed an electrodynamics with no fields at all—just direct particle interactions. "The field disappears," he said, "as nothing but bookkeeping variables."[922] He meant this quite literally. In standard physics, the field is no longer a clearly defined structure—it functions as a flexible stand-in for interactions across time-sliced frames. But when time is treated as continuous recalibration, the need for an intermediary mechanism—

however defined—falls away. What remains is the structure itself: orientation, synchronization, coherence.

Feynman also noted the strangeness of multiple models—fields, particles, and waves—describing the same reality: "There is always another way to say the same thing that doesn't look at all like the way you said it before... I don't know what it means... but maybe that is a way of defining simplicity."[923]

What appears as strangeness is actually a meaningful signal. The system is relational, not reductionist. Feynman clarified that this involved direct, time-delayed interactions between particles, excluding self-action and fields. As he put it, "the idea that a particle acts on itself... is not a necessary one—it is a sort of a silly one, as a matter of fact."[924] Instead, "there was no field. It was just that when you shook one charge, another would shake later." Interaction occurred with a delay—direct and structured, not mediated.[925]

Feynman abandoned this view under pressure from the field paradigm. Our most precise technologies confirm that spin determines structure, and alignment—not force—regulates motion. Our technologies don't operate this way by design preference—they do so because they have no choice. Precision demands adherence to the same systemic logic that governs nature. We didn't design gyroscopes to simulate planetary orientation—but that's exactly what they do.

## Magnetism

In mainstream accounts, magnetism arises from the alignment of microscopic spins—tiny angular momentum vectors associated with electrons. In ferromagnetic materials, these spins align into domains, forming macroscopic magnetic fields. This behavior exposes a deeper architecture that mainstream models often describe without fully recognizing. As Charles Kittel explains in *Introduction to Solid State Physics*:

> Domain structure always has its origin in the possibility of lowering the energy of a system by going from a saturated configuration with high magnetic energy to a domain configuration with a lower energy.[926]

The system transitions from saturation to order by reorganizing orientation.

The same principle applies on a cosmic scale. Anomalous galactic rotation curves—where stars orbit faster than expected—don't require hidden mass. They reflect the stabilizing effect of spin coherence across the orientation lattice. The spin of galaxies is not a fossil of formation. What appears as a discrepancy in gravitational models is better understood as a systemic tendency toward distributed balance. Black holes are not singularities but saturation points—regions where the coherence lattice reaches maximum orientation density. Information isn't lost, but becomes inaccessible, as relational space runs out of capacity to encode new alignment.

This coherence at the cosmic scale echoes what is already observable in condensed matter systems. Ferromagnetic domain behavior offers a material template for this logic: spontaneous organization arises from internal energy minimization, not external force. The system seeks equilibrium through alignment. As Kittel notes, "Landau and Lifshitz showed that domain structure is a natural consequence of the various contributions to the energy—exchange, anisotropy, and magnetic—of a ferromagnetic body."[927] Domain stability arises not from dominance, but from structured energetic preference—coherence shaped by anisotropic and exchange parameters. Kittel continues: "The magnetization of the crystal sees the crystal lattice through orbital overlap of the electrons: the spin interacts with the orbital motion by means of the spin-orbit coupling."[928]

The orientation of spin is not arbitrary—it is informed by the material's structural memory, operating through spin-orbit coupling and lattice symmetry. Even the anisotropy energy, as Kittel notes, defines the preferred directions of spin: "The anisotropy energy is of the order of the anisotropy constant times the thickness..."[929] This is a system holding its own configuration through feedback and spatial logic. In this light, ferromagnetic domains are not just aggregations of particles. They are relational structures—coherent, energy-optimized, and memory-retaining.

Ferromagnetic behavior offers a material, measurable prototype of the principle of spatial coherence. Magnetic domains are not just regions of aligned electrons. They resist perturbation, restores alignment after disruption, and maintains order not through interaction, but through embedded orientation logic. Researchers studying magnetism routinely refer to "hysteresis loops"—a phenomenon where a material remembers the direction of past magnetization.[930] This memory is not stored in a material substance, but in the spatial alignment of spins.[931] Framed through this coherence perspective, hysteresis expresses relational persistence—a structure that tracks its prior states and resists disorder. This is spatial memory, not field decay.

There is a bridge, if you know what to look for. What appears as hysteresis in ferromagnetism and as quasi-integrals in celestial mechanics are not isolated phenomena. They're expressions of the same coherence principle across scale. Different domains, same logic. Each system, whether atomic or astronomical, sustains alignment.

In another focus, quantum studies of magnetism offer further support. In spintronics, scientists manipulate electron spin as a unit of information, describing it as stable, non-volatile, and inherently binary—on or off, up or down.[932] Devices operate using spin orientation to perform computation and store data. What we call computation or information storage in engineered systems is simply a continuation of what nature already does through rotational coherence. Biological and physical systems are not opposites, but unique expressions of the same coherence-based architecture.

It would be wiser to learn from this structure than to reduce it into fragmented abstractions. Unlike the reactive models that dominate much of contemporary science—often driven by data without grounding—this coherence-based framework offers clarity. It reflects how functional systems behave, whether engineered or organic.

## Electromagnetism

In quantum terms, electromagnetism is governed by U(1) gauge symmetry—a formal system for maintaining directional consistency across space.[933] But this isn't just abstract algebra. It encodes a deeper principle: electromagnetism as a protocol for preserving alignment within a spatially intelligent lattice.[934] As David Tong explains, "A gauge field is the tool we use to compare fields at different points in space. It tells us how to parallel transport internal degrees of freedom from one location to another."[935] In other words, the gauge field maintains orientation relationships across space—it keeps track of how elements should align as they move through the system. This behavior isn't simply a gauge effect—it operates as a coherence-enforcing layer embedded in spatial structure itself.

Conventional models describe the results—stability, memory, and alignment—but misclassify the cause. They label these patterns "emergent" or "field-based," framing the structure as incidental rather than intentional. As David Tong puts it: "We typically do not think of gauge symmetry as a fundamental symmetry of Nature. Instead, it is a redundancy in our mathematical description."[936]

In mainstream science, "emergent" is just a fancy way of saying, 'We don't know, and we're not asking.' It is used to label what they can't explain—as if complexity just appears out of nowhere. It's an empty word, used to disguise willful ignorance as explanation.

Much of modern physics tries to describe coherence through the lens of interaction. Stable patterns—memory, alignment, structure—are interpreted as the result of particle exchanges within force fields. But these "particles" are not real in any ontological sense. What's actually observed is not a discrete object but a structured behavior—an expression of how coherence responds under pressure. The coherence is real. The particles are shadows. This entire framework misreads system-wide alignment as object-to-object forces, trying to explain orientation with the wrong glasses.

As David Tong notes in his gauge theory lectures, understanding the low-energy dynamics of strongly coupled gauge theories is "typically a very hard problem," especially since "the physical spectrum need not look anything like the fields that appear in the Lagrangian."[937] This difficulty exposes the limitations of a framework that begins with particles and tries to force coherence into it, rather than recognizing coherence as the basis of structure from the outset.

This is why even the most sophisticated field theories often collapse. Consider the difficulty of explaining confinement and mass generation in strongly coupled gauge theories—where particles predicted by the Lagrangian never appear in the observable spectrum. As Tong notes: "the physical spectrum need not look anything like the fields that appear in the Lagrangian." This isn't just a technical issue. It exposes a deeper failure: the theory cannot explain how its own structure gives rise to what we observe. The gap is routinely labeled as "emergent," but that word functions more as a shield than an answer—a way to avoid the real question: why does the structure of physical law appear coherent, ordered, and intelligible—as if shaped by the logic of a mind? That question isn't asked because it's not allowed. But the failure to resolve it keeps showing up—disguised, renamed, or ignored.

Rather than consider that the field concept itself may be the misreading, a zoo of invisible intermediaries is invented—gravitons, gluons, inflatons—to patch its failures. These are compensations not discoveries. The hard problem isn't confinement. It's interpretation. Materialist science has spent decades describing systems that behave like engineered constructs—regulated, resilient, and information-driven—yet refuses to recognize them as such. They walk like engineering, perform like engineering, and are modeled using engineered principles—but within the materialist frame, "emergence" is the mandated explanation.

Electromagnetism is traditionally framed as one of the four fundamental forces—a field associated with electric charge and mediated, in quantum theory, by photons. But when freed from materialist assumptions, its behavior does not suggest a force traversing empty space, but structured coordination across spatial relations. What's labeled as "electromagnetic interaction" is better understood as an orientation-preserving protocol embedded in the architecture of coherence.

Electromagnetism is structured through the movement and arrangement of charge. What defines a charge, beyond convention, is its relation to spin and phase. An electron's magnetic moment is a rotational identity rather than an added trait. Charge interactions unfold along field lines: directional structures shaped by spatial configuration.[938] These are alignment maps, shaped by something deeper maintaining orientation across the system.[939]

Maxwell's equations—celebrated as a milestone of 19th-century physics—describe how electric and magnetic fields evolve together. But they do more than describe—they reveal symmetry, balance, and continuous feedback. A changing electric field generates a magnetic field, and vice versa—not as a push or a pull, but as a regulatory update. This is not how a particle system behaves; it's how a coupled coherence engine operates. Every motion triggers a counter-orientation, just as gyroscopic systems stabilize through reciprocal alignment.

Electromagnetic wave propagation reveals even deeper structure.[940] These aren't mechanical waves—they move through vacuum at a fixed speed, implying that vacuum is structured.[941] The wave is made of oscillating electric and magnetic components, always perpendicular to each other and to the direction of motion. This tri-orthogonal geometry is exact, not incidental. Space doesn't transmit waves—it resolves orientation disturbances in synchrony. Classical models treat space as an inert container, hosting matter and energy. But that view imposes passivity where structure is clearly active.

Henri Poincaré, writing at the edge of classical physics, challenged this assumption directly. He argued that space is not a given—it is constructed through reasoning, shaped by the logic of coordination, and inseparable from the relationships it supports. Geometry, in his view, was not "true or false" but a language we use to describe interaction.[942] His point removes the idea of a fixed, neutral background. What we call "space" is not where phenomena occur—it's part of how they occur. Within this view, the photon isn't a traveling particle—it's a recalibration event moving through a coherent orientation system.

This understanding is echoed, perhaps unintentionally, by Hecht, who notes that "in the subatomic domain, the classical concept of a physical wave is an illusion."[943] Hecht goes further: "in a sense, the field is the particle and the particle is the field," collapsing the dualism that underlies most photon models.[944] He explains, "The contradiction of Relativity is only an apparent one, arising from the fact that although a monochromatic wave can have a speed in excess of $c$, it cannot convey information."[945] This reflects a constraint built into the structure of wave behavior. A monochromatic wave expresses a stable phase configuration, propagating through space as a continuous expression of alignment. Its properties remain fixed across time, sustained by coherence rather than content transfer. Light, in this form, operates as a patterned recalibration across a spatially ordered field.

When light meets matter, the interaction is precise. Surfaces reflect, absorb, or refract based on phase relationships, angular momentum, and resonance—all signs that light isn't just energy, but structured alignment. Polarization confirms this: it's orientation control, not just intensity or direction. As Hecht writes, "the concepts of polarization and coherence are related in a fundamental way," and the coherence time defines how long a polarization state remains stable.[946] In engineered systems like fiber optics, phase is preserved through directional consistency, not energetic confinement—further confirming that orientation, not momentum, governs propagation.

Quantum electrodynamics (QED) models electromagnetic interaction as photon exchange between particles. The math works. But it doesn't explain why this exchange preserves coherence—or why it always coordinates charge interactions with such precision. Photon exchange is just the surface view. Underneath is a deeper process: real-time alignment correction inside an intelligent spatial substrate.

Experiments like the double-slit test—where single photons form interference patterns—show that space preserves relational memory and superpositional orientation. Each photon reflects a set of configuration options, and the field responds as a whole, not as separate parts. Light doesn't choose a path. It follows embedded coherence rules.

Hecht defines a monochromatic wave as "a disturbance whose properties have been fixed for all time."[947] What he frames as a mathematical condition is, in this context, a signature of relational memory—an imprint of coherence that remains stable across propagation.

Electromagnetism is an operating logic, a distributed protocol that preserves phase, rotation, and orientation across distance and time. Every working antenna, circuit, and fiber optic cable confirms it: nature is already wired for coordinated transmission. Like magnetism, electromagnetism exposes what mainstream models miss: not blind force, but real-time communication. This is system built on principles, not particles. The perspective of Natural Technology fits what space is actually doing.

## Forces

Mainstream physics treats the strong and weak forces as short-range quantum interactions, mediated by gauge bosons—gluons for the strong force, W and Z bosons for the weak. These models describe what happens at high energies, but they don't explain why these forces behave this way—or what principle defines their structure. In this orientation-based framework, these "forces" aren't forces at all—they are localized protocols of coherence and transformation. They do not act on particles; rather, they maintain system behavior.

Quantum Chromodynamics (QCD) is the quantum field theory that describes the strong interaction—the force that binds quarks into protons, neutrons, and other hadrons.[948] The residual effects of this interaction also account for the binding of nucleons inside atomic nuclei.[949] It's called "strong" because it dominates at small scales—far outweighing electromagnetism—yet fades quickly with distance.[950] But what's actually happening? Gluons are said to carry "color charge" between quarks—but these are mathematical constructs, not physical intuitions.[951] QCD is a working model, but its mechanism is obscure, probabilistic, and nearly impossible to visualize.[952]

What's modeled as quarks exchanging gluons is more accurately understood as a tightly constrained coherence system—where relational orientations among nodes are locked into balance, and no discrete entities transfer anything. The "strong force" is simply the system's highest-order stabilization behavior—where coherence is non-negotiable and phase misalignment is energetically forbidden. The exponential drop-off of this behavior (strong at close range, nonexistent beyond) is not because the force weakens, but because outside the nucleus, the coherence logic does not apply. The region of allowable alignment ends.

This is not unlike high-speed networks or processor architectures, where components don't exchange "things"—they maintain coherence through phase alignment. If a node drifts out of sync, the system destabilizes or shuts down. There is no transfer—only enforced relational timing. Strength arises not from force but from constraint.

Even CERN's own infrastructure unintentionally mirrors this logic. Their White Rabbit Project synchronizes distributed systems with sub-nanosecond precision—not by transferring discrete units, but by enforcing strict phase coherence across nodes. The system maintains stability not through force, but through alignment. Misalignment doesn't trigger correction—it ends participation. The irony is clear: while CERN's theoretical framework models interaction as mediated by particle exchange, its technology operates on coherence without transmission.[953]

CERN didn't set out to build communication systems—they had no choice. Their experiments demanded it. And the system they built doesn't function through exchanged packets—it relies on phase-locked alignment, the very coherence logic their own particle models refuse to acknowledge. It mirrors a non-local structure that theory resists but technology makes unavoidable. It's a documented technical feature of the system.[954]

According to CERN's documentation, the White Rabbit system maintains synchronization by adjusting the phase through phase-steering, rather than manipulating the time counter value.[955] Specifically, the White Rabbit Phase-Locked Loop (WR PLL) uses a technique called Digital Dual Mixer Time Difference (DDMTD) to measure and correct phase differences between clocks.[956] This ensures that all nodes in the network are tightly phase-aligned, maintaining coherence across the system.

In conventional QCD, confinement refers to the fact that isolated quarks are never observed as free particles, no matter how much energy is applied. As Srednicki puts it, "a theory is confining if all finite-energy states are invariant under a global gauge transformation."[957] This is a nonperturbative feature of the theory. What's called "confinement" isn't a mystery in this view. It's a structural boundary. Quarks aren't standalone entities; they're not particles at all, but fractal expressions of rotational equilibrium. They don't leave the lattice because there's no coherent state for them outside it.

In standard models, the weak force explains events like beta decay, where a neutron becomes a proton through a quark-level shift.[958] Despite its name, the weak force plays a central role in such transformations. "Weak" refers to interaction strength, not significance—its effects are less powerful than the strong or electromagnetic forces but no less essential. But transformation isn't random or decaying—it's a recalibration triggered by structural imbalance.

Standard physics treats it as a stochastic event—a rare, probabilistic shift carried by massive virtual bosons. But the weak force isn't a decay mechanism. It's a regulated pathway for system-level reconfiguration. When coherence breaks—through stress, age, or disturbance—the system triggers an internal update. Or as Srednicki describes it in the context of chiral symmetry, the vacuum reorients—its internal structure shifting across spacetime.[959] It's not random, however, it's precise. And the change isn't decay. It's realignment.

What standard physics call the four fundamental forces—gravity, electromagnetism, and the strong and weak interactions—aren't separate forces to unify. They're systemic behaviors—coherence protocols embedded in space. Each reflects a rule of stability, regulation, or transformation that sustains structural order. The massive W and Z bosons aren't force carriers—they're the visible signature of intense local reconfiguration.[960] A shift in orientation so energetically dense, it briefly shows up as mass. It's not a messenger. It's a gear shift—loading the system with stress as it reorients. From this perspective:

- The strong force = internal cohesion in high-stability zones (nuclear coherence locks)

- The weak force = transformation pathways when that coherence must update

- Electromagnetism = relational calibration over longer distances

- Gravity = inertial coherence over the largest scales

None of these are true "forces" in the Newtonian sense—no pushes or pulls between discrete objects. Srednicki alludes to this shift when describing gauge fields not as mechanical forces, but as structures that enforce symmetry. "The gauge field $A_\mu(x)$ transforms in such a way that the covariant derivative $D_\mu\psi$ transforms in the same way as $\psi$ itself," ensuring local gauge invariance.[961] These are not classical forces nor tangible exchanges—they are structural rules embedded in field behavior. What physics models as force is, in this view, how the orientation lattice sustains coherence across scale.

"Gravity" reflects inertial memory and orientation logic that holds balance across spatial nodes. Mass doesn't pull—it anchors. Objects don't fall—they align. Coherence replaces curvature as the governing principle. Electromagnetism isn't the interaction of charged particles. It's a distributed coherence system—more like a gyroscopic network that regulates balance, spin logic, and phase alignment across space. Its behavior mirrors technologies that stabilize through real-time feedback: oscillators, phase-locked loops, and control circuits all follow the same principle.

Electromagnetism is often called the most well-understood of the fundamental forces—unified by Maxwell's equations and formalized in QED. But under the precision is a conceptual gap, especially regarding light. In 1865, Maxwell wrote: "The velocity of electromagnetic waves is the same as that of light… we can scarcely avoid the inference that light itself is an electromagnetic disturbance." That inference became dogma: light was no longer a distinct phenomenon, but a self-sustaining oscillation of electric and magnetic fields. Yet no physical medium was ever found.

Feynman openly admitted the unresolved core of it: "The strange theory of light and matter is called quantum electrodynamics. It has no underlying mechanism. It just works." QED is precise—but it doesn't explain wave-particle duality, how fields "know" to oscillate, or what sustains coherence across vast distances. As Feynman also noted: "Nobody ever gives a definition of a field anymore. It's just something that's there, in space, that changes."

Rather than a force moving through empty space, electromagnetism is a behavior of spatial orientation under high-frequency recalibration—more like dynamic resonance in a tightly coupled system. What's called "field strength" may simply be local orientation alignment. And the speed of light—not variable, but constant—would then reflect the system's refresh rate: a synchronized update across space, faster than any object can travel, but regulated by structural rhythm.

Maxwell once hinted at what's now largely forgotten: "There is a great probability that the interplanetary and interstellar spaces are not empty, but occupied by a material substance or body… which is the recipient of light and electromagnetic disturbances." Physics kept the motion, but discarded the medium. This framework restores the logic: light doesn't travel through emptiness—it propagates through a structured orientation lattice that maintains coherence across space.

Materialist physics calls light both wave and particle—but explains neither. Wave-particle duality is accepted not because it makes sense, but because it's mathematically useful. Feynman, describing the double-slit experiment, said: "We choose to examine a phenomenon which is impossible, absolutely impossible, to explain in any classical way... We say it's a mystery." But mystery isn't a theory. By avoiding the structure of space itself, physics has resorted to elaborate circumventions. Electromagnetic behavior is modeled as vector quantities varying across space and time, yet key features—like phase alignment, polarization, and propagation speed— are treated as givens. There's no underlying mechanism—just equations without structure.

Electromagnetism is more openly technological. It defines the behavior of electricity, magnetism, and light—all of which are embedded in every piece of modern engineering. But calling it a force hides the structure. Electromagnetic behavior arises from field alignment and phase interaction. As Eugene Hecht notes, "a time-varying magnetic field will have an electric field associated with it"—but not from a source point.[962] This is an internal update within the structure itself. These are not fields in a material sense— they are rules of configuration, unfolding through resonance.

In magnetostatics, theoretical physicist John D. Jackson describes magnetic field lines as orthogonal to equipotential surfaces, indicating that they follow structural constraints rather than exerting mechanical force: "The surfaces of the high-permeability material are approximately 'equipotentials,' and the lines of H are perpendicular to these equipotentials."[963] This is why magnetic fields aren't "forces" in the classical sense—they act as constraints on movement, directing trajectories through alignment with the surrounding field geometry. In waveguides, circuit design, and rotating magnetic systems, we already harness this relational structure. The design of these systems depends on the geometry of field propagation and boundary conditions—what Jackson shows are regulated by vector potentials and shaped magnetic surfaces.[964]

This leads to one of the most glaring misinterpretations in modern physics: the idea that light "becomes" a particle. This claim originates from the photoelectric effect and related quantum phenomena, where light interacting with matter exhibits quantized energy transfer.[965] But interpreting this as light transforming into a particle assumes the model is literal. It's not.

Photons are not objects; they are outputs of relational measurement. If the system behaves coherently, then interaction isn't a collision—it's a structural negotiation. The system adjusts to maintain alignment before anything "happens" in the classical sense.

This leads to the idea of an anticipation gauge: a structural constraint within a coherence system that adjusts local or global configuration based on potential perturbation, ensuring stability across interaction thresholds. In quantum terms, this means the system doesn't wait for an interaction to occur—it preconditions the outcome space. What appears as a discrete event (like a photon detection) is the result of the system aligning itself in advance to deliver a coherent, permissible response to the mode of engagement. It exhibits discrete effects only when filtered through a measuring system—

one already structured by expectation, by interpretation, and ultimately, by consciousness.

There is no "dual nature" in light. There is a dual nature in how we observe. Awareness resolves the system. Without a mind to observe, no properties exist: no position, no spin, no particle, no relationship. What we call the physical world is a system of relationships resolved through conscious interaction.

Our technology already reflects this structure in form, if not in awareness. In phased-array antennas, wireless networks, and optical systems, coherence only occurs when internal alignment, phase locking, and tuned feedback are built into the system. Signal strength and range don't depend on abstract vectors—they depend on structural orientation and stability. Materialists say light travels through "nothing." But nothing doesn't enforce anything. Light cannot exist as what it is—it cannot "be"—without the underlying structure. This is even supported by Krauss, who argues that "nothing" is a fluctuation within a law-bound quantum field. It has rules, symmetries, dimensionality, and energy states. In any meaningful sense, it's something.

The very fact that different types of signals (radio, microwave, optical, etc.) require and obey different structural constraints—like carrier frequency, modulation type, bandwidth, and propagation medium—implies that structure is foundational. Signal behavior is always relational to a structure. That structure may be engineered (as in phased arrays), or it may be inherent to space itself—but either way, waves don't just propagate in the abstract. They require: a frame of coherence; a constraint system; and rules of transformation that shape how information appears and moves.

Even the zero-point field—introduced to explain background energy in quantum electrodynamics—was described by Feynman and Vernon as a "sea of oscillators."[966] That's not nothing. That's structure with rules. Maxwell once proposed a mechanical model for the electromagnetic medium.[967] "By pointing out the mechanical consequences of such hypotheses," he wrote, "I hope to be of some use to those who consider the phenomena as due to the action of a medium, but are in doubt as to the relation of this hypothesis to the experimental laws already established."[968] He didn't know its nature, but he was certain space had internal structure.

In this framework, electromagnetic behavior is a recalibration process within a coherent spatial lattice.

- Wave propagation is synchronized orientation updates—like a ripple through a coordinated network.

- Polarization shows the directional constraints of alignment—like fixed axes in a physical system.

- Phase coherence reflects locked timing across spatial nodes—just like in oscillating circuits.

- The speed of light isn't the velocity of an object—it's the system's intrinsic update rate.

The photon has no rest mass because it doesn't travel through space—it pulses within it. When QED speaks of "virtual particles" mediating force, it's describing surface effects of a deeper process: interaction protocols within a dynamic orientation system.

The cosmic microwave background is usually framed as the afterglow of the Big Bang—a faint echo from a hot, dense, chaotic past, stretched thin by expansion. This view assumes order emerged from chaos, structure from undirected heat. But the CMB is the visible onset of coherence in an orientation-based universe. It is the activation moment—when relational logic came online.

The CMB marks the coherence boundary, when the orientation lattice first stabilized into global relational balance, an activation signature. Disorder gives way to structure. The CMB marks that moment—the first stable output of an orientation lattice locking into balance through internal coherence regulation. Inflation theory is ideobabble—not an explanation, but a patch after the fact. What cosmologists call a thermal fog is really the radiant trace of spatial alignment—structure not from temperature, but from orientation.

In the conventional view, high velocities—planetary spin, galactic rotation—are treated as mechanical outcomes: caused by past events or requiring extra mass, like dark matter, to hold things together. But in a coherence-based framework, speed isn't caused by force or hidden mass. It's the visible expression of alignment within an orientation lattice.

Fast motion isn't what creates stability, and it is not evidence of missing mass. It signals how tightly a system is locked into its coherent path. Inertia at speed shows the system's resistance to disruption—like a gyroscope holding its axis through conserved orientation. What looks like momentum is not energy pushing mass, but coherence maintaining direction.

Velocity isn't an independent variable—it's an outcome of stability under constraint. High-speed systems aren't moving at random; they follow paths of least resistance within a framework built to preserve balance. Galactic spin, orbital precision, even the spread of angular momentum—all of it points to orientation logic embedded in the structure of space, not force-based mechanics.

Gyroscopes, used in aerospace and navigation, hold orientation through internal spin. In gimbals or inertial navigation systems, they resist changes to their axis—not because of gravity or anchors, but because of self-sustaining rotational coherence. Like a planet or star, a gyroscope doesn't stay aligned because it's held—it stays aligned because it's already in sync with the system it's part of.[969]

Satellites and space telescopes, like Hubble, don't rely on jets or external forces to turn. They use reaction wheels—internal flywheels whose spin adjusts to transfer angular momentum and rotate the craft.[970] No mass is expelled, no force applied externally. This is a direct parallel to the coherence-preserving reorientations in this model: movement through internal rebalancing, not external pull.

Maglev trains use the repulsion and attraction between superconducting magnets and guideways to generate lift and propulsion without contact. These systems move by maintaining a delicate balance of spatial orientation. The train levitates, moves, and stabilizes through embedded coherence logic—high-speed alignment without friction. This mirrors the model: motion generated by relational orientation, not by mass-induced curvature.

Atomic clocks measure time by the frequency of microwave signals emitted when electrons change energy levels in atoms.[971] These oscillations stay stable regardless of velocity or location, even on satellites moving at thousands of kilometers per hour. GPS systems account for time dilation—not as time warping, but as a shift in oscillation rhythm due to differences in positional coherence.[972] This aligns with my view of time as recalibration speed, not a fixed dimension stretched by force.

This coherence-based framing also recasts how we understand drift. While commonly treated as sensor error or entropy, drift is a measurable breakdown in relational alignment—an indicator of how well a system remains in sync with its underlying coherence lattice.[973] As Wojciech Zurek explains, decoherence arises from entanglement with the environment, which leads to "the emergence of the preferred set of pointer states"—those most stable under interaction.[974] Drift, then, is a directional signal: a deviation from these stable, environment-selected states that marks where coherence begins to fail and recalibration may be required.

By analogy, drift reflects a similar breakdown in structured systems: a slow erosion of alignment under stress, time, or disturbance. What appears as mechanical failure may in fact be coherence degradation—a macroscopic signature of decoherence. This is the same challenge in quantum computing—not simply decoherence in the statistical sense, but phase misalignment among qubits. Stability isn't just preserving state—it's preserving relational coherence. This insight opens new paths for system design: rather than relying on brute-force correction, systems can recalibrate in real time by detecting and responding to coherence loss. Drift, then, becomes a diagnostic tool—one that correlates with spatial position, orientation, and systemic fatigue.

Gyroscopes, used throughout this work as examples of inertial memory and orientation logic, illustrate this principle well. Their gradual misalignment over time isn't just wear—it's relational drift, a structural echo of the same decoherence Zurek frames as the bridge from quantum to classical order.[975] This ancient technology embodies coherence principles that today's physics is still catching up to.

The earliest inertial systems were developed in secret—classified tools of missile guidance, submarine navigation, and spaceflight. Only in recent decades have gyro-free inertial navigation systems (GF-INS) been developed, capable of maintaining orientation through distributed sensing and feedback without relying on a central spinning mass.

Gyro-free inertial systems show that orientation and navigation don't require a centralized spinning mechanism.[976] Movement and stability arise from integral relational logic and internal recalibration. This reflects the same coherence structure I've identified across natural systems—from particles to galaxies. These systems don't need anchors to maintain direction—they sustain themselves through embedded coherence. GF-INS confirms this. It's not just a technical achievement; it's a proof-of-concept for what I call Natural Technology. And yet, even our most advanced technologies— whether gyroscopes or gyro-free—degrade over time. In contrast, coherence systems in nature can maintain alignment for billions of years.

Moreover, if time itself represents an increasing accumulation of disorder, this, too, implies that the universe began in a state of profound coherence—an initial condition characterized by extraordinary structural alignment and order. Such foundational coherence, evident from the smallest quantum systems to the largest cosmic structures, raises clear implications about intentionality or underlying design.

Order of this magnitude does not appear accidental, nor is it easily attributable to random initial conditions. Instead, it suggests an intrinsic logic or structure embedded within the universe itself, serving as a baseline from which all subsequent changes and disruptions unfold.

Recalibration is necessary because expressions of coherence—like biological systems or planetary processes—can degrade. But this degradation is not a loss of structure itself—only a breakdown in how the system maintains its relation to that structure. Coherence isn't an isolated force; it's a relational pattern within a larger logic that enables resistance, adaptation, and persistence. This is no different from engineered technologies: machines break down, but the laws that govern their function remain intact. Entropy, in this view, is not the failure of reality—it's the stress signature of coherence under constraint. What persists beneath failure is a logic too stable to collapse. The defining relationship of coherence is not stasis, but resistance. Real order isn't the absence of change, but the capacity to endure it without losing identity.

Time itself can be understood as more than just an objective measurement of change; it is an active record of drift—the progressive accumulation of recalibration errors away from that initially coherent state. The forward progression of time is the record of a system's ongoing effort to maintain coherence—and the gradual accumulation of recalibration errors as it drifts from its initially coherent state. These are the observable effects of both structural successes and subtle failures.

CHAPTER 31

# Vision for the Future

## Toward a New Science of Meaning

Throughout this work, we've challenged the assumption that technology is something humans impose onto a neutral, chaotic world. Instead, we've traced a profound claim: that human innovation arises from our embeddedness in a universe already governed by Natural Technology. The gyroscopic logic of gravity, the fine-tuning of cosmic constants, the networked communication of quantum systems—all these reveal a structure not invented by us, but inherited.

Again, our technology isn't separate from nature—it's conceived through the union of nature's logic with our reasoning, as beings of natural technology who extend its patterns through mind and will. We don't invent logic; we discover and apply it. Every advancement, from microprocessors to synthetic molecular machines, is a translation of preexisting structure. Even our brightest minds cannot introduce a logic that doesn't already exist—nor fully unravel the natural logic that does. Its structure precedes us, and exceeds us.

This is not fudge factoring. This is origin. And it is only by recognizing that origin that we can begin to understand our place—not as full creators of meaning, but as participants in a logic that precedes us. We've seen that spin, structure, and stability are not random traits but system behaviors. Forces don't push from outside—they arise from within. Time isn't a sequence of fixed moments; it's a continuous process of relational recalibration.

This model that holds when others collapse. A model that answers questions materialism cannot even ask. And a model that does not resort to invisible fields or infinite realms to make sense of what is—in plain sight.

What comes next is not dogma. It is exploration. If the universe is technological in its logic—relational, responsive, precise—then our role is not to dominate or decode it in isolation, but to learn from it as participants in a living system. This is the invitation: to move forward not as machines trapped in randomness, but as observers awakened to the architecture we were born into. We are not anomalies in a dead universe—we are interfacing with meaning.

## The School of Natural Technology and Engineering

Interdisciplinary approaches like Stanford's BIO-X program remain the exception rather than the norm in higher education. Adopting a NatTech perspective could open new avenues of inquiry. In this context, the Dover case, which dismissed the exploration of design, represents a significant misstep. Revisiting this decision could advance scientific progress and give students a more well-rounded understanding, ensuring fairness for design-oriented perspectives.

To bridge this gap between theoretical understanding and practical application, I propose the establishment of the School of Natural Technology and Engineering within universities. This institution would honor the complex systems and processes observed in nature and transform these observations into advanced technological applications. It would focus on harnessing the principles observed in natural world phenomena to pioneer innovative technologies. It is the methodological basis for a new discipline.

Our technologies hold many keys to understanding the deeper architecture of nature—not because they are novel creations, but because they echo patterns already embedded in the world around us. Gyroscopic stability, genetic coding, feedback systems—all mirror dynamics we find in natural systems. By studying how our tools function, we can reverse-engineer the logic of the universe itself. Studying the dynamic assembly and disassembly of molecular machines like kinesin and microtubules, we can derive new models for structural advancements and self-repairing materials. The curriculum will transcend traditional biotechnology, delving into natural technological principles from the quantum scale of fundamental particles to the molecular interactions within living systems, all the way to the cosmic scale of galactic structures.

The proposed school would encourage an intertwined relationship between various engineering disciplines and biological sciences. Future biologists would benefit from a strong foundation in life sciences coupled with expertise in fields such as electrical, mechanical, civil, computer, or architectural engineering. Such interdisciplinary training will enable them to unravel biological complexities and apply this knowledge in diverse contexts—from enhancing biomechanical devices to designing smarter, more responsive urban ecosystems.

While disciplines such as bioelectrical engineering, mechanical bioengineering, environmental systems engineering, and computational biology are well-established, the proposed School of Natural Technology and Engineering aims to strengthen the integration of these fields. Unlike traditional programs, this school would specifically focus on the interconnectedness of natural processes from a holistic perspective, applying this understanding across disciplines to drive innovations that are not only technologically advanced but also sustainable and ethically sound.

The objective of founding the School of Natural Technology and Engineering is to create an educational ecosystem where new pathways cross traditional academic boundaries, fostering a generation of innovators and thinkers who can approach global challenges with a mindset that is attuned to the ancient wisdom of natural designs. This school will serve as a catalyst for progress, inspiring a new wave of scientific research. We should call upon academic institutions, industry leaders, and policymakers to support this transformative initiative, paving the way for a future that harmonizes with the principles of nature's own designs.

## The Human Body

A brighter path forward should also consider the interconnectedness of humanity. Together, we have explored the relational frameworks that structures everything from quantum mechanics to cellular systems and the vast cosmos. Incidentally, parallels can be seen between the collective progression of humanity and the irreducible systems of living organisms. It is overly simplistic to reduce us to merely individual parts. Humankind has a shared cosmic trajectory and a shared cosmic purpose, underscoring our ethical responsibility. Each individual's life represents a cell in the body of humankind, a unique chapter in the larger story of humanity. How we interact affects the body as a whole.

Historically, although we have an ethical responsibility towards one another, not all contributions from the "cells" of humanity have been benevolent. Dark chapters like Nazi Germany and the era of American slavery remind us how parts of our collective body have caused harm, and these periods left lasting scars akin to debilitating diseases that undermine the well-being of the entire organism.

On the other hand, the growth and maturing of our collective body, especially in terms of technology, showcases our remarkable capacity for progress and adaptation. From the simplest tools to artificial intelligence, each technological milestone enhances our collective capabilities and efficiency. This journey from basic toolmaking to the digital age is a maturation process, mirroring how cells develop specialized functions to support the larger organism.

However, the application of Darwinian Materialism has introduced severe consequences, notably the potential devaluation of life. This worldview increasingly pervades our education system, teaching children from a young age that the universe lacks purpose, thereby diminishing their lives' perceived value. The frequent suggestion that free will does not exist and that agency is largely predetermined by external processes poses serious risks to children's existential development.

The Theory of Natural Technology rekindles the sense of purpose, hope, and childlike wonder essential for driving meaningful scientific progress. This book is not the end of the conversation—it's the invitation to a new one. The science of tomorrow must begin with deeper questions. Not "What can we invent?" but "What have we been given?" Not "What do we control?" but "What are we participating in?" Not "How can we fix theoretical problems?" but "How can we accept reality as it is and follow the evidence wherever it leads?" The Theory of Natural Technology is not merely a model of how the universe works—it is a call to become aligned with how the universe means.

Moments, which we often misinterpret as scientific serendipity, may seem like chance encounters but are instead intricately woven into the structure of existence, guiding us toward greater understanding and insight. We could call this phenomenon an *inspired gifting*. It's as if the system of nature itself has historically encouraged inquiry, ensuring that those who seek will inevitably find. This often leads to discoveries more fascinating than initially sought or imagined.

It's reasonable to perceive that the elaborate systems in the universe point to an Engineer. However, a creator lacking consciousness, intention, or the ability to hear, see, or communicate is incompatible with the beauty of mathematics, the abundance of water, the sky at dawn and dusk, biological systems, consciousness, and the countless interconnected relationships that form the story of life. To exalt and celebrate the idea of a creator incapable of listening, speaking, or thinking seems, to me, fundamentally irrational. Yet, this is the position many take today—and that's fine with me.

I honestly find the heated debates over whether the universe evolved from nothing—often taking place in prestigious universities—a bit foolish. After all, by materialist logic, not just organisms, but even these universities, their buildings, and the very debates themselves are merely inevitable products of blind evolutionary forces. That logic makes no sense to me. How can I ignore what is right in front of my eyes? Two and two are four.

# Notes

11 "Quotes Relevant to Science," University of Connecticut, accessed August 17, 2024, https://media.pluto.psy.uconn.edu/291quotess05.html.

2 Stephen C. Meyer, *Signature in the Cell: DNA and the Evidence for Intelligent Design* (New York: HarperOne, 2009).

3 Years ago, I had a strong intuition that, despite Hitchens' remarkable intellect, he might have been intoxicated or possibly hungover during debates, especially during his most famous debate with William Lane Craig. Not wanting to rely solely on a hunch, I decided to research his drinking habits, particularly during public debates. After further investigation, it seems my intuition was likely correct. Sources from those who knew him, including his ex-wife Carol Blue and Hitchens himself, revealed that alcohol was a significant part of his lifestyle and could have influenced his behavior during debates. Blue, who was married to Hitchens, observed that although he was an alcoholic, he functioned professionally at a high level. She mentioned that while he didn't always appear drunk, there were times when it seemed he might have been intoxicated. As Blue put it, "Once in a while, it seems like he might be drunk. Aside from that, even though he's obviously an alcoholic, he functions at a really high level and he doesn't act like a drunk." This observation aligns with my intuition that Hitchens could have been under the influence during his debates, including the one with William Lane Craig. Carol Blue, quoted in "He Knew He Was Right," The New Yorker, October 16, 2006, https://www.newyorker.com/magazine/2006/10/16/he-knew-he-was-right-2.

4 In his 2003 Vanity Fair article, "Living Proof," Hitchens discusses how alcohol was deeply ingrained in his daily life. He reflects on how alcohol, despite its negative effects, never seemed to interfere with his professional abilities. Hitchens writes, "Some drank to meet a deadline, and some drank to give themselves an excuse to miss one. The latter crew had a tendency to check out prematurely. When the late Murray Kempton was asked by a copyboy how much longer it would be until his column was ready, Kempton held up a bottle and jovially said, 'About an inch.'" This commentary on alcohol as a constant companion could help explain Hitchens' behavior in debates, suggesting that his drinking habits might have affected his performance in debates, including the one with William Lane Craig.

5 Richard L. Doty, "Human Pheromones: Do They Exist?" in *Neurobiology of Chemical Communication*, ed. Carla Mucignat-Caretta (Boca Raton, FL: CRC Press/Taylor & Francis, 2014), chap. 19, accessed November 21, 2024, https://www.ncbi.nlm.nih.gov/books/NBK200980/.

6 Wikipedia, s.v. "Technology Transfer," last edited June 11, 2024, https://en.wikipedia.org/wiki/Technology_transfer.

7 "What Is Intelligent Design?" *Intelligent Design*, accessed January 22, 2025, https://intelligentdesign.org/whatisid/.

8 Giulio Ruffini, "Information, Complexity, Brains and Reality (Kolmogorov Manifesto)," *arXiv* preprint arXiv:0704.1147 (2007): 10, accessed January 8, 2024, https://arxiv.org/pdf/0704.1147.pdf.

9 "Agnosticism," Wikipedia, last modified December 21, 2024, https://en.wikipedia.org/wiki/Agnosticism.

10 Huxley, Thomas Henry. *Science and Morals*. 1886. In Collected Essays IX: Evolution & Ethics and Other Essays. http://aleph0.clarku.edu/huxley/CE9/S-M.html.

11 Michael Behe, "A (R)evolutionary Biologist," MichaelBehe.com, accessed November 5, 2024, https://michaelbehe.com/.

12 American Civil Liberties Union, "Frequently Asked Questions about Intelligent Design," last modified September 16, 2005, accessed February 11, 2024, https://www.aclu.org/documents/frequently-asked-questions-about-intelligent-design.

13 Michael J. Behe, "A Mousetrap Defended: Response to Critics," Discovery Institute, July 31, 2000, https://www.discovery.org/a/446/.

14 "Nobel Laureates Urge Kansas to Reject Intelligent Design," NBC News, September 16, 2005, https://www.nbcnews.com/id/wbna9368495.

15 Ibid.

16 Patricia Bauer, "Flying Spaghetti Monster," *Encyclopaedia Britannica*, accessed January 9, 2025, https://www.britannica.com/topic/Flying-Spaghetti-Monster#ref1254227.

17 Kathleen Craig, "Passion of the Spaghetti Monster," Wired, December 22, 2005, https://www.wired.com/2005/12/passion-of-the-spaghetti-monster/.

18 Andrei Linde, "A Brief History of the Multiverse," arXiv preprint, arXiv:1512.01203v3 [hep-th], December 8, 2017, Stanford Institute for Theoretical Physics and Department of Physics, Stanford University, Stanford, CA.

19 Ibid., 2.

20 "Parallel Universes Explained: Brian Greene & Joe Rogan Discuss the Mind-Bending Possibilities," Science Clips, December 21, 2023, YouTube video, 0:00–0:59, https://www.youtube.com/watch?v=Opig-w3dBTw.

21 As mentioned, Greene frequently discusses the classical interpretations of wave collapse (Copenhagen theory) and alternates between Many-Worlds Interpretation (MWI) and multiverse theory, often extrapolating between the two without providing sufficient context for listeners. This blending creates the impression that these distinct theories are interconnected or stem "naturally" from the same mathematical foundation. In two different videos referenced below, Greene conflates these ideas in ways that could mislead audiences about their respective implications.

  In one explanation, Greene begins with MWI, describing how quantum equations suggest that all possible outcomes of a quantum event occur in separate universes. For example, an electron's potential positions—left or right—each correspond to a branching universe. This interpretation assumes that all universes share the same laws of physics but differ only in the outcomes of quantum events. Greene then transitions to a broader claim, stating, "The gods…see that there are…many universes out there. And basically, anything that's allowed by the laws of quantum physics is represented in this menagerie of universes." While consistent with MWI, this phrasing starts to blur the distinction between quantum branching and the multiverse theories of cosmology.

In another discussion, Greene expands n this idea, stating, "There have to be Realms out there that duplicate ours as well. Many can be different, but there have to be versions of this reality that are also instantiated." Here, Greene introduces the idea of variation among universes, implying that not all universes are identical or governed by the same outcomes. This aligns more closely with multiverse theories derived from eternal inflation or string theory, where universes may have fundamentally different physical constants or laws. By weaving these concepts together, Greene creates the impression that quantum mechanics inherently predicts a multiverse containing universes with varying laws of physics—a concept not supported by MWI, which operates exclusively within the framework of shared physical laws.

This conflation arises because Greene does not clarify the distinct origins and implications of these theories. MWI is not a direct consequence of the mathematics of quantum mechanics; it is an interpretive framework that posits a Darwinian-like branching structure of universes governed by the same fundamental laws. In contrast, multiverse theories stem from cosmology and often require additional physical assumptions, such as those in inflationary models or string theory. By presenting these frameworks as a unified whole, Greene, possibly, misleads his audience into thinking they describe a single, cohesive narrative, when in reality, they are distinct hypotheses addressing different questions in physics. References:

"Brian Greene Challenges Parallel Universes," Pangburn, January 7, 2019, YouTube video, 0:00–1:56, https://www.youtube.com/watch?v=y-Qv9LRlKV8.

"Parallel Universes Explained: Brian Greene & Joe Rogan Discuss the Mind-Bending Possibilities," Science Clips, December 21, 2023, YouTube video, 0:00–0:59, https://www.youtube.com/watch?v=Opig-w3dBTw.

22 Y. Chang, B. L. Carroll, and J. Liu, "Structural Basis of Bacterial Flagellar Motor Rotation and Switching," *Trends in Microbiology* 29, no. 11 (November 2021): 1024-1033, https://doi.org/10.1016/j.tim.2021.03.009.

23 Bruce Simpson, *Chrysle's Gas Turbine Cars: Jet Engine Technologies for Interested Amateurs*, last updated May 11, 2002, © 2001–2009, accessed February 8, 2025, https://www.aardvark.co.nz/pjet/chrysler.shtml.

24 "Reduction drive." *Wikipedia*. Last modified October 22, 2024. Accessed February 8, 2025. https://en.wikipedia.org/wiki/Reduction_drive.

25 Mike McNessor, "Below the Hood: Chrysler A831 Turbine," Hemmings, March 22, 2024, https://www.hemmings.com/stories/belowthehood-chrysler-a831-turbine/.

26 J. Tan, L. Zhang, X. Zhou, et al., "Structural Basis of the Bacterial Flagellar Motor Rotational Switching," *Cell Research* 34 (2024): 788–801, https://doi.org/10.1038/s41422-024-01017-z.

27 G. N. Cohen-Ben-Lulu et al., "The Bacterial Flagellar Switch Complex Is Getting More Complex, *The EMBO Journal* 27, no. 7 (2008): 1134–1144, https://doi.org/10.1038/emboj.2008.48.

28 T. Yamaguchi, F. Makino, T. Miyata, et al., "Structure of the molecular bushing of the bacterial flagellar motor," *Nature Communications* 12 (2021): 4469, https://doi.org/10.1038/s41467-021-24715-3.

29 Ibid.

30 Maria Santiveri et al., "Structure and Function of Stator Units of the Bacterial Flagellar Motor," *Cell* 183, no. 1 (2020): 244-257, https://doi.org/10.1016/j.cell.2020.08.016.

31 Ibid.

32 Ibid.

33 T. Yamaguchi, F. Makino, T. Miyata, et al., "Structure of the molecular bushing of the bacterial flagellar motor."

34 Ibid.

35 Ibid.

36 David Graeber and David Wengrow, *The Dawn of Everything: A New History of Humanity* (New York: Farrar, Straus and Giroux, 2021), Kindle edition, loc. 190.

37 Christopher S. Henshilwood et al., "Emergence of Modern Human Behavior: Middle Stone Age Engravings from South Africa," *Science* 295, no. 5558 (January 10, 2002): 1278–1280, https://doi.org/10.1126/science.1067575. These engravings provide evidence of early symbolic thought, suggesting that anatomically modern humans possessed advanced cognitive abilities far earlier than previously assumed.

38 W. W. Rouse Ball, *A Short Account of the History of Mathematics*, 4th ed. (1908), "Sir Isaac Newton (1642–1727)," https://www.maths.tcd.ie/pub/HistMath/People/Newton/RouseBall/RB_Newton.html.

39 Josh H. Wimpenny et al., "Cognitive Processes Associated with Sequential Tool Use in New Caledonian Crows," *PLoS One* 4, no. 8 (2009): e6471, https://doi.org/10.1371/journal.pone.0006471.

40 Christian Rutz, Sarah J. Jelbert, Barbara C. Klump, and James J. H. St Clair, "New Caledonian Crows Plan for Specific Future Tool Use," *Proceedings of the Royal Society B* 287, no. 1938 (November 2020): 1–7, https://doi.org/10.1098/rspb.2020.1490.

41 Ibid.

42 L. Sprague de Camp, *The Ancient Engineers* (Garden City, NY: Doubleday, 1963).

43 David Graeber and David Wengrow, *The Dawn of Everything*, loc. 176.

44 Ibid., loc 186.

45 Ibid.

46 Ibid.

47 Ibid.

48 Ibid., 379.

49 Kevin Kelly, *What Technology Wants* (New York: Viking Penguin, 2010), 15.

50 Andreas Kontokanis, "Etymology of words related to Technology," U Speak Greek, accessed June 11, 2024, https://uspeakgreek.com/category/technology/page/3/.

51 Ibid.

52 Ibid.

53 Ibid.

54 Ibid.

55 E. Montuschi, "Order of man, order of nature: Francis Bacon's idea of a 'dominion' over nature," Order: God's, Man's and Nature's, ISSN 2045-5577, accessed June 11, 2024, https://iris.unive.it/retrieve/e4239ddb-3008-7180-e053-3705fe0a3322/MontuschiBacon.pdf

56 C. B. Christensen, "Karl Marx on Technology, Economy and Society," May 27, 1994, CB Christensen, http://www.cbchristensen.net/courses/technology/karl-marx-on-technology-economy-and-society/.

57 Martin Heidegger, *The Question Concerning Technology and Other Essays*, trans. William Lovitt (New York & London: Garland Publishing, Inc., 1977).

58 "Guide to Heidegger's 'The Question Concerning Technology'," Criticalink: Martin Heidegger: *The Question Concerning Technology*, accessed June 11, 2024, https://english.hawaii.edu/criticalink/heidegger/guide1.html.

59 Ibid.

60 Ibid., xx.

61 Don Ihde and Evan Selinger, eds., *Chasing Technoscience: Matrix for Materiality* (Bloomington & Indianapolis: Indiana University Press, 2003).

62 Irena Avsenik Nabergoj, "Ethics and Information Technology: A Phenomenological Perspective," *Stanford Encyclopedia of Philosophy*, last modified February 14, 2020, https://plato.stanford.edu/entries/ethics-it-phenomenology/.

63 Bruno Latour, "Reassembling the Social: An Introduction to Actor-Network-Theory," (PDF file, last accessed August 11, 2024), http://www.bruno-latour.fr/sites/default/files/P-67%20ACTOR-NETWORK.pdf.

---

64 Jacques Ellul, *The Technological Society* (New York: Vintage Books, 1964).

65 Ibid., 212-213.

66 Ibid., Note to the Reader.

67 Ibid. 299-301.

68 Kevin Kelly, *What Technology Wants.*

69 Ibid., 46-50.

70 Stephen C. Meyer, *Return of the God Hypothesis: Three Scientific Discoveries That Reveal the Mind Behind the Universe* (HarperAudio, 2021), Chapter 22, discussing the low entropy of the universe under the Big Bang model, audiobook, https://www.everand.com/audiobook/635394419/Return-of-the-God-Hypothesis-Three-Scientific-Discoveries-That-Reveal-the-Mind-Behind-the-Universe.

71 Luisa Alba-Lois, Ph.D., and Claudia Segal-Kischinevzky, M.Sc., "Yeast Fermentation and the Making of Beer," Nature Education (Facultad de Ciencias, Universidad Nacional Autonoma de Mexico), 2010, accessed September 19, 2024, https://www.nature.com/scitable/topicpage/yeast-fermentation-and-the-making-of-beer-14372813/.

72 "Egyptian Wine," *Wikipedia*, last modified January 18, 2025, https://en.wikipedia.org/wiki/Egyptian_wine.

73 Peter Damerow, "Sumerian Beer: The Origins of Brewing Technology in Ancient Mesopotamia," *Cuneiform Digital Library Journal* 2012:2, January 22, 2012, https://cdli.mpiwg-berlin.mpg.de/articles/cdlj/2012-2.

74 "History of Tortilla," *History of Bread*, accessed January 23, 2025, https://www.historyofbread.com/bread-history/history-of-tortilla/.

75 Peter Damerow, "Sumerian Beer: The Origins of Brewing Technology in Ancient Mesopotamia."

76 Chomsky, Noam. *Syntactic Structures*. The Hague: Mouton, 1957.

77 Noam Chomsky, *The Minimalist Program* (Cambridge, MA: MIT Press, 1995).

78 Ibid.

79 Ibid.

80 Ibid.

81 Daniel Everett has controversially argued that the Pirahã language lacks recursion, challenging Chomsky's claim that recursion is a universal feature of human language. However, Everett's findings remain highly disputed, with critics arguing that his conclusions are based on limited and anecdotal evidence rather than rigorous linguistic analysis. See Daniel Everett, *Cultural Constraints on Grammar and Cognition in Pirahã: Another Look at the Design Features of Human Language*, Current Anthropology 46, no. 4 (2005): 621–646.

82 Ibid.

83 H. Takahashi, K. Takahashi, and F. C. Liu, "FOXP Genes, Neural Development, Speech and Language Disorders," in *Madame Curie Bioscience Database* [Internet] (Austin, TX: Landes Bioscience, 2000–2013), https://www.ncbi.nlm.nih.gov/books/NBK7023/.

84 "FOXP2-Related Speech and Language Disorder," *MedlinePlus*, last updated April 18, 2023, https://medlineplus.gov/genetics/condition/foxp2-related-speech-and-language-disorder/.

85 Joseph S. Perkell, *Brain Mechanisms for Hearing and Speech*, Harvard-MIT Division of Health Sciences and Technology, HST.722J, Fall 2005, https://ocw.mit.edu/courses/hst-722j-brain-mechanisms-for-hearing-and-speech-fall-2005/170ec54f67d650df07afc52712cfc75d_6_mot_con_sp_per.pdf.

86 Ibid.

87 "Serial Order," ScienceDirect, accessed March 20, 2025, https://www.sciencedirect.com/topics/engineering/serial-order. ScienceDirect indexes this topic under engineering, a classification that reflects a broader understanding in cognitive science, neuroscience, and robotics. Serial order *problems* arise in fields requiring precise sequencing and hierarchical control, such as speech production, motor control, artificial intelligence, and industrial automation. The reason it is often categorized under engineering is that serial order relies on structured control systems, similar to those found in designed mechanisms rather than simple biological reflexes. This further complicates evolutionary explanations, as natural selection cannot produce intricate, multi-level control systems all at once.

88 "Étienne Bonnot de Condillac," in *The Stanford Encyclopedia of Philosophy*, Fall 2020 Edition, ed. Edward N. Zalta, section 6, "Signs and the 'Language of Action'," https://plato.stanford.edu/entries/condillac/.

89 Ibid.

90 Michael Corballis, *From Hand to Mouth: The Origins of Language* (Princeton, NJ: Princeton University Press, 2002).

91 David McNeill, *Gesture and Thought* (Chicago: University of Chicago Press, 2005).

92 "Language Acquisition Device (LAD)," Shiken AI, accessed January 13, 2025, https://shiken.ai/english-language-topics/language-acquisition-device-lad.

93 "Introduction to Grammar in Theory of Computation," GeeksforGeeks, accessed January 12, 2025, https://www.geeksforgeeks.org/introduction-to-grammar-in-theory-of-computation/.

94 Ibid.

95 Ibid.

96 Stephen J. Pyne, *Fire: A Brief History* (Seattle: University of Washington Press, 2001).

97 H. K. Pae, "The Emergence of Written Language: From Numeracy to Literacy," in *Script Effects as the Hidden Drive of the Mind, Cognition, and Culture*, Literacy Studies 21 (2020): 25, https://doi.org/10.1007/978-3-030-55152-0_2.

98 Peter Rowley-Conwy, "The Concept of Prehistory and the Invention of the Terms 'Prehistoric' and 'Prehistorian:' The Scandinavian Origin, 1833–1850," *European Journal of Archaeology* (2006), https://www.academia.edu/49730833/The_Concept_of_Prehistory_and_the_Invention_of_the_Terms_Prehistoric_and_Prehistorian_The_Scandinavian_Origin_1833_1850.

99 Commodore E. S. Clark Jr., "The Evolution of the Sextant," *Proceedings* 62, no. 11/405 (November 1936), https://www.usni.org/magazines/proceedings/1936/november/evolution-sextant.

100 Richard J. Cox, "The Information Age and History: Looking Backward to See Us," University of Pittsburgh, accessed September 19, 2024, https://d-scholarship.pitt.edu/2698/1/r_cox_1.html.

101 Dmitry Solenov, Jan Brieler, and Jeffrey F. Scherrer, "The Potential of Quantum Computing and Machine Learning to Advance Clinical Research and Change the Practice of Medicine," Missouri Medicine 115, no. 5 (September-October 2018): 463-467, PMID: 30385997; PMCID: PMC6205278.

102 Gokul Alex, "Quantum Chaos and the Digital Universe: Tower of Babel 2.0?" LinkedIn Pulse, November 7, 2014, https://www.linkedin.com/pulse/20141107071141-23034297-quantum-chaos-and-the-digital-universe-tower-of-babel-2-0/.

103 CBS News. 'Prosthetics breakthroughs: 2006 to now.' CBS News, July 16, 2023. Accessed January 12, 2024. https://www.cbsnews.com/news/prosthetics-breakthroughs-2006-to-now-60-minutes-2023-07-16/.

104 Hardwick, Senia. "Computers and the Human Brain." Howard Brain Sciences Foundation. Last modified March 30, 2020. Accessed January 13, 2024. https://www.brainsciences.org/computers-and-the-human-brain.

105 Craig Rusbult, "What is a World View?" American Scientific Affiliation, accessed February 4, 2024, https://www.asa3.org/ASA/education/views/index.html. Quote: "Your worldview includes your answers for a wide range of questions: What are humans, why we are here, and what is our purpose in life? What are your goals for life? What aspects of life (and actions of life) are more important, and less important? When you make decisions about using time – and thus life – what are your values and priorities? What can we know, and how? and with how much certainty? Does reality include only matter/energy, or is there more?"

106 Grammarist. "Familiarity breeds contempt." Accessed January 13, 2024. https://grammarist.com/proverb/familiarity-breeds-contempt/.

107 "Information," in Cambridge Dictionary, accessed November 3, 2023, https://dictionary.cambridge.org/us/dictionary/english/information.

108 Peter A. Corning et al., eds., Evolution "On Purpose", 15. "Their dedication to natural selection led neo-Darwinists like Coyne to deny well-documented sensory, cognitive, computational, and decision-making abilities in living organisms. In doing so, they discounted the discoveries about intricately

organized biological soft matter that molecular genetics revealed to exist and to exercise sensory, regulatory, and computational processes in even the simplest living organisms (Bray, 2009; Regolin & Vallortigara, 2021)."

109 Ruffini, "Information, Complexity, Brains and Reality," 15, accessed January 8, 2024.

110 Fredrik Ege Abrahamsen, Yun Ai, and Michael Cheffena, "Communication Technologies for Smart Grid: A Comprehensive Survey," arXiv:2103.11657v1 [eess.SP], March 22, 2021.

111 Richard Dawkins, *The Blind Watchmaker* (1st American ed.; New York: Norton, 1986), 133.

112 Lemmon, M. A., & Schlessinger, J. "Cell Signaling by Receptor Tyrosine Kinases." *Cell*, 141(7), 1117-1134. doi:10.1016/j.cell.2010.06.011.

113 Divya Chauhan and Jawahar Thakur, "Data Mining Techniques for Weather Prediction: A Review," *International Journal on Recent and Innovation Trends in Computing and Communication* 2, no. 8 (2014): 2184-2189.

114 Richard Dawkins, *The Ancestor's Tale: A Pilgrimage to the Dawn of Evolution* (Boston: Houghton Mifflin Harcourt, 2004), accessed via Goodreads, https://www.goodreads.com/quotes/11048729-the-universe-could-so-easily-have-remained-lifeless-and-simple.

115 Jewish Learning Institute. "The Parting of the Red Sea - Natural or Miraculous?" YouTube video, 6:00. January 13, 2022. Accessed March 9, 2024. https://www.youtube.com/watch?v=Nc2F3-Hlnm8. Note: This video is not the original referenced but is exemplary of the narrative recalled.

116 Ibid.

117 Ibid.

118 Ark Encounter."Governor Beshear, Ark Encounter Announce Plans to Build Full-Scale Noah's Ark." Accessed March 9, 2024. https://arkencounter.com/press/governor-beshear-ark-encounter-announce-plans-build-full-scale-noahs-ark. Note: This article was adapted from a news release recently distributed to the media.

119 Ken Ham, "Get Answers: Animals on the Ark," Answers in Genesis, March 20, 2018, https://answersingenesis.org/blogs/ken-ham/2018/03/20/get-answers-animals-ark/. In this blog, Ham estimates that about 1,400 kinds, or around 6,700 individual animals, were taken on the Ark, noting that this number could be lower.

120 Ibid.

121 S.J. Theroux, *A Most Improbable Story: The Evolution of the Universe, Life, and Humankind*, 1st ed. (CRC Press, 2022), 8, https://doi.org/10.1201/9781003270294. Theroux quotes Max Tegmark: "some of the fine-tuning (of the universe) appears extreme enough to be quite embarrassing—for example, we need to tune the dark energy to about 123 decimal places to make habitable galaxies. To me, an unexplained coincidence can be a tell-tale sign of a gap in our scientific understanding. Dismissing it

by saying 'We just got lucky—now stop looking for an explanation!' is not only unsatisfactory, but is also tantamount to ignoring a potentially crucial clue."

122 Ibid.

123 Scientists at NASA's Imagine the Universe. 'When Did the Universe Begin?' NASA. Accessed March 9, 2024. https://imagine.gsfc.nasa.gov/educators/elements/imagine/02.html. This resource includes the statement: 'Scientists have traced the expansion back to a time when the entire Universe was smaller than an atom.

124   *Stanford    Encyclopedia    of    Philosophy*, 'Fine-Tuning,'    accessed    March    9,    2024, https://plato.stanford.edu/entries/fine-tuning/#DoesImprFineTuniUnavCallForResp. This entry explores the philosophical implications of the universe's apparent fine-tuning for the existence of life, examining various theories and responses to this phenomenon.

125 For more information on cosmic inflation, see Chapter 9, "Cosmic Dawning."

126 Adam Frank, "Our standard model of cosmology is both a triumph and a jumbled mess," Big Think, April 20, 2023, https://bigthink.com/13-8/standard-model-cosmology-jumbled-mess/. In the article, Frank observes: 'CMB looked the same no matter where astronomers directed their gaze... Either a miracle occurred in the early Universe, or the cosmic microwave background's smoothness had a deeper explanation... What they came up with in the 1980s was inflation.'

127 "Falsifiability." ScienceDirect Topics, https://www.sciencedirect.com/topics/mathematics/falsifiability.

128 Krauss, Lawrence M. A Universe from Nothing: Why There Is Something Rather Than Nothing. New York: Free Press, 2012.

129 Anja Steinbauer, "News: April/May 2016," Philosophy Now, accessed February 19, 2024, https://philosophynow.org/issues/113/News_April_May_2016. Vilenkin is quoted as stating, "It is said that an argument is what convinces reasonable men and a proof is what it takes to convince even an unreasonable man. With the proof now in place, cosmologists can no longer hide behind the possibility of a past-eternal universe. There is no escape: they have to face the problem of a cosmic beginning."

130 Ibid.

131 Vilenkin, "The Beginning of the Universe."

132 Anja Steinbauer, "News: April/May 2016," Philosophy Now.

133 Ibid.

134 U.S. Department of Energy, Office of Scientific and Technical Information. "The Manhattan Project: An Interactive History." The Atomic Bomb: The Manhattan Project (2016). Accessed June 23, 2024. https://www.osti.gov/opennet/manhattan-project-history/Science/Atom/atom.html.

135 ZEUS Collaboration. "Limits on the Effective Quark Radius from Inclusive Ep Scattering at HERA." *Physics Letters B* 757 (2016): 468-472. https://doi.org/10.1016/j.physletb.2016.04.007.

136 David Christian, *Maps of Time: An Introduction to Big History* (Berkeley: University of California Press, 2004), 22–23.

137 R.H. Luecke, W.D. Wosilait, and J.F. Young, "Mathematical Modeling of Human Embryonic and Fetal Growth Rates," *Growth Development and Aging* 63, no. 1-2 (Spring-Summer 1999): 49-59, PMID: 10885857: "Early embryonic growth immediately following fertilization is exponential; i.e., one cell goes to 2, then 4, then 8... etc., with essentially no decrease in relative growth rate."

138 Ibid.

139 Scott F. Gilbert, "An Introduction to Early Developmental Processes," in *Developmental Biology*, 6th ed. (Sunderland, MA: Sinauer Associates, 2000), accessed April 12, 2024, https://www.ncbi.nlm.nih.gov/books/NBK9992/. "Meanwhile, the cleavage of nuclei occurs at a rapid rate never seen again (not even in tumor cells). A frog egg, for example, can divide into 37,000 cells in just 43 hours. Mitosis in cleavage-stage Drosophila embryos occurs every 10 minutes for over 2 hours and in just 12 hours forms some 50,000 cells."

140 Paul Davies, *The Goldilocks Enigma: Why Is the Universe Just Right for Life?* (Boston: Houghton Mifflin, 2006). Paul Davies tackles the question of why the universe appears to be finely tuned for the emergence and sustenance of life. He examines various explanations, from the anthropic principle to the possibility of a multiverse, providing insights into the fundamental laws and physical constants that make our universe habitable. This book explores how these conditions might have originated from the singularity, addressing the "just right" conditions.

141 Ibid.

142 Harvard-Smithsonian Center for Astrophysics, "Dark Energy and Dark Matter," Harvard-Smithsonian Center for Astrophysics, accessed May 20, 2024, https://www.cfa.harvard.edu/research/topic/dark-energy-and-dark-matter.

143 Stephen C. Meyer, About "Signature in the Cell," Signature in the Cell, accessed April 12, 2024, https://signatureinthecell.com/about-the-book/.

144 Ibid.

145 Seth Shostak, "SETI and Intelligent Design," Space.com, published December 1, 2005, https://www.space.com/1826-seti-intelligent-design.html.

146 "Speed of Light," CERN, https://home.cern/tags/speed-light.

147 Roger Penrose, *The Road to Reality: A Complete Guide to the Laws of the Universe* (New York: Alfred A. Knopf, 2004), 390.

148 John D. Norton, "The World's Quickest Derivation of $E = mc^2$," Department of History and Philosophy of Science, University of Pittsburgh, accessed January 7, 2025,

https://sites.pitt.edu/~jdnorton/teaching/HPS_0410/chapters_June_6_2024/E%3Dmcsquared/proof.html.

149 Christopher S. Baird, "Why Is the Speed of Light a Random Finite Number?" The Top 50 Science Questions with Surprising Answers (West Texas A&M University, February 7, 2024), https://www.wtamu.edu/~cbaird/sq/2024/02/07/why-is-the-speed-of-light-a-random-finite-number/.

150 Ibid.

151 "What Is Refresh Rate for TVs, Monitors, and Projectors: 60Hz vs 120Hz and More," Best Buy Blog, https://blog.bestbuy.ca/tv-audio/what-is-refresh-rate-for-tvs-monitors-and-projectors-60hz-vs-120hz-and-more.

152 Eline Technology, "Definition: Ultra HD/4K," https://elinetechnology.com/definition/ultra-hd-4k/.

153 Christopher S. Baird, "Why Is the Speed of Light a Random Finite Number?," Emphasis added.

154 Martin Rees, *Just Six Numbers: The Deep Forces That Shape the Universe* (New York: Basic Books, 2000), 47.

155 Penrose, *The Road to Reality*, 399-401.

156 Ibid.

157 John D. Barrow, "Outer Space: Are the Constants of Nature Really Constant?" *Plus Magazine*, June 1, 2009. https://plus.maths.org/content/outer-space-are-constants-nature-really-constant

158 Penrose, *The Road to Reality*, 399-401.

159 "What is Bandwidth?," T-Mobile, "In computer networking, bandwidth refers to the amount of data a network can transmit over a connection in a given amount of time." https://www.t-mobile.com/home-internet/the-signal/internet-devices/what-is-bandwidth.

160 Ibid.

161 Harley H. McAdams and Adam Arkin, "Stochastic Mechanisms in Gene Expression," *PNAS* 94, no. 3 (February 4, 1997): 814-819, https://doi.org/10.1073/pnas.94.3.814. Quoted in Ibid., "In cellular regulatory networks, genetic activity is controlled by molecular signals that determine when and how often a given gene is transcribed. In genetically controlled pathways, the protein product encoded by one gene often regulates expression of other genes. The time delay, after activation of the first promoter, to reach an effective level to control the next promoter depends on the rate of protein accumulation. We have analyzed the chemical reactions controlling transcript initiation and translation termination in a single such 'genetically coupled' link as a precursor to modeling networks constructed from many such links. Simulation of the processes of gene expression shows that proteins are produced from an activated promoter in short bursts of variable numbers of proteins that occur at random time intervals. As a result, there can be large differences in the time between successive events in regulatory cascades across a cell population."

162 "The Horizon Problem," ResearchGate (2019), accessed May 3, 2024, https://www.researchgate.net/publication/334857975_The_Horizon_Problem.

163 "Critical Density," Swinburne University of Technology, accessed July 9, 2024, https://astronomy.swin.edu.au/cosmos/C/Critical+Density.

164 Ibid.

165 Alan H. Guth, "Inflationary Universe: A Possible Solution to the Horizon and Flatness Problems," *Physical Review D* 23, no. 2 (1981): 347-356.

166 "Cosmic Inflation Theory Faces Challenges," Scientific American, accessed July 9, 2024, https://www.scientificamerican.com/article/cosmic-inflation-theory-faces-challenges/.

167 "Section 1: Parametric Equations," 18.022: Calculus of Several Variables, accessed August 8, 2024, https://math.mit.edu/~djk/18_022/chapter03/section01.html.

168 Brad Lemley and Larry Fink, "Guth's Grand Guess: Most People Really Want to Know Where We Came From. We Have Evidence. We No Longer Have to Rely on Stories We Were Told When We Were Young," Discover Magazine, April 1, 2002, https://www.discovermagazine.com/the-sciences/guths-grand-guess.

169 Wikipedia contributors, "Cosmogony," Wikipedia, The Free Encyclopedia, last modified September 22, 2023, https://en.wikipedia.org/wiki/Cosmogony. Wikipedia defines cosmogony as any model concerning the origin of the cosmos or universe, dividing it into two categories: scientific theories and myths.

170 Brad Lemley and Larry Fink, "Guth's Grand Guess: Most People Really Want to Know Where We Came From. We Have Evidence. We No Longer Have to Rely on Stories We Were Told When We Were Young."

171 William J. Wolf. "Cosmological Inflation and Meta-Empirical Theory Assessment." *Studies in History and Philosophy of Science* 103 (February 2024): 146–158. https://doi.org/10.1016/j.shpsa.2023.12.006.

172 Paul J. Steinhardt, "The Inflation Debate," Scientific American, April 1, 2011, https://www.scientificamerican.com/article/the-inflation-debate/.

173 Ibid.

174 Ibid.

175 Brad Lemley and Larry Fink, "Guth's Grand Guess."

176 "Supernatural," Wikipedia, last modified November 1, 2024, https://en.wikipedia.org/wiki/Supernatural.

177 Paul J. Steinhardt, "The Inflation Debate," Note: Steinhardt refers to the components of the multiverse as multi "islands," a term he uses to emphasize the distinct, separate scenarios that can arise in an eternally inflating universe.

178 Ibid.

179 Ibid.

180 "What Happened in the Early Universe?," Harvard-Smithsonian Center for Astrophysics, accessed September 20, 2024, https://www.cfa.harvard.edu/big-questions/what-happened-early-universe.

181 Brian Greene, "Is our universe the only universe?" TED video, 21:45, April 23, 2012, https://www.youtube.com/watch?v=bf7BXwVeyWw.

182 *"Physical Constant,"* New World Encyclopedia, last modified October 9, 2020, https://www.newworldencyclopedia.org/entry/Physical_constant.

183 "Calibration Instruments, Standards and Reference Sources," *GlobalSpec*, accessed March 24, 2025, https://www.globalspec.com/productfinder/test_measurement_equipment/calibration_instruments_standards_reference_sources.

184 Ibid.

185 Zeeya Merali, "The Biggest Crisis in Astronomy," *John Templeton Foundation*, accessed January 25, 2025, https://www.templeton.org/news/the-biggest-crisis-in-astronomy.

186 For more detailed information on the importance of processor speed in computing devices, see Intel's article, "CPU Clock Speed: A Guide to Understanding Processor Speed," Intel, accessed February 22, 2024, https://www.intel.com/content/www/us/en/gaming/resources/cpu-clock-speed.html.

187 Martin Rees, *Just Six Numbers*.

188 National Institute of Standards and Technology, "Newtonian Constant of Gravitation G," accessed May 4, 2024, https://physics.nist.gov/cgi-bin/cuu/Value?bg.

189 National Institute of Standards and Technology, "Fine-Structure Constant," accessed May 4, 2024, https://physics.nist.gov/cuu/Constants/alpha.html.

190 Ibid.

191 Rees, *Just Six Numbers*.

192 "Specs," TFT Central, accessed February 22, 2025, https://tftcentral.co.uk/specs-2.

193 WhatIs.com, "Milliampere-hour (mAh)," TechTarget, accessed June 23, 2024, https://www.techtarget.com/whatis/definition/milliampere-hour-mAh.

194 "What Is an ARM Processor?," Lenovo, accessed February 22, 2025, https://www.lenovo.com/us/en/glossary/arm-

processor/?orgRef=https%253A%252F%252Fwww.google.com%252F&srsltid=AfmBOopdO1psl 4IFIRwJBeV_mrHxRO3fJRLu0143-OF5euNV1BwMN64b.

195 John D. Barrow, *The Constants of Nature* (Pantheon, 2002).

196 Martin Rees, *Just Six Numbers*, 96-97, 154-155.

197 Gribbin and Rees, *Cosmic Coincidences*, "if gravity were stronger, stars would be smaller and would run through their life cycles more quickly—perhaps so quickly that there would be no time for intelligent life to evolve on any planets orbiting those stars."

198 John Gribbin and Martin Rees, *Cosmic Coincidences: Dark Matter, Mankind, and Anthropic Cosmology* (New York: Bantam Books, 1989), "If the Universe had been expanding too rapidly, the clouds would have been spread thin and pulled apart before gravity could dominate, even on a local scale, and make them collapse into galaxies and stars."

199 "Lambda Leads the Way," Stanford News, accessed February 22, 2024, https://news.stanford.edu/2018/09/13/lambda-leads-way/: "Equally perplexing, lambda's tiny value lay just within the narrow range able to support life. If it were much larger, the universe would expand too quickly for galaxies and stars to form; much smaller, and creation would collapse back into a point."

200 Ibid.

201 Carl Sagan, *Cosmos* (Random House, 1980), "The hypersphere is expanding from a point, like a four-dimensional balloon being inflated, creating in every instant more space in the universe. Sometime after the expansion begins, galaxies condense and are carried outward on the surface of the hypersphere. There are astronomers in each galaxy, and the light they see is also trapped on the curved surface of the hypersphere. As the sphere expands, an astronomer in any galaxy will think all the other galaxies are running away from him."

202 Jason Lisle, "Exploring the Findings of the James Webb Telescope," Denver Society of Creation, YouTube video, 1:16:38-1:17:04, May 2, 2024, https://www.youtube.com/watch?v=EOx7tbSLGk0.

203 Brian Greene, "The Hidden Reality" (New York: Alfred A. Knopf, 2011), 281: "The data also allowed the researchers to fix the numerical value of the cosmological constant—the amount of dark energy suffusing space. Expressing the result in terms of an equivalent amount of mass, as is conventional among physicists (using $E = mc2$ in the less familiar form, $m = E/c2$), the researchers showed that the supernova data required a cosmological constant of just under $10^{-29}$ grams in every cubic centimeter."

204 M. Hanson, "Lecture 23: The Cosmological Constant," Department of Physics, University of Cincinnati, accessed February 22, 2024, https://homepages.uc.edu/~hansonmm/ASTRO/LECTURENOTES/W03/Lec23/Page8.html.

205 John D. Barrow, *The Constants of Nature* (Vintage Books, 2003).

206 Gribbin and Rees, *Cosmic Coincidences*, "The Universe must have expanded, and be expanding, neither too fast nor too slow, but at just the 'right' rate to allow elements to be cooked in stars."

207 Ibid.

208 Rees, *Just Six Numbers*.

209 Ibid.

210 Rees, *Just Six Numbers*, 76, 90.

211 "Transmission Control Protocol/Internet Protocol," in IBM Knowledge Center, IBM, accessed February 22, 2024, https://www.ibm.com/docs/en/aix/7.2?topic=management-transmission-control-protocolinternet-protocol.

212 "Introduction to TCP/IP," IONOS Digitalguide, https://www.ionos.com/digitalguide/server/know-how/introduction-to-tcpip/: "TCIP/IP forms the backbone of the Internet: Without these protocols, we wouldn't be able to surf the Web."

213 John F. Donoghue, "Random values of the cosmological constant," *Journal of High Energy Physics* JHEP08 (2000) 022 (Sep. 18, 2000), doi: 10.1088/1126-6708/2000/08/022, 1. "Donoghue discusses the potential for different values of the cosmological constant ($\Lambda$) in various domains of the universe, stating: 'The cosmological constant is a parameter that can take on different values in different domains of the universe, with an assumed cosmological evolution such that we live entirely within a single domain.' This observation parallels the concept of distinct network domains interconnected by the TCP/IP protocol, emphasizing the importance of standards or parameters in governing the behavior and functionality of complex systems, whether in the vast expanse of the cosmos or within the interconnected network of the internet."

214 "Network Bridge Explained," Study-CCNA, accessed April 10, 2024, https://study-ccna.com/network-bridge-explained/. This source provides a detailed explanation of the concept of network bridging in computer networking. It outlines how network bridges function to connect multiple network segments, facilitating communication and data transfer as though they were a single network. The article delves into the technical aspects of how bridges use MAC addresses to forward traffic between different segments.

215 V. M. Mostepanenko and V. M. Frolov, eds., *Cosmological Constant: The Weight of the Vacuum* (Springer, 2017).

216 Barrow., and Frank J. Tipler. *The Anthropic Cosmological Principle*.

217 NASA, *Universe 101: What's Driving the Expansion?*, accessed February 22, 2025, https://map.gsfc.nasa.gov/universe/uni_accel.html. "Einstein first proposed the cosmological constant (not to be confused with the Hubble Constant) usually symbolized by the Greek letter 'lambda' ($\Lambda$), as a mathematical fix to the theory of general relativity. In its simplest form, general relativity predicted that the universe must either expand or contract. Einstein thought the universe was static, so he added this new term to stop the expansion."

218 Ibid., "Very recently it has become practical for astronomers to observe very bright rare stars called supernova in an effort to measure how much the universal expansion has slowed over the last few billion years. Surprisingly, the results of these observations indicate that the universal expansion is speeding up, or accelerating! While these results should be considered preliminary, they raise the possibility that the universe contains a bizarre form of matter or energy that is, in effect, gravitationally repulsive. The cosmological constant is an example of this type of energy."

219 Julian De Vuyst, "A Natural Introduction to Fine-Tuning," *Department of Physics & Astronomy*, Ghent University, December 10, 2020, https://arxiv.org/html/2012.05617v2.

220 Ibid.

221 Martin Rees, "New Page," accessed April 10, 2024, https://www.martinrees.uk/new-page.

222 Rees, *Just Six Numbers*.

223 Ibid.

224 Ibid.

225 Buckminster Fuller first introduced the concept of the "Knowledge Doubling Curve," suggesting that human knowledge doubled every century by 1900 and every 25 years by 1945. More recent estimates by IBM and industry analysts suggest certain domains, such as medical knowledge, have reached doubling rates as fast as every 73 days. See IBM, "What will we make of this moment?" *2013 IBM Annual Report* (2013), https://www.ibm.com/annualreport/2013/bin/assets/2013_ibm_annual.pdf?utm_source; and Buckminster Fuller, *Critical Path* (New York: St. Martin's Press, 1981).

226 Vern Bender, "The Universe's Operating System Runs from the Quantum Level," Vern Bender Blog, July 19, 2022, https://vernbender.com/28483-2/.

227 Internet of Things (IoT) - Oracle," accessed February 25, 2024, https://www.oracle.com/internet-of-things/what-is-iot/.

228 IBM, "Internet of Things," accessed May 4, 2024, https://www.ibm.com/topics/internet-of-things.

229 Ibid.

230 Ibid.

231 Ibid.

232 IBM, "Internet of Things," https://www.ibm.com/topics/internet-of-things. The article outlines several benefits of IoT for businesses, including improved efficiency through automation and optimized processes, which enhance productivity and reduce downtime; data-driven decision-making, where the vast data collected by IoT devices facilitates better-informed strategic decisions based on insights into customer behavior, market trends, and operational performance; and significant cost savings through reduced manual processes and enhanced sustainability practices.

233 Ibid.

234 "Delivery to Consumers," U.S. Energy Information Administration (EIA)," accessed February 25, 2024, https://www.eia.gov/energyexplained/electricity/delivery-to-consumers.php.

235 "Once stepped up to the appropriate voltage, the power is then placed on the transmission system which consists of lines and poles… Once the power reaches its delivery point, it goes through a step-down (or reduction of voltage) process at switching stations," in "How Power is Delivered to Your Home," Central Alabama Electric Cooperative, accessed February 25, 2024, https://caec.coop/electric-service/how-power-is-delivered-to-your-home/.

236 "The Big Misconception About Electricity," Technology.org, November 20, 2021, https://www.technology.org/2021/11/20/the-big-misconception-about-electricity/.

237 Ibid.

238 "Understanding the Role of Transformer Substations in Power Distribution," UTB Transformers, May 15, 2024, https://utbtransformers.com/understanding-the-role-of-transformer-substations-in-power-distribution/.

239 Ibid.

240 Edraw Content Team, *Network Topology Examples & Templates*, Edraw Software, accessed March 22, 2025, https://www.edrawsoft.com/topology-diagram-example.html.

241 M. D. Fricker, L. L. M. Heaton, N. S. Jones, and L. Boddy, "The Mycelium as a Network," *Microbiology Spectrum* 5, no. 3 (2017), https://doi.org/10.1128/microbiolspec.FUNK-0033-2017.

242 "Network Topologies, Protocols and Layers – AQA," *BBC Bitesize*, accessed March 22, 2025, https://www.bbc.co.uk/bitesize/guides/z666pbk/revision/2.

243 "Cosmic Microwave Background," Center for Astrophysics, Harvard & Smithsonian, accessed February 25, 2024, https://www.cfa.harvard.edu/research/topic/cosmic-microwave-background.

244 "9.1. Base Load Energy Sustainability," *EME 807: Technologies for Sustainability Systems*, Penn State University, accessed March 22, 2025, https://www.e-education.psu.edu/eme807/node/667.

245 "It would seem an awful waste of space. To design a whole universe requiring over 100 billion galaxies, each containing 100 billion stars lasting over 13 billion years, just to allow the evolution of one species on one planet less than a million years ago, seems like a remarkably inefficient design," in Lawrence Krauss, "Cosmology without Design," Inference Review, accessed February 25, 2024, https://inference-review.com/article/cosmology-without-design.

246 Martin Greenwald, "How the Universe Got Its Magnetic Field," *MIT News*, May 25, 2022, https://news.mit.edu/2022/how-universe-got-its-magnetic-field-0525.

247 Verschuur, *The Invisible Universe*, 109: "There will be magnetic fields that are literally tied to the black hole because some of the gas will have plunged into the hole and dragged the magnetic fields with it. Those magnetic fields at first remain connected to the gas outside and will rapidly wind up."

248 "How Maglev Trains Work," HowStuffWorks, accessed February 22, 2024, https://science.howstuffworks.com/transport/engines-equipment/maglev-train.htm.

249 Ibid.

250 Martin Greenwald, "How the Universe Got Its Magnetic Field," *MIT News*, May 25, 2022, https://news.mit.edu/2022/how-universe-got-its-magnetic-field-0525.

251 P. Heinzel and U. Anzer, "Prominence fine structures in a magnetic equilibrium: Two-dimensional models with multilevel radiative transfer," *Astronomy & Astrophysics* 375 (2001): 1082–1090, https://doi.org/10.1051/0004-6361:20010926.

252 "Magnetic Beads: A Simple Guide," Cytiva Life Sciences, accessed May 30, 2024, https://www.cytivalifesciences.com/en/us/news-center/magnetic-beads-a-simple-guide-10001.

253 Venkataragavalu Sivagnanam, "Microfluidic Immunoassays Based on Self-Assembled Magnetic Bead Patterns and Time-Resolved Luminescence Detection" (PhD diss., Lausanne: EPFL, 2010), 8-9, https://doi.org/10.5075/epfl-thesis-4644.

254 Ibid.

255 Ibid., 111.

256 Ibid., 100.

257 Cambridge Dictionary, s.v. "system," accessed September 26, 2024, https://dictionary.cambridge.org/us/dictionary/english/system.

258 NASA, The Balance of Power in the Earth-Sun System, accessed October 5, 2024, https://www.nasa.gov/wp-content/uploads/2015/03/135642main_balance_trifold21.pdf.

259 Edward Dolnick, *The Clockwork Universe: Isaac Newton, the Royal Society, and the Birth of the Modern World* (New York: HarperCollins, 2011).

260 Bernard Schutz, "Gravity from the Ground Up: An Introductory Guide to Gravity and General Relativity" (Cambridge: Cambridge University Press, 2003), 9–10.

261 "Aristotle and the Science of Motion," Rutgers University, accessed February 22, 2025, https://www.physics.rutgers.edu/~croft/aristotle.htm.

262 For an in-depth exploration of Venus, including its orbit, appearance in our skies, the eight-year cycle, its significance in mythology, and its portrayal in art and literature, see: Guy Ottewell, Venus, a Longer View (Universal Workshop, 2020), accessed October 5, 2024, https://www.amazon.com/Venus-longer-view-Guy-Ottewell/dp/0934546819#detailBullets_feature_div.

263 A. C. Correia and J. Laskar, "Mercury's Capture into the 3/2 Spin-Orbit Resonance as a Result of Its Chaotic Dynamics," *Nature* 429, no. 6994 (June 24, 2004): 848–50, https://doi.org/10.1038/nature02609.

264 "Kepler did not believe this 'music' to be audible, but felt that it could nevertheless be heard by the soul," Wikipedia, s.v. "Musica universalis," last modified September 11, 2024, https://en.wikipedia.org/wiki/Musica_universalis.

265 National Museum of the American Indian, "Maya Skies," accessed October 5, 2024, https://americanindian.si.edu/nk360/informational/maya-skies.

266 OpenStax, "Formation of the Solar System," in Astronomy 2e, accessed October 5, 2024, https://openstax.org/books/astronomy-2e/pages/14-3-formation-of-the-solar-system.

267 OpenStax, "Formation of the Solar System," in Astronomy 2e, accessed October 5, 2024, https://openstax.org/books/astronomy-2e/pages/14-3-formation-of-the-solar-system.

268 OpenStax, "Deeply Branching Bacteria," in Microbiology, accessed October 5, 2024, https://openstax.org/books/microbiology/pages/4-5-deeply-branching-bacteria.

269 "Last Universal Common Ancestor," Wikipedia, last modified January 5, 2025, https://en.wikipedia.org/wiki/Last_universal_common_ancestor.

270 OpenStax, "Star Formation," in Astronomy 2e, accessed October 5, 2024, https://openstax.org/books/astronomy-2e/pages/21-1-star-formation.

271 OpenStax, "The Formation of the Galaxy," in Astronomy 2e, accessed October 5, 2024, https://openstax.org/books/astronomy-2e/pages/25-6-the-formation-of-the-galaxy.

272 Ibid.

273 NASA, "Our Sun: Facts," NASA Science Solar System Exploration, accessed September 27, 2024, https://science.nasa.gov/sun/facts/.

274 Philip J. Armitage explains, "Gravitational instability is thus likely to occur early, when protoplanetary disks are still massive." He discusses the need for "external forces" like "irradiation" or other factors to influence disk dynamics, supporting the argument that gravitational collapse requires more complex forces (Philip J. Armitage, *Dynamics of Protoplanetary Disks, Annual Review of Astronomy and Astrophysics* 49 (2011): 8-9, 34, https://doi.org/10.48550/arXiv.1011.1496).

275 Piero Madau and Mark Dickinson, "Cosmic Star-Formation History," *Annual Review of Astronomy and Astrophysics* 52 (2014): 3, https://doi.org/10.48550/arXiv.1403.0007. The authors explain that "the key idea of standard cosmological scenarios is that primordial density fluctuations grow by gravitational instability driven by cold, collisionless dark matter, leading to a 'bottom-up'" formation process. However, they do not adequately address how the denser regions of gas first emerged, reflecting an assumption embedded within the standard model.

276 Daniele Sorini, John A. Peacock, and Lucas Lombriser, "The Impact of the Cosmological Constant on Past and Future Star Formation," Monthly Notices of the Royal Astronomical Society, preprint, November 13, 2024, https://arxiv.org/abs/2411.07301.

277 University of Oregon, "Chapter 19: The Formation of Planetary Systems," accessed October 5, 2024, https://pages.uoregon.edu/jimbrau/astr122/Notes/Chapter19.html.

278 Harri A. T. Vanhala and Alan P. Boss, "Injection of Radioactivities into the Forming Solar System," (Department of Terrestrial Magnetism, Carnegie Institution of Washington, 2001), https://doi.org/10.48550/arXiv.astro-ph/0111180.

279 Highline College, "Formation of the Solar System," accessed October 5, 2024, https://people.highline.edu/iglozman/classes/astronotes/solsys_form.htm.

280 Telemachos Ch. Mouschovias, "Magnetic Braking, Ambipolar Diffusion, Cloud Cores, and Star Formation: Natural Length Scales and Protostellar Masses," The Astrophysical Journal 373 (May 20, 1991): 169-186, https://ui.adsabs.harvard.edu/abs/1991ApJ...373..169M/abstract.

281 Ibid.

282 Ibid.

283 Ibid.

284 NASA, "How Do Planets Form?" https://science.nasa.gov/exoplanets/how-do-planets-form/.

285 Smithsonian Center for Astrophysics, "Intergalactic Medium," accessed September 29, 2024, https://pweb.cfa.harvard.edu/research/topic/intergalactic-medium.

286 John Chambers, "Terrestrial Planet Formation, Carnegie Institution for Science," accessed October 5, 2024, https://websites.pmc.ucsc.edu/~fnimmo/eart290q_11/chambers_exoplanets.pdf, 2.

287 Tuition Physics, "4 Examples of Static Electricity in Our Daily Lives," January 2022, https://tuitionphysics.com/jan-2022/4-examples-of-static-electricity-in-our-daily-lives/.

288 John Chambers, "Terrestrial Planet Formation" (Washington, DC: Carnegie Institution for Science), 5. "Numerical simulations show that in the absence of turbulence, μm-sized dust grains aggregate into m-sized bodies in ∼104 orbital periods for plausible sticking probabilities. Growth mainly occurs when large aggregates sweep up small ones. Without turbulence, similarly-sized objects have low relative velocities, so mutual collisions are not disruptive, even though they may not lead to growth."

289 Philip J. Armitage notes, "The conditions for planetesimal formation are sensitive to the level of intrinsic disk turbulence." He indicates that turbulence could either help or hinder the clumping process and suggests that "dead zones" may play a role in particle accumulation, but adds that these processes are still not fully understood (Philip J. Armitage, Dynamics of Protoplanetary Disks, Annual Review of Astronomy and Astrophysics 49 (2011): 195-236, https://doi.org/10.48550/arXiv.1011.1496).

290 John Chambers, "Terrestrial Planet Formation," 2.

291 Anders Johansen et al., "A Pebble Accretion Model for the Formation of the Terrestrial Planets in the Solar System."

292 John Chambers, "Terrestrial Planet Formation."

293 Ibid., 4. "In a turbulent nebula, meter-sized objects collide with one another at substantial speeds, probably leading to fragmentation rather than growth. Meter-sized particles also have short drift lifetimes, severely limiting the amount of time available to grow into larger objects. These difficulties suggest that growth may stall when objects reach ~ 1 m in diameter, a problem referred to as the 'meter-size barrier'"

294 While the term "pebbles" typically refers to centimeter to decimeter scale objects in cosmological studies, some scientists use the term to describe much larger objects, up to the size of meters. This usage, though somewhat logical given the vast scale of cosmic structures, may lead to confusion or misinterpretation of the barriers in planetary formation theories, such as the critical meter size barrier for accretion processes.

295 John Chambers, Terrestrial Planet Formation.

296 American Astronomical Society, *Extreme Solar Systems 4*, id. 317.18, *Bulletin of the American Astronomical Society* 51, no. 6 (August 2019). "The meter-size barrier is a persistent problem in current planet formation models, where particles on the order of a meter in size fail to grow because they either fragment or drift into the star."

297 Ibid., 5.

298 Ibid., "Although it is currently unclear how planetesimals formed, it is assumed they did so in large numbers at an early stage in the solar nebula."

299 Ibid., 5-10.

300 Ibid., "Although it is currently unclear how planetesimals formed, it is assumed they did so in large numbers at an early stage in the solar nebula."

301 Ibid., 10. "The Moon probably formed as a result of an oblique impact between a Mars-mass embryo and Earth towards the end of its accretion. Numerical simulations show that if the two bodies were already differentiated, their iron cores would coalesce;" 12, "Planetary embryos the size of Mars or larger are likely to melt and differentiate as a result of the kinetic energy released during impacts with other large bodies (Tonks and Melosh 1992). Iron and other siderophile elements sink to the centre of these bodies, leaving a silicate-mantle surrounding a metal-rich core."

302 Ibid.

303 Stephen A. Nelson, "Meteorites, Impacts, and Mass Extinction," Tulane University, last updated April 27, 2018, https://www2.tulane.edu/~sanelson/Natural_Disasters/impacts.htm.

304 Ibid.

305 Ibid.

306 NASA, "Our Sun: Facts."

307 Note: Despite the detailed discussions on processes such as gravitational instability and external triggers like supernova shockwaves, none of the papers I reviewed—whether Madau and Dickinson (2014), Kuiper (1951), or Vanhala and Boss (2002)—address the first cause of collapse in the early universe. Each assumes pre-existing conditions, such as denser regions of gas or external forces, without explaining the origin of these initial states. This reliance on speculative initial conditions reflects the broader conflict in mainstream cosmology to account for the first triggers of star formation, undermining the theory that builds upon this foundational assumption.

308 Ibid.

309 Brian Cox, *Wonders of the Solar System*, 110–111. "The Pillars of Creation exemplify how stars form in these vast clouds, with sprinkles of heavier elements scattered throughout," indicating the ongoing process of stellar formation from enriched interstellar matter.

310 Lawrence Wright, "Orwell on Writing: 'Clarity Is the Remedy,'" *NPR*, September 22, 2006, https://www.npr.org/2006/09/22/6124822/orwell-on-writing-clarity-is-the-remedy.

311 Kelly E. Miller et al., "The Composition of Saturn's Rings," arXiv, November 29, 2023, https://doi.org/10.48550/arXiv.2311.17344, 6.

312 Ibid., 7.

313 Ibid., "The Composition of Saturn's Rings," 6. Following this reference, I argue that the concept of quasi-integrals of motion provides a significant theoretical framework for understanding the long-term dynamics and stability of Saturn's rings. Despite the apparent chaos and complexity of the rings, characterized by varying densities, particle sizes, and external influences, these quasi-integrals help maintain a certain degree of order and predictability. They function as nearly conserved quantities that, despite slight variations due to interactions within the rings or with external bodies, manage to sustain the overall structure and behavior of the rings over extended periods.

314 S. F. Singer, "How Did Venus Lose Its Angular Momentum?," *Science* 170, no. 3963 (1970): 1196-1198, https://doi.org/10.1126/science.170.3963.1196.

315 Scafetta, Nicola. "The complex planetary synchronization structure of the solar system." *arXiv* preprint (2014), https://arxiv.org/abs/1405.0193.

316 Renu Malhotra, "Orbital Resonances and Chaos in the Solar System," in *Solar System Formation and Evolution*, ed. D. Lazzaro et al. (ASP Conference Series, Vol. 149, 1998), https://www.lpl.arizona.edu/~renu/malhotra_preprints/rio97.pdf.

317 Giuseppe Pucacco, "Dynamical Stability of the Laplace Resonance," *arXiv*:2410.12768v1 [astro-ph.EP], October 16, 2024, https://arxiv.org/abs/2410.12768.

318 J. J. O'Connor and E. F. Robertson, "Pierre-Simon Laplace," MacTutor History of Mathematics Archive, last updated January 1999, https://mathshistory.st-andrews.ac.uk/Biographies/Laplace/.

319 Laura Canil and Michael Schirber, "Why Is the Solar System So Stable?" *Observatoire de Paris - PSL*, published May 4, 2023, last updated April 3, 2024, https://www.observatoiredeparis.psl.eu/pourquoi-le-systeme-solaire.html?lang=en.

320 Ibid.

321 Ibid.

322 "These symmetries are broken only by weak resonances, leading to the existence of quasi-integrals of motion that are shown to relate to the smallest Lyapunov exponents. Strong evidence emerges that they effectively constrain the chaotic diffusion of the orbits, playing a crucial role in the statistical stability over the Solar System lifetime," Federico Mogavero, Nam H. Hoang, and Jacques Laskar, "Timescales of Chaos in the Inner Solar System: Lyapunov Spectrum and Quasi-Integrals of Motion," Physical Review X 13, no. 2 (2023): 1, https://doi.org/10.1103/PhysRevX.13.021018.

323 Federico Mogavero, Nam H. Hoang, and Jacques Laskar, "Timescales of Chaos in the Inner Solar System: Lyapunov Spectrum and Quasi-Integrals of Motion," *Physical Review X* 13, no. 2 (2023), 17, https://doi.org/10.1103/PhysRevX.13.021018.

324 Ibid., 2-17

325 Ibid.

326 "Quasi-local 'conserved quantities' (i.e., in the situation when the gravitating system has a non-vanishing energy flux)... Applications include energy-momentum and angular momentum at spatial and null infinity, asymptotically anti-deSitter spacetimes, and thermodynamics of the isolated horizons." Roh S. Tung, "Quasi-Local Conserved Quantities," in The Ninth Marcel Grossmann Meeting: On Recent Developments in Theoretical and Experimental General Relativity, Gravitation and Relativistic Field Theories (Singapore: World Scientific, 2002), 1597–1598.

327 Sabine Hossenfelder, Sabine Hossenfelder: Physicist & Author, accessed October 6, 2024, https://sabinehossenfelder.com/.

328 Sabine Hossenfelder, YouTube Channel, accessed October 6, 2024, https://www.youtube.com/channel/UC1yNl2E66ZzKApQdRuTQ4tw.

329 Sabine Hossenfelder, "This is Why Physics is Dying," YouTube video, 6:22–8:30, October 5, 2024, https://www.youtube.com/watch?v=cBIvSGLkwJY&t=336s.

330 Christopher Sykes, "HD Feynman: FUN TO IMAGINE complete (with optional Chinese subtitles)", YouTube, August 24, 2020, https://www.youtube.com/watch?v=nYg6jzotiAc, 55:04–55:55.

331 Zygmunt Kowalik and John L. Luick, *Modern Theory and Practice of Tide Analysis and Tidal Power* (Austides Consulting, n.d.), 8–9, https://uaf.edu/cfos/files/research-projects/people/kowalik/Book2019_tides.pdf.

332 Ibid.

333 Ibid.

334 OpenStax, Astronomy 2e (Houston: OpenStax, 2016), https://openstax.org/books/astronomy-2e/pages/9-4-the-origin-of-the-moon.

335 D.U. Wise, "Origin of the Moon by Fission," in The Earth-Moon System, ed. B.G. Marsden and A.G.W. Cameron (Boston, MA: Springer, 1966), 289-311, https://doi.org/10.1007/978-1-4684-8401-4_14.

336 Ibid.

337 Alice Friedemann, "Energy Skeptic," Energy Skeptic, accessed July 2, 2024, https://energyskeptic.com/page/2/.

338 Ibid.

339 "Hypothesis for Moon Formation," HyperPhysics, accessed February 26, 2024, http://hyperphysics.phy-astr.gsu.edu/hbase/Astro/moonform.html.

340 "Question 38," StarChild, NASA, accessed February 26, 2024, https://starchild.gsfc.nasa.gov/docs/StarChild/questions/question38.html.

341 Ibid.

342 Hartmann, William K., Ron Miller, and Pamela G. Conrad. "The Origin of the Earth and Moon." In The Grand Tour: A Traveler's Guide to the Solar System, 3rd ed., 102-136. New York: Copernicus Books, 2005.

343 Ibid.

344 "How does the Moon affect life on Earth?" Natural History Museum, accessed February 26, 2024, https://www.nhm.ac.uk/discover/how-does-the-moon-affect-life-on-earth.html.

345 "The Moon's Orbit and Rotation," NASA, accessed February 26, 2024, https://moon.nasa.gov/resources/429/the-moons-orbit-and-rotation/.

346 "Why Does One Side of the Moon Always Face Earth?" New Space Economy, February 16, 2025, https://newspaceeconomy.ca/2025/02/16/one-side-of-the-moon-always-faces-earth/.

347 "How Does the Moon Affect Tides and Weather?" The Old Farmer's Almanac, accessed February 26, 2024, https://www.almanac.com/how-does-moon-affect-tides-and-weather.

348 JoEllen McBride, "Stirring the Ocean: How the Moon Mixes Things Up Beneath the Waves," The Pipette Pen, June 23, 2015, accessed February 26, 2024, https://www.thepipettepen.com/stirring-the-ocean-how-the-moon-mixes-things-up-beneath-the-waves/. Note: This article describes the role of internal waves in the ocean, generated by tidal forces, in mixing nutrient-rich water and trapping CO2 at deeper levels. These processes, compared to stirring creamer in coffee, are crucial for marine ecosystems and the global carbon cycle. The study's computational models offer insights into how the Moon's gravitational influence extends far beyond surface tides, deeply impacting marine life and carbon storage in the ocean depths.

349 Robert Lea, "Understanding Geosynchronous Orbit," Space.com, accessed February 26, 2024, https://www.space.com/29222-geosynchronous-orbit.html.

350 Ibid.

351 Ibid.

352 "Webb's Sunshield," NASA Science, accessed February 23, 2025, https://science.nasa.gov/mission/webb/webbs-sunshield/.

353 Kirill Vankov and Anatoli Vankov, *Dark Matter in Galaxies and Lambda-CDM Universe* (2023), accessed March 16, 2025, https://hal.science/hal-04055449v4/file/hal-04055449v4.pdf.

354 Rees, *Just Six Numbers*, 73-74.

355 NASA Science. "Mapping the Cosmic Web." *Hubble Space Telescope Science Highlights*. Accessed March 24, 2025. https://science.nasa.gov/mission/hubble/science/science-highlights/mapping-the-cosmic-web/.

356 "Anonymous Functions," PHP Manual, accessed February 22, 2024, https://www.php.net/manual/en/functions.anonymous.php.

357 "Alonzo Church," The Stanford Encyclopedia of Philosophy (Spring 2021 Edition), ed. Edward N. Zalta, accessed February 22, 2024, https://plato.stanford.edu/entries/church/.

358 "Lambda Calculus in Functional Programming Paradigm," FasterCapital, accessed February 22, 2024, https://fastercapital.com/topics/lambda-calculus-in-functional-programming-paradigm.html.

359 Rees, *Just Six Numbers* 73-83.

360 Rees, *Just Six Numbers*, 74-76. The book emphasizes that without dark matter, galaxies would not be stable but would disintegrate. Galaxies are effectively held together by dark matter, making them much larger and heavier than previously thought. This is also true for entire clusters of galaxies (Page 75). Dark matter, though not directly observable, exerts a significant gravitational pull, influencing the structure and behavior of visible matter in galaxies and galaxy clusters (Page 76).

361 Ibid.

362 I. Ciufolini et al., "An Improved Test of the General Relativistic Effect of Frame-Dragging Using the LARES and LAGEOS Satellites," *European Physical Journal C* 79 (2019), https://doi.org/10.1140/epjc/s10052-019-7386-z.

363 Julio F. Navarro, Carlos S. Frenk, and Simon D. M. White, "The Structure of Cold Dark Matter Halos," *Astrophysical Journal* 462 (1996): 563, https://doi.org/10.1086/177173.

364 *The Invisible Man* (1933), *IMDb*, accessed February 22, 2024, https://www.imdb.com/title/tt0024184/.

365 Rees, *Just Six Numbers*, 74-76.

366 Brian Greene, *The Hidden Reality*.

367 Yannick Joye and Jan Willem Bolderdijk, "An Exploratory Study into the Effects of Extraordinary Nature on Emotions, Mood, and Prosociality," *Frontiers in Psychology* 5 (2015): 1577, https://doi.org/10.3389/fpsyg.2014.01577.

368 A.P.M. Baede, E. Ahlonsou, Y. Ding, and D. Schimel, "The Climate System: an Overview," in Climate Change 2001: The Scientific Basis, Contribution of Working Group I to the Third Assessment Report of the Intergovernmental Panel on Climate Change, eds. J.T. Houghton et al. (Cambridge, UK: Cambridge University Press, 2001), accessed March 11, 2024, https://www.ipcc.ch/site/assets/uploads/2018/03/TAR-01.pdf. This chapter from the IPCC's Third Assessment Report provides a comprehensive overview of the Earth's climate system, including its various components and the interactions among them.

369 Ibid.

370 NASA Science Space Place, 'Weather on Other Planets,' accessed March 10, 2024, https://spaceplace.nasa.gov/weather-on-other-planets/en/. This webpage provides information on the various weather phenomena observed on different planets in our solar system.

371 Matt Williams, "Mars compared to Earth," Phys.org, December 7, 2015, https://phys.org/news/2015-12-mars-earth.html. This article provides a comparative analysis of Mars and Earth, discussing their environmental, geological, and atmospheric differences and similarities.

372 "Mass Movement," Interactive Mapping of Mars Tutorial, University of Western Ontario, accessed March 12, 2024, https://imars.uwo.ca/tutorial/mass-movement.

373 Alexander T. Basilevsky and James W. Head, "The Surface of Venus," *Reports on Progress in Physics* 66, no. 10 (2003): 1699, https://doi.org/10.1088/0034-4885/66/10/R04.

374 Ibid.

375 "Hubble Tracks Jupiter's Stormy Weather," HubbleSite News Releases, March 14, 2024, Release ID: 2024-009, https://hubblesite.org/contents/news-releases/2024/news-2024-009. This release details the Hubble Space Telescope's latest observations of Jupiter, capturing dynamic weather patterns and storms. It emphasizes the Great Red Spot and other atmospheric phenomena.

376 "Hubble Tracks Jupiter's Stormy Weather," HubbleSite News Releases, March 14, 2024, Release ID: 2024-009, https://hubblesite.org/contents/news-releases/2024/news-2024-009.

377 Ibid.

378 Ibid.

379 "Mercury: Facts," NASA Science, accessed March 12, 2024, https://science.nasa.gov/mercury/facts.

380 "Saturn: Facts," NASA Science, accessed March 12, 2024, https://science.nasa.gov/saturn/facts/.

381 Deborah Byrd, "Seasons of Uranus, a Sideways World," EarthSky, accessed March 12, 2024, https://earthsky.org/space/seasons-of-uranus-strange-sideways-world.

382 Patrick Pester, "Neptune: The farthest planet from our sun," Live Science, May 24, 2022, https://www.livescience.com/neptune.

383 "Global Atmospheric Circulation," NOAA Jetstream, accessed March 12, 2024, https://www.noaa.gov/jetstream/global/global-atmospheric-circulations.

384 "The Water Cycle on Earth," NOAA Education, accessed March 12, 2024, https://www.noaa.gov/education/resource-collections/freshwater/water-cycle.

385 Ibid.

386 Ibid.

387 Daniel Binkley, Tom Sisk, Carol Chambers, Judy Springer, and William Block, "The Role of Old-Growth Forests in Frequent-Fire Landscapes," Ecology and Society 12, no. 2 (2007), http://www.jstor.org/stable/26267895.

388 "How Do Hurricanes Impact the Deep Ocean?," NOAA Office of Ocean Exploration and Research, accessed March 15, 2024, https://oceanexplorer.noaa.gov/facts/hurricane-impact.html. Note: The article outlines the benefits of hurricanes on deep ocean ecosystems.

389 Michael Arthur and Demian Saffer, "Nutrient Supply to Floodplains," in EARTH 111: Water: Science and Society, Pennsylvania State University, accessed March 15, 2024, https://www.e-education.psu.edu/earth111/node/823.

390 A.I. Dounis and C. Caraiscos, "Advanced Control Systems Engineering for Energy and Comfort Management in a Building Environment—A Review," Renewable and Sustainable Energy Reviews 13, nos. 6–7 (2009): 1246-1261, https://doi.org/10.1016/j.rser.2008.09.015.

391 Meera Narvekar and Priyanca Fargose, "Daily Weather Forecasting using Artificial Neural Network," D.J. Sanghvi College of Engineering, accessed March 12, 2024, https://www.e-education.psu.edu/earth111/node/823. Note: This article provides insight into the use of artificial neural networks for weather forecasting, drawing a parallel between the complexity of weather systems and networked processes. It illustrates how interconnected elements in weather systems can be modeled and analyzed similarly to a neural network, emphasizing the intricate nature of atmospheric interactions.

392 Ibid.

393 Ibid.

394 Ibid.

395 Ibid.

396 Ibid.

397 Ibid.

398 NASA Science Editorial Team, "Earth's Magnetosphere: Protecting Our Planet from Harmful Space Energy," NASA Science, August 3, 2021, https://science.nasa.gov/science-research/earth-science/earths-magnetosphere-protecting-our-planet-from-harmful-space-energy/.

399 Radiation Protection Systems, "Radiation Protection Systems," accessed February 23, 2025, https://radprosys.com/.

400 Ibid.

401 Colin Stuart, "Earth's Mysterious Core," Science Focus, August 21, 2023, accessed June 15, 2024, https://www.sciencefocus.com/planet-earth/earths-mysterious-core.

402 Yang and Song, "Multidecadal Variation of the Earth's Inner-Core Rotation," *Journal of Geophysical Research*, 2022.

403 Stuart, "Earth's Mysterious Core."

404 Julien Aubert and Prof. Christopher Finlay, "Supercomputer Simulations of Earth's Magnetic Field," Technical University of Denmark, 2019.

405 Stuart, "Earth's Mysterious Core."

406 The Editors of Encyclopaedia Britannica, "Rayleigh scattering," Encyclopedia Britannica, October 27, 2023, accessed March 2, 2024, https://www.britannica.com/science/Rayleigh-scattering.

407 Alan Buis, "The Atmosphere: Earth's Security Blanket," NASA, October 2, 2019, accessed February 26, 2024, https://climate.nasa.gov/news/2914/the-atmosphere-earths-security-blanket/.

408 Frontiers in Psychology, "Perceived Restorativeness of Visible Sky: Examining the Role of Perceived Openness and Spatial Boundaries," *Frontiers in Psychology*, October 27, 2022, https://doi.org/10.3389/fpsyg.2022.932507.

409 "Skychology: What Is It and Does It Really Work?" Health News, accessed March 2, 2024, https://healthnews.com/mental-health/self-care-and-therapy/skychology-what-is-it-and-does-it-really-work/.

410 "The Enchanting Beauty of the Sunset," Vocal Media, accessed March 2, 2024, https://vocal.media/earth/the-enchanting-beauty-of-the-sunset-zb7y0v1g.

411 "Planets' Green Lungs: The Most Fascinating Forests," BBVA OpenMind, accessed March 2, 2024, https://www.bbvaopenmind.com/en/science/environment/planets-green-lungs-most-fascinating-forests-e/.

412 "Planets' Green Lungs: The Most Fascinating Forests," BBVA OpenMind, accessed March 2, 2024, https://www.bbvaopenmind.com/en/science/environment/planets-green-lungs-most-fascinating-

forests-e/. Quote: "Even five minutes around trees or in green spaces may improve health. Think of it as a prescription with no negative side effects that's also free."

413 Ibid.

414 Ibid.

415 Ibid.

416 Miki Huynh, "Clues of Earth's Early Rise of Oxygen," NASA Astrobiology, March 05, 2019, accessed March 2, 2024, https://astrobiology.nasa.gov/nai/articles/2019/3/5/clues-of-earths-early-rise-of-oxygen/index.html.

417 S.L. Olson, E.W. Schwieterman, C.T. Reinhard, and T.W. Lyons, "Earth: Atmospheric Evolution of a Habitable Planet," arXiv preprint, arXiv:1803.05967 (2018), https://arxiv.org/abs/1803.05967.

418 "What is Ozone?," EPA, accessed July 2, 2024, https://www.epa.gov/ozone-pollution-and-your-patients-health/what-ozone

419 J.B. West, "The Strange History of Atmospheric Oxygen," *Physiological Reports* 10, no. 6 (March 2022): e15214, doi:10.14814/phy2.15214, PMID: 35347882, PMCID: PMC8960603. West notes the unique privilege of air-breathing organisms on Earth, emphasizing that humans and most other animals are the only known living creatures in the universe with an unlimited oxygen supply. This abundance is attributed to photosynthesis, a process West describes as "one of the greatest miracles of nature" for its ability to release oxygen from water using sunlight energy.

420 Alan Buis, "The Atmosphere: Earth's Security Blanket."

421 Ibid.

422 J.B. West, "The Strange History of Atmospheric Oxygen," *Physiological Reports* 10, no. 6 (March 2022): e15214, doi:10.14814/phy2.15214, PMID: 35347882, PMCID: PMC8960603. West notes the unique privilege of air-breathing organisms on Earth, emphasizing that humans and most other animals are the only known living creatures in the universe with an unlimited oxygen supply. This abundance is attributed to photosynthesis, a process West describes as "one of the greatest miracles of nature" for its ability to release oxygen from water using sunlight energy.

423 Jimmy Kwon, Jothsna Kethar, and Rajagopal Appavu, "Skin Cancer: The Ozone Layer and UV Radiation," *Journal of Student Research* 11, no. 4 (2022), https://doi.org/10.47611/jsrhs.v11i4.3836.

424 "Brooklyn College Professor Helps Researchers Prove Water Has Multiple Liquid States," CUNY, November 20, 2020, https://www1.cuny.edu/mu/forum/2020/11/20/brooklyn-college-professor-helps-researchers-prove-water-has-multiple-liquid-states/.

425 "Origin of Water on Earth," Harvard University, accessed March 2, 2024, https://courses.seas.harvard.edu/climate/eli/Courses/EPS281r/Sources/Origin-of-oceans/1-Wikipedia-Origin-of-water-on-Earth.pdf.

426 Ibid.

427 "Ocean Worlds," NASA, accessed June 15, 2024, https://www.nasa.gov/specials/ocean-worlds/

428 "How Much Natural Water Is There?" U.S. Geological Survey (USGS), accessed March 2, 2024, https://www.usgs.gov/faqs/how-much-natural-water-there.

429 "Origin of Water on Earth," Harvard University.

430 "How Much Natural Water Is There?".

431 "Comet" *Ask an Earth and Space Scientist*, Arizona State University, accessed February 20, 2025, https://askanearthspacescientist.asu.edu/comets.

432 "How Much Natural Water Is There?," U.S. Geological Survey (USGS).

433 "Earth May Have Kept Its Own Water, Rather than Getting It from Asteroids".

434 Michael Marshall, "Dinosaur-Killing Chicxulub Asteroid Formed in Solar System's Outer Reaches," *Nature*, August 15, 2024, https://www.nature.com/articles/d41586-024-02647-4.

435 C. B. Senel, P. Kaskes, O. Temel, et al., "Chicxulub Impact Winter Sustained by Fine Silicate Dust," *Nature Geoscience* 16 (2023): 1033–1040, https://doi.org/10.1038/s41561-023-01290-4.

436 "Earth May Have Kept Its Own Water, Rather than Getting It from Asteroids," Science.org, accessed March 2, 2024, https://www.science.org/content/article/earth-may-have-kept-its-own-water-rather-getting-it-asteroids. Note: This article presents findings that significantly challenge the theory that Earth's water originated from carbonaceous chondrites, as their isotopic signatures do not align with those found in Earth's water. A key quote from the article, "The carbonaceous chondrites don't really work," underscores this disparity, further weakening the hypothesis that comets or these asteroids were the primary sources of Earth's water.

437 NASA, "Comets," NASA Space Place, accessed June 15, 2024, https://spaceplace.nasa.gov/comets/en/.

438 "Earth May Have Kept Its Own Water, Rather than Getting It from Asteroids".

439 Wikipedia, s.v. "Late Heavy Bombardment," last edited May 26, 2024, https://en.wikipedia.org/wiki/Late_Heavy_Bombardment.

440 "Impacts," Lunar and Planetary Institute, accessed March 2, 2024, https://www.lpi.usra.edu/publications/slidesets/impacts/. Note: The page discusses the effects of large impact events, including the generation of extreme pressures and temperatures that can vaporize meteorites or melt them, mixing them with the target rocks.

441 Ibid.

442 "Earlier Water on Earth? Oldest Rock Suggests Hospitable Young Planet," National Science Foundation, January 10, 2001, accessed March 2, 2024, https://www.nsf.gov/od/lpa/news/press/01/pr0102.htm.

443 Ibid.

444 "Ocean Worlds," NASA.

445 Denton, Michael. *The Wonder of Water: Water's Profound Fitness for Life on Earth and Mankind*. Illustrated edition. Discovery Institute Press, 2017. ISBN 1936599473, 9781936599479.

446 "Goldilocks Zone," NASA Exoplanet Exploration, last updated December 15, 2022, accessed February 26, 2024, https://exoplanets.nasa.gov/resources/323/goldilocks-zone/.

447 Thomas Henry Huxley, *Autobiography*, accessed January 26, 2025, http://aleph0.clarku.edu/huxley/CE1/AutoB.html.

448 "Early studies showed the psychological effects of interpersonal touch (e.g.7), but recent studies have started to assess its physiological effects10. For example, such positive physical contact as hugging and massages from partners reduce cortisol, increase oxytocin, and lower systolic blood pressure in stressful situations." Sumioka H, Nakae A, Kanai R, Ishiguro H. "Huggable communication medium decreases cortisol levels." Scientific Reports. 2013 Oct 23;3:3034. doi: 10.1038/srep03034. PMID: 24150186; PMCID: PMC3805974.

449 "Excavator Manipulator Arm," Pearson Engineering, accessed April 3, 2023, https://www.pearson-eng.com/product/excavator-manipulator-arm/.

450 Kyung Myun Moon, Jinsu Kim, Young Seong, Byung-Chang Suh, Kyongseok Kang, Hyungbae Kwon Choe, and Kiyoung Kim, "Proprioception, the Regulator of Motor Function," *BMB Reports* 54, no. 8 (August 2021): 393-402, https://doi.org/10.5483/BMBRep.2021.54.8.052.

451 Leah Schwed, "Learning the Five Senses," The Spark Innovations, January 23, 2023, https://thesparkinnovations.com/blogs/blog/learning-the-five-senses.

452 John L. Sobiesk and Sunil Munakomi, "Anatomy, Head and Neck, Nasal Cavity," in *StatPearls* (Treasure Island (FL): StatPearls Publishing, 2024), last updated July 24, 2023, https://www.ncbi.nlm.nih.gov/books/NBK544232/.

453 Ibid.

454 Ibid.

455 Ibid.

456 Nicola A. Britton, Niels A. J. Blom, and Alessandro Rizzo, "The Immunological Functions of the Human Nasal Mucosa," *Nature Reviews Immunology* 20, no. 10 (2020): 579-593, https://www.nature.com/articles/s41385-020-00359-2.

457 Ibid.

458 Ibid.

459 "What is a HEPA filter?" U.S. Environmental Protection Agency, last updated March 5, 2024, https://www.epa.gov/indoor-air-quality-iaq/what-hepa-filter.

460 Sobiesk and Munakomi, "Anatomy, Head and Neck, Nasal Cavity."

461 Ibid.

462 Ibid.

463 Ibid.

464 Ibid.

465 "The Nasal Cavity and Paranasal Sinuses," Canadian Cancer Society, last reviewed January 2020, https://cancer.ca/en/cancer-information/cancer-types/nasal-and-paranasal-sinus/what-is-nasal-cavity-and-paranasal-sinus-cancer/the-nasal-cavity-and-paranasal-sinuses.

466 Liron Oren, Amelie Kummer, and Stephen Boyce, "Understanding Nasal Emission During Speech Production: A Review of Types, Terminology, and Causality," *Cleft Palate Craniofacial Journal* 57, no. 1 (January 2020): 123-126, doi: 10.1177/1055665619858873, published online July 1, 2019, PMID: 31262198, PMCID: PMC9153061.

467 Ibid.

468 Ibid.

469 Ibid.

470 "Breathing Part 2," On The Go Physical Therapy, last modified October 12, 2019, https://www.onthegophysicaltherapy.com/blog/2019/10/12/breathing-part-2.

471 "Ear Infections and Hearing Loss," Mayo Clinic, accessed May 25, 2024, https://www.mayoclinic.org/diseases-conditions/hearing-loss/in-depth/ear-infections/art-20546801.

472 "Ear Anatomy: Outer Ear," University of Texas Health Science Center at Houston, accessed August 12, 2024, https://med.uth.edu/orl/online-ear-disease-photo-book/chapter-3-ear-anatomy/ear-anatomy-outer-ear/.

473 Ibid.

474 Ibid.

475 "The Power of Hearing," Physics World, May 2, 2002, https://physicsworld.com/a/the-power-of-hearing/.

476 Ibid.

477 Ibid.

478 Ibid.

479 Ibid.

480 Joseph Casale, Taylor Browne, IV Murray, et al., "Physiology, Vestibular System," in *StatPearls* (Treasure Island (FL): StatPearls Publishing, 2024), last updated May 1, 2023, https://www.ncbi.nlm.nih.gov/books/NBK532978/.

481 Ibid.

482 Ibid.

483 Ibid.

484 Ibid.

485 Casale, Browne, Murray, et al., "Physiology, Vestibular System."

486 Ibid.

487 Ibid.

488 Ibid.

489 Adelbert W. Bronkhorst, "The Cocktail-Party Problem Revisited: Early Processing and Selection of Multi-Talker Speech," *Attention, Perception, & Psychophysics* 77, no. 5 (July 2015): 1465-1487, doi: 10.3758/s13414-015-0882-9, PMID: 25828463, PMCID: PMC4469089.

490 Ibid.

491 U. Saxena, B. P. Singh, S. B. R. Kumar, G. Chacko, and K. N. S. V. Bharath, "Acoustic Reflexes in Individuals Having Hyperacusis of the Auditory Origin," *Indian Journal of Otolaryngology and Head & Neck Surgery* 72, no. 4 (2020): 497-502, doi:https://doi.org/10.1007/s12070-020-02002-9.

492 "What Is Speech Recognition?" IBM, accessed May 25, 2024, https://www.ibm.com/topics/speech-recognition.

493 "Acoustic Reflex," Wikipedia, last modified March 8, 2024, https://en.wikipedia.org/wiki/Acoustic_reflex#cite_note-7.

494 Ibid.

495 Ibid.

496 Ibid.

497 Ibid.

498 "The Future of Prosthetics," Medical Device Network, accessed April 3, 2023, https://www.medicaldevice-network.com/features/future-prosthetics/.

499 Ibid.

500 "Bionics," Etymonline, accessed July 2, 2024, https://www.etymonline.com/word/bionics.

501 "Bionics," Wikipedia, last modified March 8, 2024, https://en.wikipedia.org/wiki/Bionics.

502 The Six Million Dollar Man (television series), IMDb, https://www.imdb.com/title/tt0071054/.

503 "The Six Million Dollar Man," Wikipedia, last modified June 29, 2024,
     https://en.wikipedia.org/wiki/The_Six_Million_Dollar_Man

504 Ibid.

505 "The Six Million Dollar Man," IMDb, accessed July 2, 2024,
     https://www.imdb.com/title/tt0073965/.

506 "The Bionic Woman," Wikipedia, last edited on June 18, 2024,
     https://en.wikipedia.org/wiki/The_Bionic_Woman.

507 "Protein Folding: The Good, The Bad, and The Ugly," Science in the News, Harvard University,
     February 28, 2010, accessed March 31, 2023, https://sitn.hms.harvard.edu/flash/2010/issue65/.
     Quote: "the world of proteins is a fascinating one, full of molecules with such intricate shapes and
     precise functions that they seem almost fanciful."

508 Ibid

509 Ibid.

510 Ibid.

511 Ibid.

512 Ibid.

513 Naama Barkai and Stan Leibler, "Robustness in simple biochemical networks," Nature 387, no. 6636
     (1997): 913-917, Abstract, explaining the concept of robust adaptation in biochemical networks,
     particularly in bacterial chemotaxis, and emphasizing the importance of robustness for proper
     functioning. Available at: https://www.nature.com/articles/43199

514 Ibid.

515 Denis Noble, The Music of Life Sourcebook (2020), accessed May 28, 2024,
     https://www.denisnoble.com/wp-content/uploads/2020/11/The-Music-of-Life-sourcebook.pdf.

516 "DNA as a Revolutionary Storage Medium," Wyss Institute, accessed March 17, 2024,
     https://wyss.harvard.edu/news/save-it-in-dna/, The text notes: George Church, Ph.D., and his
     team demonstrated the remarkable storage capacity of DNA, converting a 5.27 megabit book into
     DNA. This technique illustrates DNA's potential to store immense amounts of data, estimated at 215
     petabytes per gram. To contextualize, this means approximately 36 million copies of a 6GB movie
     could be stored in just one gram of DNA. This pioneering research underscores DNA's potential as a
     highly efficient storage medium in the face of exponentially growing global digital data.

517 Denis Noble, *The Music of Life Sourcebook.*

518 Thomas A. Kunkel and Katarzyna Bebenek, "DNA Replication Fidelity," *Annual Review of Biochemistry* 69 (2000): 497-529, https://doi.org/10.1146/annurev.biochem.69.1.497. This review article comprehensively discusses DNA replication fidelity, highlighting the importance of hydrogen bonding, base pair geometry, and conformational changes in DNA polymerases. It highlights how these factors contribute to genome stability, error rates, and the nuanced mechanisms underlying genetic integrity and variability.

519 Catherine L. Jopling, "Stop that nonsense!" eLife 3 (September 9, 2014): e04300, doi: 10.7554/eLife.04300, PMID: 25205670, PMCID: PMC4155323.

520 T. A. Brown, *Genomes*, 2nd ed. (Oxford: Wiley-Liss, 2002), chap. 14, "Mutation, Repair and Recombination," available from https://www.ncbi.nlm.nih.gov/books/NBK21114/.

521 Mateusz Kciuk, Barbara Marciniak, Mariusz Mojzych, and Renata Kontek, "Focus on UV-Induced DNA Damage and Repair-Disease Relevance and Protective Strategies," *International Journal of Molecular Sciences* 21, no. 19 (October 1, 2020): 7264, doi:10.3390/ijms21197264, PMID: 33019598, PMCID: PMC7582305.

522 Ibid.

523 "DNA: The Language of Life," Epic of Evolution, accessed March 17, 2024, https://epicofevolution.com/biological-evolution/dna-language-of-life. This article delves into the intricate structure of DNA, portraying it as the fundamental language of life, with a focus on its coding system of four chemical bases and the formation of codons.

524 Denis Noble, *The Music of Life Sourcebook.*

525 Alberts et al., *Molecular Biology of the Cell*, 6th Edition, Chapter 1: Eukaryotic cells compartmentalize DNA in the nucleus as a central repository of genetic information.

526 Alberts, Bruce, Alexander Johnson, Julian Lewis, et al., *Molecular Biology of the Cell*, 4th ed. (New York: Garland Science, 2002), "From DNA to RNA," https://www.ncbi.nlm.nih.gov/books/NBK26887/.

527 Michaela Frye, Bryan T. Harada, Mikaela Behm, and Chuan He, "RNA Modifications Modulate Gene Expression During Development," *Science* 361, no. 6409 (2018): 1346-1349, https://doi.org/10.1126/science.aau1646.

528 Alberts et al., Chapter 6: "Selective gene expression in different cell types ensures functional diversity."

529 Lee, H. T., Oh S., Ro, D. H., Yoo, H., and Kwon, Y. W. "The Key Role of DNA Methylation and Histone Acetylation in Epigenetics of Atherosclerosis." *Journal of Lipid and Atherosclerosis* 9, no. 3 (2020): 419-434. https://doi.org/10.12997/jla.2020.9.3.419.

530 Patterson and Hennessy, Computer Organization and Design: The Hardware/Software Interface, 5th Edition, 490.

531 Ibid., 402.

532 Alberts, Bruce, Alexander Johnson, Julian Lewis, et al., *Molecular Biology of the Cell*, 4th ed. (New York: Garland Science, 2002), "From DNA to RNA," https://www.ncbi.nlm.nih.gov/books/NBK26887/.

533 Patterson and Hennessy, Computer Organization and Design: The Hardware/Software Interface, 377.

534 Ibid., c38-c39.

535 Ille, A. M., H. Lamont, and M. B. Mathews, "The Central Dogma Revisited: Insights from Protein Synthesis, CRISPR, and Beyond," *Wiley Interdisciplinary Reviews: RNA* 13, no. 5 (2022): e1718, https://doi.org/10.1002/wrna.1718.

536 Patterson and Hennessy, Computer Organization and Design: The Hardware/Software Interface, 5th Edition, c38-c39.

537 Patterson and Hennessy, Computer Organization and Design: The Hardware/Software Interface, 423.

538 S. Ravindran, "Barbara McClintock and the Discovery of Jumping Genes," *Proceedings of the National Academy of Sciences of the United States of America* 109, no. 50 (December 11, 2012): 20198–99, https://doi.org/10.1073/pnas.1219372109.

539 Ibid.

540 "Transposons: The Jumping Genes," Nature Education, accessed May 29, 2024, https://www.nature.com/scitable/topicpage/transposons-the-jumping-genes-518/.

541 M. Ricci, V. Peona, E. Guichard, et al., "Transposable Elements Activity is Positively Related to Rate of Speciation in Mammals," *Journal of Molecular Evolution* 86, no. 6 (2018): 303–310, https://doi.org/10.1007/s00239-018-9847-7.

542 Keith R. Oliver and Wayne K. Greene, "Transposable Elements: Powerful Facilitators of Evolution," *BioEssays* 31, no. 7 (2009): 703-14, https://doi.org/10.1002/bies.200800219.

543 Ana M. Soto et al., "Toward a Theory of Organisms: Three Founding Principles in Search of a Useful Integration," *Progress in Biophysics and Molecular Biology* 122, no. 1 (October 2016): 77–82, https://doi.org/10.1016/j.pbiomolbio.2016.07.006.

544 Elena Casacuberta and Josefa González, "The Impact of Transposable Elements in Environmental Adaptation," *Molecular Ecology* 22 (2022): 1503-1517.

545 Ibid.

546 Ibid.

547 Ibid.

548 Ibid.

549 Ibid.

550 Ibid.

551 Ibid.

552 Imakawa, K., et al. "Endogenous Retroviruses and Placental Evolution, Development, and Diversity." *Cells* 11, no. 15 (2022): 2458. https://doi.org/10.3390/cells11152458.

553 N. Jansz and G. J. Faulkner, "Endogenous Retroviruses in the Origins and Treatment of Cancer," *Genome Biology* 22 (2021): 147, https://doi.org/10.1186/s13059-021-02357-4.

554 Ibid.

555 Imakawa, K., et al. "Endogenous Retroviruses and Placental Evolution, Development, and Diversity."

556 "Code Reusability and Modularity," OutSystems, accessed May 29, 2024, https://success.outsystems.com/documentation/10/developing_an_application/reuse_and_refactor/code_reusability_and_modularity/.

557 Lenovo, "Dynamic Link Library," accessed May 29, 2024, https://www.lenovo.com/us/en/glossary/dynamic-link-library/.

558 Lenovo, "Plug and Play (PnP)," accessed May 29, 2024, https://www.lenovo.com/us/en/glossary/pnp/.

559 "Dynamic Data Structures," JavaTpoint, "The main use case for which the Dynamic Data Structures are defined is to easily facilitate the change in the size of the data structure at the runtime without hindering the other operations that are associated with that data structure before increasing or decreasing the size of the data structure," accessed May 29, 2024, https://www.javatpoint.com/dynamic-data-structure.

560 "What is Linked List?" GeeksforGeeks, accessed May 29, 2024, https://www.geeksforgeeks.org/what-is-linked-list/.

561 Casacuberta and González, "The Impact of Transposable Elements in Environmental Adaptation."

562 "BSTs in Java: A Comprehensive Guide," GeeksProgramming, accessed May 30, 2024, https://geeksprogramming.com/bsts-java-comprehensive-guide/.

563 "Genetic Algorithm," ScienceDirect, accessed May 30, 2024, https://www.sciencedirect.com/topics/engineering/genetic-algorithm.

564 Note: While genetic algorithms are said to be modeled after the concept of natural selection, it's important to distinguish that engineered selection involves human input and predefined criteria, unlike Darwinian natural selection, which is theorized to be an unguided process that led to all the diversity on Earth with no initial information or goal.

565 K. Prasad, R.S. Cross, and M.R. Jenkins, "Synthetic Biology, Genetic Circuits and Machine Learning: A New Age of Cancer Therapy," *Molecular Oncology* 17, no. 6 (June 2023): 946-949, https://doi.org/10.1002/1878-0261.13420.

566 Julio Pulecio, Nikhil Verma, Eva Mejía-Ramírez, Danwei Huangfu, and Angel Raya, "CRISPR/Cas9-Based Engineering of the Epigenome," *Cell Stem Cell* 21, no. 4 (October 5, 2017): 431-447, https://doi.org/10.1016/j.stem.2017.09.006.

567 John Archibald Wheeler, *"Information, Physics, Quantum: The Search for Links,"* in *Proceedings of the Third International Symposium on Foundations of Quantum Mechanics in the Light of New Technology*, ed. S. Toyoda (Tokyo: Physical Society of Japan, 1989), 354-368.

568 Warner Bros. *The Jetsons*. Accessed February 29, 2024. https://www.warnerbros.com/tv/jetsons.

569 "History of Quantum Mechanics," Wikipedia, last edited May 8, 2024, https://en.wikipedia.org/wiki/History_of_quantum_mechanics.

570 "Quantum Information Science," NIH Data Science, accessed June 24, 2024, https://datascience.nih.gov/quantum-information-science.

571 "Lecture 1," in Quantum Physics I, Spring 2013, MIT OpenCourseWare, accessed June 24, 2024, https://ocw.mit.edu/courses/8-04-quantum-physics-i-spring-2013/resources/lecture-1/.

572 "WHAT IS INFORMATICS?," The University of Edinburgh, accessed June 28, 2024, https://www.ed.ac.uk/files/atoms/files/what20is20informatics.pdf.

573 MIT OpenCourseWare, "Chapter 3," in *Quantum Physics III*, Spring 2016, accessed June 24, 2024, https://ocw.mit.edu/courses/8-06-quantum-physics-iii-spring-2016/resources/mit8_06s16_chap3/.

574 University of Illinois, PHYS 419: Space, Time, and Matter Lecture Notes, Department of Physics, University of Illinois at Urbana-Champaign, accessed February 11, 2025, https://courses.physics.illinois.edu/phys419/sp2011/lectures/L20r.html.

575 Erwin Schrödinger, *Nature and the Greeks* (Cambridge: Cambridge University Press, 1954). Schrödinger writes, "The world is given to me only once, not one existing and one perceived. Subject and object are only one." This suggests he saw reality and consciousness as interconnected, aligning with interpretations that consider consciousness a fundamental aspect of existence rather than a mere byproduct of physical processes.

576 Carlo Rovelli, *"Consciousness Is Irrelevant to Quantum Mechanics,"* interview by IAI News, July 19, 2022, https://iai.tv/articles/consciousness-is-irrelevant-to-quantum-mechanics-auid-2187.

577 Massimiliano Sassoli de Bianchi, The Observer Effect, arXiv:1109.3536v3 [quant-ph], last revised June 18, 2012. To appear in *Foundations of Science*. "That light is able to exert a pressure, and therefore

disturb the objects it illuminates, was well-known also before the advent of quantum physics. However, it wasn't considered as a problem, as it was also believed that it was always possible, at least in principle, to arbitrarily reduce the intensity of the light source and diminish its effective disturbance, thus allowing one to observe whatever entity—macroscopic or microscopic—without sensibly affecting its condition." https://arxiv.org/abs/1109.3536.

578 Ibid.

579 Henry P. Stapp, Quantum Theory and the Role of Mind in Nature.

580 Niels Bohr, "Discussion with Einstein on Epistemological Problems in Atomic Physics," in *Albert Einstein: Philosopher-Scientist*, ed. Paul Arthur Schilpp (Evanston: Library of Living Phil Aarij Hussaan, Generation of Adaptive Pedagogical Scenarios in Serious Games (PhD diss., Université Claude Bernard - Lyon I, 2012), 97–99, https://theses.hal.science/tel-02868764/file/TH2012HussaanAarij.pdf.osophers, 1949), 201-241.

581 Henry P. Stapp, *Quantum Theory and the Role of Mind in Nature* (Lawrence Berkeley National Laboratory, March 8, 2001), 6-8, *arXiv*:quant-ph/0103043v1.

582 David J. Chalmers and Kelvin J. McQueen, "Consciousness and the Collapse of the Wave Function," in *Consciousness and Quantum Mechanics*, ed. Shan Gao (Oxford: Oxford University Press, 2022).

583 Raoni Wohnrath Arroyo, Lauro de Matos Nunes Filho, and Frederik Moreira dos Santos, "Towards a Process-Based Approach to Consciousness and Collapse in Quantum Mechanics," *Manuscrito 47*, no. 1 (March 2024): 6, DOI:10.1590/0100-6045.2024.v47n1.ra.

584 Six Jonathan, "Frustum Culling," *LearnOpenGL*, September 2021, https://learnopengl.com/Guest-Articles/2021/Scene/Frustum-Culling.

585 Aarij Hussaan, *Generation of Adaptive Pedagogical Scenarios in Serious Games* (PhD diss., Université Claude Bernard - Lyon I, 2012), 97–99, https://theses.hal.science/tel-02868764/file/TH2012HussaanAarij.pdf.

586 MIT OpenCourseWare, "Lecture 3," in *Quantum Physics I*, Spring 2013, accessed June 24, 2024, https:/ocw.mit.edu/courses/8-04-quantum-physics-i-spring-2013/ resources/lecture-3/.

587 Harald Atmanspacher and Hans Primas, "The Hidden Side of Wolfgang Pauli: An Eminent Physicist's Extraordinary Encounter with Depth Psychology," *Journal of Consciousness Studies* 3, no. 2 (1996): 112-26.

588 Erwin Schrödinger, Nature and the Greeks.

589 Erwin Schrödinger, *My View of the World* (Cambridge: Cambridge University Press, 1961), 21-22, https://archive.org/details/myviewofworld0000schr/page/22/mode/2up?q=obviously+

590 Erwin Schrödinger, *Mind and Matter* (Cambridge: Cambridge University Press, 1958), 129, http://strangebeautiful.com/other-texts/schrodinger-what-is-life-mind-matter-auto-sketches.pdf.

591 Erwin Schrödinger, *My View of the World*, 67-68

592 Ibid.

593 Ibid.

594 Ibid.

595 David Bohm, *Wholeness and the Implicate Order* (London: Routledge, 1980)

596 "The School Wellness Project," *Little Flower Yoga*, accessed February 15, 2025, https://www.littlefloweryoga.com/the-school-wellness-project/.

597 "Yoga & Mindfulness Teacher Preparation Program (YMTP²)," YMTPP, accessed February 15, 2025, https://www.ymtpp.org/.

598 Raoni Wohnrath Arroyo, Lauro de Matos Nunes Filho, and Frederik Moreira dos Santos, "Towards a Process-Based Approach to Consciousness and Collapse in Quantum Mechanics," 8.

599 X. Ma et al., "Quantum Erasure with Causally Disconnected Choice," Proceedings of the National Academy of Sciences of the United States of America 110, no. 4 (2013): 1221–1226, https://doi.org/10.1073/pnas.1213201110.

600 Ibid.

601 *Stanford Encyclopedia of Philosophy*, s.v. "Quantum Decoherence," last modified January 23, 2025, https://plato.stanford.edu/entries/qm-decoherence/.

602 Sean Carroll, "Why the Many-Worlds Formulation of Quantum Mechanics Is Probably Correct," *Preposterous Universe* (blog), June 30, 2014, https://www.preposterousuniverse.com/blog/2014/06/30/why-the-many-worlds-formulation-of-quantum-mechanics-is-probably-correct/.

603 Hugh Everett, "'Relative State' Formulation of Quantum Mechanics," *Reviews of Modern Physics* 29, no. 3 (July 1957): 454–462, https://doi.org/10.1103/RevModPhys.29.454.

604 Stephen Boughn, "Making Sense of the Many Worlds Interpretation," *arXiv* preprint, arXiv:2303.11249 [physics.hist-ph], Department of Physics, Princeton University, and Departments of Astronomy and Physics, Haverford College, accessed, https://arxiv.org/pdf/2303.11249,1.

605 Ibid., 1-2.

606 Shan Gao, "Understanding Branching in the Many-Worlds Interpretation of Quantum Mechanics," Research Center for Philosophy of Science and Technology, Shanxi University, July 22, 2024, https://philsci-archive.pitt.edu/23724/1/branching%202024%20v9.pdf.

607 Brian Greene and Sean M. Carroll, World Science Festival, May 1, 2024, 8:33-32:55, https://www.youtube.com/@WorldScienceFestival. Note: In this YouTube discussion, moderated by Brian Greene with Sean M. Carroll providing responses, several quantum interpretations are referenced, but distinctions between them are often glossed over, presenting a seemingly unified narrative that misrepresents the unique approaches and implications of each theory. This lack of

caution is emblematic of a tendency in these discussions of quantum mechanics, where different theories are often blended to maintain an appearance of coherence. Key examples from the discussion include:

Carroll initially describes GRW theory, which proposes spontaneous particle collapse to explain stable, definite positions in macroscopic objects. He then shifts to the MWI, which avoids collapse entirely by suggesting that each quantum event creates a separate reality. Moving between GRW and MWI without clearly distinguishing their distinct approaches can give the impression that these theories address the measurement problem in similar ways, despite their fundamentally different mechanisms.

Carroll addresses the issue of probability in MWI, noting that "every outcome is happening" across branches, which conflicts with classical probability, where only one outcome occurs. This blending of classical and quantum probability creates a misleading sense of coherence, skating around a central tension in MWI related to defining likelihood in a framework where all outcomes exist.

Carroll references the measurement problem and the role of the observer in causing collapse. In the MWI, measurement is redefined as a physical result of branching rather than collapse. In contrast, the Copenhagen Interpretation explicitly involves the observer or measurement as the trigger for wave function collapse. GRW theory, however, proposes that collapse occurs spontaneously, requiring no observation or interaction. Blending these perspectives glosses over how each theory differently interprets the observer's influence (or lack thereof) on quantum states, downplaying essential distinctions.

Carroll describes MWI as deterministic, following the Schrödinger equation without uncertainty, while also introducing "subjective probability" to explain branching outcomes. This blending of determinism and subjective probability is misleading, suggesting that probability in MWI is straightforward when, in reality, it represents a major interpretive challenge.

Carroll's use of metaphor, such as comparing branching to throwing a dart at a dartboard that then "splits," oversimplifies the differences between GRW, MWI, and hidden variable theories. This merging of metaphor and technical explanation makes these theories appear more compatible than they are, contributing to an oversimplified narrative.

608 Alan H. Guth, Eternal Inflation and Its Implications, February 22, 2007, https://arxiv.org/abs/hep-th/0702178. Guth discusses the role of eternal inflation within theoretical physics, stating, "Eternal inflation then has potentially a direct impact on fundamental physics, since it can provide a mechanism to populate the landscape of string vacua" (p. 10). This suggests that eternal inflation was developed as a theoretical extension rather than an inherent feature of inflationary theory itself. Furthermore, Guth acknowledges the unobservable nature of pocket universes, noting, "Although the infinity of pocket universes produced by eternal inflation are unobservable, it is argued that eternal inflation has real consequences in terms of the way that predictions are extracted from theoretical models" (p. 1). This highlights that the multiverse is not a testable scientific claim but rather a framework derived from theoretical considerations. Additionally, Guth recognizes the difficulties in defining probabilities within an eternally inflating universe, stating, "However, as soon as one attempts to define probabilities in an eternally inflating spacetime, one discovers ambiguities" (p. 11). The ambiguity in defining

probabilities within eternal inflation underscores the conceptual challenges associated with this framework, further indicating that it is not grounded in empirical evidence.

609 Ibid.

610 Ibid.

611 See Chapter 1.

612 Sean M. Carroll, "Sean Carroll Explains: What Is the Many-Worlds Interpretation?" *The Science and Cocktails Foundation*, January 8, 2020, YouTube video, 0:50–1:00, https://www.youtube.com/watch?v=t_xqNDII--Q.

613 Sean M. Carroll; Addressing the quantum measurement problem. *Physics Today* 1 July 2022; 75 (7): 62–63. https://doi.org/10.1063/PT.3.5046

614 Ricardo Karam, "Schrödinger's Original Struggles with a Complex Wave Function," *American Journal of Physics* 88, no. 6 (June 1, 2020): 433–438, https://doi.org/10.1119/10.0000852.

615 Sean M. Carroll, "Sean Carroll Explains: What Is the Many-Worlds Interpretation?", 1:10–1:30

616 Ibid., 1:30–5:34.

617 Ibid.

618 Ibid.

619 Ibid.

620 Hugh Everett, "'Relative State' Formulation of Quantum Mechanics."

621 Ibid., 6.

622 Hugh Everett III, *The Many-Worlds Interpretation of Quantum Mechanics: A Fundamental Exposition*, ed. Bryce S. DeWitt and Neill Graham (Princeton, NJ: Princeton University Press, 1973).

623 Stanford Encyclopedia of Philosophy, *The Many-Worlds Interpretation of Quantum Mechanics*, last modified June 3, 2021, https://plato.stanford.edu/entries/qm-manyworlds/.

624 Sean M. Carroll, "The Multiverse Is Real. Just Not in the Way You Think It Is. | Sean Carroll," *Big Think*, December 9, 2022, YouTube video, 0:00–9:28, https://www.youtube.com/watch?v=2bZi3Xm9tJE&t=222s.

625 Ibid.

626 Ibid.

627 Ibid.

628 Ibid.

629 Ibid.

630 Ibid.

631 ScienceDirect, "Theory of Everything." Accessed January 13, 2024.
   https://www.sciencedirect.com/topics/psychology/theory-of-everything.

632 Hawking, Stephen, 1942-2018. 2017. *A Brief History of Time*. New York, Bantam Books.

633 Hawking, Stephen, and Leonard Mlodinow. *The Grand Design*. Bantam Books, 2010, p. 180.

634 Austin L. Hughes, "The Folly of Scientism," The New Atlantis, Fall 2012, accessed February 4, 2024,
   https://www.thenewatlantis.com/publications/the-folly-of-scientism. Quote: "Scientists ought to
   be able to recognize how often philosophical issues arise in their work...practicing scientists, like all
   people, are prone to philosophical errors...scientists can be prone to errors of elementary logic, and
   these can often go undetected by the peer review process and have a major impact on the literature —
   for instance, confusing correlation and causation, or confusing implication with a biconditional.
   Philosophy can provide a way of understanding and correcting such errors. It addresses a largely
   distinct set of questions that natural science alone cannot answer, but that must be answered for natural
   science to be properly conducted."

635 Ludwig Wittgenstein, quoted in John C. Lennox, *Gunning for God: Why the New Atheists are Missing the
   Target* (Oxford: Lion Books, 2011), 228.

636 Newton's Laws of Motion." Glen Research Center, NASA.

637 Ma, W. "The Essence of Life." Biol Direct 11, no. 1 (September 26, 2016): 49. doi:
   10.1186/s13062-016-0150-5. PMID: 27671203; PMCID: PMC5037589. In this article, Ma states,
   "Although biology has achieved great successes in recent years, we have not got a clear idea on 'what
   is life?'."

638 Meyer, Stephen C. "Stephen Meyer On Intelligent Design And The Return Of The God Hypothesis."
   Interview. Hoover Institution. April 6, 2021. "So you can't explain the origin of the universe as the
   result of a material cause, because it's matter and energy themselves that come into existence at that
   point before which there was no matter to do the causing." Accessed January 11, 2024.
   https://www.hoover.org/research/stephen-meyer-intelligent-design-and-return-god-hypothesis-1.

639 Sean Carroll, *"Complexity in the Universe,"* 15th Annual Biard Lecture, CCAPP OSU, March 6, 2025,
   YouTube video, 19:17–19:50, https://www.youtube.com/watch?v=Kr_S-vXdu_I&t=153s.

640 Ibid.

641 Henry P. Stapp, Quantum Mechanical Approach to the Connection Between Mind and Brain
   (Berkeley, CA: Lawrence Berkeley National Laboratory, 2024), 1-14, https://www-
   physics.lbl.gov/~stapp/QMA.pdf.

642 "Richard P. Feynman," National Science Foundation,
   https://www.nsf.gov/news/special_reports/medalofscience50/feynman.jsp.

643 Richard P. Feynman, *The Relation of Science and Religion*, lecture delivered at the Caltech YMCA Lunch Forum, May 2, 1956, accessed April 4, 2025, https://calteches.library.caltech.edu/49/2/Religion.htm.

644 Plus.maths.org. 'Schrödinger's equation — what is it?' Accessed February 29, 2024. https://plus.maths.org/content/schrodinger-1.

645 "Erwin Schrödinger – Facts," NobelPrize.org, accessed June 20, 2024, https://www.nobelprize.org/prizes/physics/1933/schrodinger/facts/.

646 Plus.maths.org. 'Schrödinger's equation — what is it?'.

647 Ibid.

648 Josh Twist, "Synchronizing Multiple Nodes in Microsoft Azure," *MSDN Magazine*, November 2010, Vol. 25, No. 11, https://learn.microsoft.com/en-us/archive/msdn-magazine/2010/november/msdn-magazine-cloud-computing-synchronizing-multiple-nodes-in-microsoft-azure.

649 Yale University, "The Rise of Quantum Computing: Deciphering the Future," accessed June 26, 2024, https://campuspress.yale.edu/ledger/the-rise-of-quantum-computing-deciphering-the-future/.

650 "The Inner Meaning of the Story of the Tower of Babel," Influx Divine, accessed June 24, 2024, https://influxdivine.com/the-inner-meaning-of-the-story-of-the-tower-of-babel/.

651 IBM, "Quantum Cryptography," accessed June 26, 2024, https://www.ibm.com/topics/quantum-cryptography.

652 Pal, S., Bhattacharya, M., Lee, S. S., & Chakraborty, C. "Quantum Computing in the Next-Generation Computational Biology Landscape: From Protein Folding to Molecular Dynamics," *Molecular Biotechnology* 66, no. 2 (February 2024): 163-178. https://doi.org/10.1007/s12033-023-00765-4. Erratum in Molecular Biotechnology, September 28, 2023. https://doi.org/10.1007/s12033-023-00881-1. PMID: 37244882; PMCID: PMC10224669.

653 Aslam, N., Zhou, H., Urbach, E. K., et al. "Quantum Sensors for Biomedical Applications," *Nature Reviews Physics* 5 (2023): 157-169. https://doi.org/10.1038/s42254-023-00558-3.

654 Ibid.

655 X. Fan, T. G. Myers, B. A. D. Sukra, and G. Gabrielse, "Measurement of the Electron Magnetic Moment," Physical Review Letters 130, no. 7 (2023): 071801, https://link.aps.org/doi/10.1103/PhysRevLett.130.071801.

656 Ibid.

657 The Quantum Mechanic, "Scientists Propose DNA as Perfect Quantum Computer, Paving Way for Advances in Medicine," Quantum Zeitgeist, January 23, 2024,

https://quantumzeitgeist.com/scientists-propose-dna-as-perfect-quantum-computer-paving-way-for-advances-in-medicine/.

658 Ibid.

659 Ibid.

660 Cuevas-Zuviría, B., Fer, E., Adam, Z. R., et al.

"The Modular Biochemical Reaction Network Structure of Cellular Translation." *npj Systems Biology and Applications* 9 (2023): 52. https://doi.org/10.1038/s41540-023-00315-3.

661 Hill, D.P., D'Eustachio, P., Berardini, T.Z., Mungall, C.J., Renedo, N., & Blake, J.A. "Modeling biochemical pathways in the gene ontology." Database, 2016, baw126. doi:10.1093/database/baw126. Available at NCBI.

662 McFadden, J., and Al-Khalili, J. "The Origins of Quantum Biology." *Proceedings of the Royal Society A: Mathematical, Physical and Engineering Sciences*, 474 (2018): 20180674. doi:10.1098/rspa.2018.0674.

663 "Quantum Physics Proposes a New Way to Study Biology—and the Results Could Revolutionize Our Understanding of How Life Works," CNSI Blog, UCLA, May 15, 2023, accessed June 24, 2024, https://cnsi.ucla.edu/blog/2023/05/15/may-15-2023-quantum-physics-proposes-a-new-way-to-study-biology-and-the-results-could-revolutionize-our-understanding-of-how-life-works/.

664 Ibid.

665 Ibid.

666 Gregory D. Scholes, "Quantum-Coherent Electronic Energy Transfer: Did Nature Think of It First?" *The Journal of Physical Chemistry Letters* 1, no. 1 (2010): 2–8, https://doi.org/10.1021/jz900062f.

667 Alessandro Marrocu et al., "Psychiatric Risks for Worsened Mental Health After Psychedelic Use," *Journal of Psychopharmacology* 38, no. 3 (March 2024): 225–235, https://doi.org/10.1177/02698811241232548.

668 "Regulation of Body Processes." In *Biology*, OpenStax, last modified March 6, 2016. Accessed March 28, 2024. https://opentextbc.ca/biology/chapter/18-3-regulation-of-body-processes/.

669 George M. Cooper, *The Cell: A Molecular Approach*, 2nd ed. (Sunderland, MA: Sinauer Associates, 2000), https://www.ncbi.nlm.nih.gov/books/NBK9953/.

670 Sushil K. Ghosh, "Camillo Golgi (1843-1926): Scientist Extraordinaire and Pioneer Figure of Modern Neurology," *Anat Cell Biol* 53, no. 4 (December 31, 2020): 385–392, https://doi.org/10.5115/acb.20.196.

671 He, M., X. Zhou, and X. Wang. "Glycosylation: Mechanisms, Biological Functions and Clinical Implications." *Signal Transduction and Targeted Therapy* 9 (2024): 194. https://doi.org/10.1038/s41392-024-01886-1.

672 Pamela Stanley, "Golgi Glycosylation." *Cold Spring Harbor Perspectives in Biology* 3, no. 4 (April 1, 2011): a005199. https://doi.org/10.1101/cshperspect.a005199.

673 Richard Dawkins, *The Greatest Show on Earth: The Evidence for Evolution* (New York: Free Press, 2009), 333.

674 Petrucci, Ralph H., et al. "12.6: Crystal Structures." In General Chemistry: Principles and Modern Applications. LibreTexts. Accessed April 2, 2024. https://chem.libretexts.org/Bookshelves/General_Chemistry/Map%3A_General_Chemistry_(Petrucci_et_al.)/12%3A_Intermolecular_Forces%3A_Liquids_And_Solids/12.6%3A_Crystal_Structures.

675 National Snow and Ice Data Center, "Why Snow Matters," University of Colorado Boulder, accessed December 18, 2024, https://nsidc.org/learn/parts-cryosphere/snow/why-snow-matters.

676 Ibid.

677 Apple. "Design: Make the Result Beautiful. And the Effort Invisible." Apple Careers. Accessed March 2, 2025. https://www.apple.com/careers/us/design.html.

678 Nurettin Sezer and Muammer Koç, "A Comprehensive Review on the State-of-the-Art of Piezoelectric Energy Harvesting," *Nano Energy* 80 (2021): 105567, https://doi.org/10.1016/j.nanoen.2020.105567.

679 Leah Burrows, "For Darwin's finches, beak shape goes beyond evolution," Harvard John A. Paulson School of Engineering and Applied Sciences, November 12, 2021, accessed April 2, 2024, https://seas.harvard.edu/news/2021/11/darwins-finches-beak-shape-goes-beyond-evolution.

680 Sangeet Lamichhaney, "Adaptive evolution in Darwin's Finches," Sangeet Lamichhaney - Harvard University, accessed April 2, 2024, https://scholar.harvard.edu/sangeet/adaptive-evolution-darwins-finches.

681 Peter Grant and B. Rosemary Grant, "Speciation Undone," *Nature* 507, no. 7491 (2014): 178–179, https://doi.org/10.1038/507178b.

682 Morgan Kelly, Office of Communications, "Gene behind 'evolution in action' in Darwin's finches identified," Princeton University, April 21, 2016, accessed April 2, 2024, https://www.princeton.edu/news/2016/04/21/gene-behind-evolution-action-darwins-finches-identified.

683 A.B. Byrne, P. Arts, T.T. Ha, et al., "Genomic Autopsy to Identify Underlying Causes of Pregnancy Loss and Perinatal Death," *Nature Medicine* 29 (2023): 180–189, https://doi.org/10.1038/s41591-022-02142-1. The study indicates that large chromosomal abnormalities, such as autosomal trisomies and copy number variants (CNVs), are responsible for approximately 25–30% of congenital abnormalities linked to fetal and neonatal loss. Additionally, monogenic disorders account for around 5% of cases, suggesting that random genetic alterations at these stages often result in outcomes incompatible with life. Due to the limitations of standard genetic testing, around 70% of congenital

447

abnormality-related deaths lack a fully understood genetic basis, highlighting the lethality of unregulated genomic changes in early development.

684 "Junk DNA," Wikipedia, last modified March 23, 2024, accessed April 2, 2024, https://en.wikipedia.org/wiki/Junk_DNA.

685 Yale Medicine. "Junk No More." Accessed April 3, 2024. https://medicine.yale.edu/news/yale-medicine-magazine/article/junk-no-more/. Note: This article highlights that the term "junk DNA" has been misleading, as scientists have known for years that large portions of the human genome that don't produce proteins are nonetheless active.

686 William Hathaway, "Junk No More," Yale Medicine Magazine, Winter 2017, https://medicine.yale.edu/news/yale-medicine-magazine/article/junk-no-more/.

687 Ibid.

688 Marcella Cesana et al., "A Long Noncoding RNA Controls Muscle Differentiation," *Nature* 487, no. 7407 (2012): 234–238, https://doi.org/10.1038/nature11247.

689 Elliott Tyler A. and Gregory T. Ryan, "What's in a genome? The C-value enigma and the evolution of eukaryotic genome content," *Philosophical Transactions of the Royal Society B* 370 (2015): 20140331, doi: 10.1098/rstb.2014.0331.

690 Ibid.

691 Gregory, T. Ryan. "Macroevolution, Hierarchy Theory, and the C-Value Enigma." *Paleobiology* 30, no. 2 (2004): 179–202. http://www.jstor.org/stable/4096842.

692 W. M. Hahn and G. A. Wray, "The g-value paradox," *Evolution & Development* 4, no. 2 (March-April 2002): 73-75, doi: 10.1046/j.1525-142x.2002.01069.x.

693 Researchers at the University of Washington and in China released the genome sequence of rice, demonstrating that humans do not possess significantly more protein-coding genes than this simpler organism. See "Researchers at the University of Washington and in China Release Genome Sequence of Rice," University of Washington News, April 5, 2002, https://www.washington.edu/news/2002/04/05/researchers-at-the-university-of-washington-and-in-china-release-genome-sequence-of-rice/.

694 Y. Liu and Q. Chen, "150 Years of Darwin's Theory of Intercellular Flow of Hereditary Information," *Nature Reviews Molecular Cell Biology* 19, no. 12 (December 2018): 749-750, https://doi.org/10.1038/s41580-018-0072-4.

695 "John Mattick," University of New South Wales, https://www.unsw.edu.au/staff/john-mattick.

696 John S. Mattick, "A Kuhnian revolution in molecular biology: Most genes in complex organisms express regulatory RNAs," *BioEssays* [Open Access], first published 15 June 2023, https://doi.org/10.1002/bies.202300080.

697 Ibid.

698 Ibid.

699 Ibid.

700 Ibid.

701 TechTarget. "Redundant." SearchStorage.
https://www.techtarget.com/searchstorage/definition/redundant

702 Ibid.

703 "System Adaptability," SEBoK Wiki, https://sebokwiki.org/wiki/System_Adaptability (accessed April 3, 2024).

704 LinkedIn. "A Comprehensive Guide to System Design Patterns for Software Engineers." Accessed April 3, 2024. https://www.linkedin.com/pulse/comprehensive-guide-system-design-patterns-software-engineers/.

705 Codium. "Programming Logic." Accessed April 3, 2024.
https://www.codium.ai/glossary/programming-logic/.

706 Refactoring Guru. "Structural Patterns." Accessed April 3, 2024. https://refactoring.guru/design-patterns/structural-patterns.

707 Tashiro, Satoshi, Nishihara, Yuki, Kugou, Kazuto, Ohta, Kunihiro, & Kanoh, Junko. "Subtelomeres Constitute a Safeguard for Gene Expression and Chromosome Homeostasis." *Nucleic Acids Research* 45, no. 18 (2017): 10333-10349. doi: 10.1093/nar/gkx780.

708 Ibid.

709 *Eugene V. Koonin*, "Darwinian Evolution in the Light of Genomics," *Nucleic Acids Research* 37, no. 4 (2009): 1011-1034, doi: 10.1093/nar/gkp089.

710 Denis Noble, *The Music of Life Sourcebook* (Oxford: Oxford University Press, 2006), 25.

711 P. Legeza, G. W. Britz, A. Shah, et al., "Impact of network performance on remote robotic-assisted endovascular interventions in porcine model," *Journal of Robotic Surgery* 16 (2022): 29–35, https://doi.org/10.1007/s11701-021-01196-6.

712 Tamara Bonaci et al., *To Make a Robot Secure: An Experimental Analysis of Cyber Security Threats Against Teleoperated Surgical Robots* (University of Washington, Department of Electrical and Computer Engineering, May 12, 2015), https://wp.ece.uw.edu/wp-content/uploads/sites/25/2014/05/arXiv_April_2015.pdf.

713 Philip Ball, *How Life Works*, 2nd ed. (New York: Random House, 2023), accessed June 26, 2024, https://www.everand.com/read/677609700/How-Life-Works-A-User-s-Guide-to-the-New-Biology.

714 Ibid., 484.

715 Ibid., 50.

716 Ibid., 14.

717 *Closer to Truth*. Hosted by Robert Lawrence Kuhn. https://closertotruth.com/episodes/.

718 Ibid., 444.

719 Ibid.

720 Ibid., 280.

721 Ibid., 131.

722 Ibid., 131.

723 Ibid., 130.

724 Ibid., 130.

725 Ibid., 13-16.

726 Ibid., 14. Emphasis added.

727 Ibid.

728 Ibid., 15.

729 Ibid.

730 Ibid., 27.

731 Ibid., 15.

732 Ibid., 23.

733 G. Gramelsberger, "Synthetic Morphology: A Vision of Engineering Biological Form," *Journal of the History of Biology* 53 (2020): 295–309, https://doi.org/10.1007/s10739-020-09601-w.

734 Ibid., 560.

735 Ibid., 562.

736 Ibid., 13.

737 Ibid., 48.

738 Ibid., 48.

739 Ibid.

740 Ibid., 479.

741 Ibid., 480-483.

742 Ibid.

743 Ibid., 483.

744 National Air and Space Museum, 'The Wright Brothers', accessed February 29, 2024,
https://airandspace.si.edu/explore/stories/wright-brothers.

745 Louis Anslow, "Haters Gonna Hate," Big Think, October 18, 2023,
https://bigthink.com/pessimists-archive/air-space-flight-impossible/.

746 Ibid.

747 Ibid.

748 Dave English, "Great Aviation Quotes | Predictions," GreatAviationQuotes.com, accessed June 15,
2024, https://www.aviationquotations.com/predictionquotes.html.

749 Ibid.

750 John Wilkins, "John Wilkins and the Royal Society," accessed August 28, 2024,
https://www.cl.cam.ac.uk/~rja14/wilkins/wilkins.html.

751 Charles Lee Lewis, "Imaginative Aeronautics," *Proceedings, U.S. Naval Institute,* April 1936,
https://www.usni.org/magazines/proceedings/1936/april/imaginative-aeronautics.

752 G. Casey Cassidy, Mankind's Fascination With Flight, Yale-New Haven Teachers Institute, 1990,
https://teachersinstitute.yale.edu/curriculum/units/1990/7/90.07.03.x.html.

753 Mark Crawford, "7 Self-Healing Materials Expand the Limits of Engineering Design," ASME,
November 22, 2022, https://www.asme.org/topics-resources/content/7-self-healing-materials-
expand-the-limits-of-engineering-design.

754 Bagus Putra Muljadi, "Let's Mimic Termite Nests to Keep Human Buildings Cool," The Conversation,
May 6, 2019, accessed August 28, 2024, https://phys.org/news/2019-05-mimic-termite-human-
cool.html.

755 "Inspired by Cheetahs, Researchers Build Fastest Soft Robots Yet," National Science Foundation
(NSF), accessed August 28, 2024, https://new.nsf.gov/news/inspired-cheetahs-researchers-build-
fastest-soft.

756 Sarah Jane Alger, "The Mimic Octopus: Master of Disguise," Accumulating Glitches, Scitable by
Nature Education, October 28, 2013, https://www.nature.com/scitable/blog/accumulating-
glitches/the_mimic_octopus_master_of/.

757 Norbert Wiener, *Cybernetics: Or Control and Communication in the Animal and the Machine*, reissue of the 1961 second edition (Cambridge, MA: MIT Press, 2019), https://mitpress.mit.edu/9780262537841/cybernetics-or-control-and-communication-in-the-animal-and-the-machine/.

758 George Washington University. "George Gamow." The Department of Physics. https://physics.columbian.gwu.edu/george-gamow.

759 "Genentech: Our Founders," *Genentech*, accessed June 20, 2024, https://www.gene.com/about-us/leadership/our-founders.

760 Awad, A., W. Pang, D. Lusseau, et al., "A Survey on *Physarum polycephalum* Intelligent Foraging Behaviour and Bio-Inspired Applications," *Artificial Intelligence Review* 56 (2023): 1–26. https://doi.org/10.1007/s10462-021-10112-1.

761 Ibid.

762 Ibid.

763 Janine Benyus, "Biomimicry: Innovation Inspired By Nature," Biomimicry 3.8, accessed June 20, 2024, https://biomimicry.net/product/janine-book/.

764 Ibid.

765 "Janine Benyus," Biomimicry.org, accessed May 12, 2024, https://biomimicry.org/janine-benyus/.

766 Biomimicry Institute. "What Is Biomimicry?" Biomimicry Institute. Accessed August 25, 2024. https://biomimicry.net/what-is-biomimicry/.

767 Janine Benyus, quoted on Bookey, accessed August 30, 2024, https://www.bookey.app/quote-author/janine-benyus.

768 Ibid.

769 "Janine Benyus on 3.8 Billion-Year-Old Solutions to Today's Design Challenges," Appalachian Today, accessed June 20, 2024, https://today.appstate.edu/2018/08/15/benyus.

770 "Psalm 100," King James Version, Bible Hub, accessed May 12, 2024, https://biblehub.com/kjv/psalms/100.htm.

771 Peter A. Corning et al., eds., Evolution "On Purpose." Materialist authors of this book state the following: "In other words, over the course of several million years, the human species in effect invented itself through an entrepreneurial process involving gradual cultural innovations that changed our ancestors' relationship to the environment—and to one another."

772 Hawking, Stephen, and Leonard Mlodinow. The Grand Design. Bantam Books, 2010, p. 180.

773 Bio-X, Stanford University, accessed August 30, 2024, https://biox.stanford.edu/.

774 Bio-X: About, Stanford University, accessed August 30, 2024, https://biox.stanford.edu/about.

775 Michael Levitt, Stanford Medicine, accessed August 30, 2024, https://med.stanford.edu/profiles/Michael_Levitt.

776 "The Science Behind Michael Levitt's Nobel Prize," Stanford Medicine, October 9, 2013, accessed August 30, 2024, https://med.stanford.edu/news/all-news/2013/10/the-science-behind-michael-levitts-nobel-prize.html.

777 Michael Levitt, "Studied Physics, Masters Biology, Won Nobel in Chemistry | Endgame #193 (Luminaries)," interview by Gita Wirjawan, YouTube video, 42:29-49:56, August 2, 2024, https://www.youtube.com/watch?v=Pg4w-rwTCXE.

778 Alan D. Attie et al., "Defending Science Education Against Intelligent Design: A Call to Action," *Journal of Clinical Investigation* 116, no. 5 (May 1, 2006): 1134–1138, https://www.jci.org/articles/view/28449.

779 "The Enigma of the Cambrian Explosion," in *Rare Earth*, ed. by Editors (Springer, New York, NY, 2000), https://doi.org/10.1007/0-387-21848-3_7.

780 Ancient origins of multicellular life. *Nature* 533, 441 (2016). https://doi.org/10.1038/533441b

781 C. F. Demoulin, Y. J. Lara, L. Cornet, et al., "Cyanobacteria Evolution: Insight from the Fossil Record," *Free Radical Biology and Medicine* 140 (2019): 206-223, https://doi.org/10.1016/j.freeradbiomed.2019.05.007.

782 Natural History Museum, "The Cambrian Explosion was far shorter than thought," last modified February 2019, https://www.nhm.ac.uk/discover/news/2019/february/the-cambrian-explosion-was-far-shorter-than-thought.html.

783 J. Y. Chen, "The Sudden Appearance of Diverse Animal Body Plans During the Cambrian Explosion," *The International Journal of Developmental Biology* 53, no. 5-6 (2009): 733–751, https://doi.org/10.1387/ijdb.072513cj.

784 Ibid.

785 Max Planck Institute for Evolutionary Anthropology, "The oldest Upper Paleolithic Homo sapiens in Europe," Max Planck Gesellschaft, last modified May 5, 2021, https://www.mpg.de/14778668/0505-evan-019609-the-oldest-upper-paleolithic-homo-sapiens-in-europe.

786 "Cambrian Explosion," Wikipedia, last modified January 25, 2025, https://en.wikipedia.org/wiki/Cambrian_explosion.

787 L. R. Kump, "The Rise of Atmospheric Oxygen," *Nature* 451, no. 7176 (2008): 277–278, https://doi.org/10.1038/nature06587.

788 Carolina Ortiz-Guerrero, "Three Pulses of Breaths Toward Three Evolutionary Shifts," *Communications Earth & Environment* (2024), https://doi.org/10.1038/s43247-024-01321-x.

The evidence used to support the so-called Great Oxidation Event (GOE) and Neoproterozoic Oxygenation Event (NOE) is far from conclusive and, in many cases, contradictory. Metal deposits, isotopic shifts, and layered sediments—often cited as indicators of atmospheric oxygenation—are just as easily explained by geological activity rather than a biologically driven oxygenation process. Volcanic eruptions, hydrothermal activity, and rapid sediment deposition can produce oxidation signatures in a matter of days or months, not millions of years. Even sulfur and carbon isotope variations, commonly interpreted as signs of ancient oxygenation, can result from tectonic upheavals, deep-sea circulation shifts, or transient chemical changes rather than a gradual increase in atmospheric oxygen.

Further undermining the NOE, new research (Ortiz-Guerrero 2024; Kaiho et al. 2024) shows that supposed oxygen levels fluctuated wildly, with alternating oxygen-rich and oxygen-poor phases rather than a steady increase. If complex life depended on stable oxygenation, how did it persist through multiple anoxic phases? If organisms thrived despite repeated oxygen crashes, then why is oxygen still assumed to be the driving force behind evolution? Even mainstream researchers admit that other potential primary controls such as nutrient availability and geological factors may have played a major role—meaning that oxygenation is not the clear-cut cause of biological complexity that it is often made out to be.

This pattern of revision exposes a deeper issue: The GOE and NOE are not established facts but retrofitted explanations designed to sustain evolutionary assumptions. Rather than accepting that the fossil record fails to align with gradualist models, evolutionists continuously reshape oxygenation theories to account for missing evidence. If the markers used to justify these events can be explained by catastrophic geological processes rather than a controlled biological progression, then the GOE and NOE cease to be necessary, and their role in evolution collapses.

789 J. Olejarz, Y. Iwasa, A. H. Knoll, et al., "The Great Oxygenation Event as a Consequence of Ecological Dynamics Modulated by Planetary Change," *Nature Communications* 12, no. 3985 (2021), https://doi.org/10.1038/s41467-021-23286-7.

790 Nick Lane, *Oxygen: The Molecule that Made the World* (Oxford: Oxford University Press, 2002), 19–20, accessed via Internet Archive, https://archive.org/details/oxygenmoleculeth0000lane.

791 Thomas Servais et al., "The Great Ordovician Biodiversification Event (GOBE): The Palaeoecological Dimension," *Palaeogeography, Palaeoclimatology, Palaeoecology* 294 (2010): 99–119, https://doi.org/10.1016/j.palaeo.2010.05.031.

792 International Commission on Stratigraphy, *International Chronostratigraphic Chart, Version 2023/09*, September 2023, https://stratigraphy.org/ICSchart/ChronostratChart2023-09.pdf.

793 T. Yamasaki and Y. Kobayashi, "Evolving Dispersal Ability Causes Rapid Adaptive Radiation," *Scientific Reports* 14 (2024): 15734, https://doi.org/10.1038/s41598-024-66435-w.

794 Charles Darwin, The Origin of Species, accessed April 4, 2024, https://www.goodreads.com/quotes/344545-if-it-could-be-demonstrated-that-any-complex-organ-existed, "If it could be demonstrated that any complex organ existed, which could not possibly

have been formed by numerous, successive, slight modifications, my theory would absolutely break down. But I can find no such case," (Charles Darwin).

795 Charles Darwin, The Origin of Species, accessed April 4, 2024, https://www.goodreads.com/quotes/344545-if-it-could-be-demonstrated-that-any-complex-organ-existed, "If it could be demonstrated that any complex organ existed, which could not possibly have been formed by numerous, successive, slight modifications, my theory would absolutely break down. But I can find no such case," (Charles Darwin).

796 Niles Eldredge and Stephen Jay Gould, "Punctuated Equilibria: An Alternative to Phyletic Gradualism," in Models in Paleobiology, ed. Thomas J. M. Schopf (San Francisco: Freeman, Cooper & Co., 1972), 98. Eldredge and Gould emphasize that the absence of transitional fossils in the fossil record does not necessarily imply that gradual evolution did not occur. Instead, they argue that the fossil record may be incomplete or biased, and that the pattern of punctuated equilibrium better fits the available data.

797 Stephen Jay Gould, The Structure of Evolutionary Theory (Cambridge, MA: Belknap Press of Harvard University Press, 2002).

798 Ibid., 61.

799 Eugenie C. Scott, Evolution vs. Creationism: An Introduction (Berkeley: University of California Press, 2004, 3.

800 Stephen Jay Gould, The Structure of Evolutionary Theory.

801 Ibid.

802 Ibid.

803 Ibid. Emphasis added.

804 Ibid.

805 Ibid.

806 U.S. Environmental Protection Agency, "Confirmation Bias," accessed April 4, 2024, https://cfpub.epa.gov/si/si_public_record_Report.cfm?Lab=CESER&dirEntryId=350904#. Quoted: "Human thought processes are not perfect. We face cognitive errors daily. Confirmation bias is a common error committed by people, and contrary to the thoughts of many, scientists are not immune to committing this same error. It's a tendency to believe that you are right, disregarding things that conflict with your ideas. Confirmation bias is reinforced with time. As scientists, we want the public to trust science, scientists and the data they put forth because they have confidence that scientists are intelligent, honest and follow procedures that ensures [sic] validity of the data and results. However, scientists are human and equally capable of succumbing to faulty thinking processes as everyone else."

807 Charles Darwin, *The Origin of Species*, accessed April 4, 2024, https://www.goodreads.com/quotes/344545-if-it-could-be-demonstrated-that-any-complex-organ-existed.

808 "0," *Wikipedia: The Free Encyclopedia*, https://en.wikipedia.org/wiki/0.

809 Jun Chen et al., "Hunting for Beneficial Mutations: Conditioning on SIFT Scores When Estimating the Distribution of Fitness Effect of New Mutations," *Genome Biology and Evolution* 14, no. 1 (January 2022): evab151, https://doi.org/10.1093/gbe/evab151.

810 Ibid.

811 Stephen Meyer, "Stephen Meyer Explains Neo-Darwinism's False Beliefs-Part 2."

812 Geoffrey M. Cooper, *The Cell: A Molecular Approach*, 2nd ed. (Sunderland, MA: Sinauer Associates, 2000), "The Development and Causes of Cancer," https://www.ncbi.nlm.nih.gov/books/NBK9963/.

813 Eugenie C. Scott, *Evolution vs. Creationism: An Introduction* (Berkeley: University of California Press, 2004), 43.

814 Charles Darwin, "The Foundations of The Origin of Species: Two Essays Written in 1842 and 1844," Darwin Online, https://darwin-online.org.uk/content/frameset?itemID=F1556&viewtype=text&pageseq=1. Note: In this text, Darwin acknowledges the challenge posed by the lack of evidence for intermediate forms in the fossil record, stating, "This want of evidence of the past existence of almost infinitely numerous intermediate forms, is, I conceive, much the weightiest difficulty on the theory of common descent; but I must think that this is due to ignorance necessarily resulting from the imperfection of all geological records."

815 Purva Variyar, "Extinction – The Mainstay of Life on Earth," Wildlife Conservation Trust, accessed April 3, 2023, https://www.wildlifeconservationtrust.org/extinction-the-mainstay-of-life-on-earth/. Quote from Pranay Lal, Indica – A Deep Natural History of The Indian Subcontinent: "It is hard to believe that since life began nearly 3.8 billion years ago, 99 percent of all living things that have ever lived on Earth have become extinct. What we see today is just a minuscule 1 percent of all life that has ever lived on Earth. In fact, it is almost a miracle that life continues to exist at all!"

816 Mary Pickard Winsor, "Darwin's Dark Matter: Utter Extinction," *Annals of Science* 80, no. 4 (2023): 357-389, https://doi.org/10.1080/00033790.2023.2194889.

817 Ruth Wilson, "Evolution Is Not Science!" The Virginian-Pilot, April 11, 1996, accessed May 23, 2024, https://scholar.lib.vt.edu/VA-news/VA-Pilot/issues/1996/vp960411/04110004.htm.

818 National Academy of Sciences (US); Fitch, W. M., and F. J. Ayala, editors. *Tempo and Mode in Evolution: Genetics and Paleontology 50 Years After Simpson*. Washington, DC: National Academies Press (US), 1995. "The Role of Extinction in Evolution." https://www.ncbi.nlm.nih.gov/books/NBK232212/.

819 Ibid.

820 Ibid.

821 Matthew C. Fitzpatrick et al., "Environmental and historical imprints on beta diversity: insights from variation in rates of species turnover along gradients," Royal Society Publishing, October 7, 2013, https://doi.org/10.1098/rspb.2013.1201. In this article, the authors discuss a common approach for analyzing geographical variation in biodiversity, highlighting the underlying assumptions about the rate of species turnover along environmental and geographical gradients.

822 Anthony Barnosky et al., "Has the Earth's sixth mass extinction already arrived?," *Nature* 471 (2011): 51–57, https://doi.org/10.1038/nature09678.

823 Casey Luskin, "A Slightly Technical Introduction to Intelligent Design," Intelligent Design, accessed July 13, 2024, https://intelligentdesign.org/articles/a-slightly-technical-introduction-to-intelligent-design/.

824 "Most products are incremental improvements on a previous version. Many parts in a new product can be reused from previous work … On average … engineers spend about 33% of their time on non value-added work such as searching for existing parts within the company that can be reused, or even recreating an existing part that cannot be found in a timely manner. If a manufacturer can reduce the time spent searching for or designing parts, the savings multiply." SIMULIA Blog, accessed July 13, 2024, https://blog.3ds.com/brands/simulia/four-problems-solved-by-intelligent-parts-reuse-standardization-and-sourcing/.

825 "Design Standardization: Definition and Best Practices," Engineering 101, accessed July 13, 2024, https://drawer.caddi.com/blog/design-standardization.

826 Ibid.

827 Ibid.

828 Ibid.

829 Ibid.

830 Ibid.

831 J.K. Rowling, *Harry Potter series* (London: Bloomsbury Publishing, 1997-2007).

832 Pranay Lal, "It is hard to believe that since life began nearly 3.8 billion years ago, 99 percent of all living things that have ever lived on Earth have become extinct. What we see today is just a minuscule 1 per cent of all life that has ever lived on Earth. In fact, it is almost a miracle that life continues to exist at all!" Wildlife Conservation Trust, accessed July 11, 2024, https://www.wildlifeconservationtrust.org/extinction-the-mainstay-of-life-on-earth/.

833 Alice Park, "After Decades of Work, the Human Genome Is Finally Complete," Time, March 31, 2022, https://time.com/6163452/human-genome-fully-sequenced/.

834 Michael Behe, "A (R)evolutionary Biologist," https://michaelbehe.com/.

835 Charles Darwin, "Notebook B: Transmutation of Species (1837–1838)," 1837, Manuscripts, Cambridge University Library, Darwin Online, https://en.m.wikipedia.org/wiki/File:Darwin_Tree_1837.png.

836 Kevin Blake, "On the Origins of 'The March of Progress'," Washington University ProSPER, December 17, 2018, https://sites.wustl.edu/prosper/on-the-origins-of-the-march-of-progress/.

837 Ibid.

838 Ibid.

839 Ibid.

840  Ernst Mayr, *The Growth of Biological Thought: Diversity, Evolution, and Inheritance* (Cambridge, MA: Harvard University Press, 1982). Mayr details the development of Neo-Darwinism and describes how the incorporation of Mendelian genetics created challenges in reconciling genetic mechanisms with Darwin's original framework.

841 Kimura, M. (1968). "Evolutionary rate at the molecular level." *Nature*, 217(5129), 624-626. Motoo Kimura's neutral theory of molecular evolution proposed that most genetic changes at the molecular level are neither beneficial nor harmful but neutral. This theory was largely developed in response to the lack of observable transitional forms in the fossil record, shifting the focus from morphology to unseen molecular changes.

842 J. S. Taylor and J. Raes, "Duplication and Divergence: The Evolution of New Genes and Old Ideas," *Annual     Review     of     Genetics*     38     (2004):     615–643, https://doi.org/10.1146/annurev.genet.38.072902.092831.
This paper discusses how gene duplication is proposed as a mechanism for evolutionary innovation. However, it also acknowledges challenges, including how newly duplicated genes acquire novel functions rather than being lost or becoming nonfunctional.

843 Sanford, J., Brewer, W., Smith, F. et al. *The waiting time problem in a model hominin population.* *Theoretical Biology and Medical Modelling* 12, 18 (2015). https://doi.org/10.1186/s12976-015-0016-z. This study models the time required for specific mutations to arise and become fixed in a hominin population, demonstrating that even under favorable assumptions, the waiting times for multiple coordinated mutations exceed feasible evolutionary timelines.

844 P. O'Donald, "Haldane's Dilemma and the Rate of Natural Selection," *Nature* 221, no. 5181 (1969): 815–816, https://doi.org/10.1038/221815a0. O'Donald discusses the limitations imposed by Haldane's Dilemma on the rate of beneficial mutation fixation, highlighting the mathematical constraints that challenge the feasibility of large-scale evolutionary changes within the available timeframes.

845 "The Model T," Ford Motor Company, accessed April 6, 2024, https://corporate.ford.com/articles/history/the-model-t.html.

846 "Cybertruck," *Tesla.com*, accessed April 6, 2024, https://www.tesla.com/cybertruck.

847 W. Ford Doolittle and Eric Bapteste, "Pattern Pluralism and the Tree of Life Hypothesis," *Proceedings of the National Academy of Sciences of the United States of America* 104, no. 7 (February 13, 2007): 2043–2049, https://doi.org/10.1073/pnas.0610699104.

848 Ibid.

849 Ibid.

850 Ibid.

851 Ibid.

852 Patrick Keeling and Jeffrey D. Palmer, "Horizontal Gene Transfer in Eukaryotic Evolution," *Nature Reviews Genetics* 9 (August 2008): 605–618, https://doi.org/10.1038/nrg2386.

853 "Scalable, Reliable, and Secure Mesh Networking," Bluetooth, accessed August 30, 2024, https://www.bluetooth.com/learn-about-bluetooth/feature-enhancements/mesh/.

854 Internet of Things (IoT), Oracle.

855 "Near Field Communication (NFC)," TechTarget, accessed August 30, 2024, https://www.techtarget.com/searchmobilecomputing/definition/Near-Field-Communication. Near Field Communication allows for close-proximity data exchange between devices, demonstrating how unrelated devices can interact and share information when in close contact, similar to how lateral gene transfer facilitates genetic exchange between unrelated species in nature.

856 C. Gilbert and R. Cordaux, "Viruses as Vectors of Horizontal Transfer of Genetic Material in Eukaryotes," *Current Opinion in Virology* 25 (August 2017): 16-22, https://doi.org/10.1016/j.coviro.2017.06.005. This study discusses the role of viruses in facilitating lateral gene transfer in eukaryotes, a process that challenges the traditional tree-like structure of evolutionary relationships by introducing genetic material across species boundaries.

857 Charles Darwin, *On the Origin of Species*, 4th ed. (London: John Murray, 1866), 61, accessed January 26, 2025, https://darwin-online.org.uk/Variorum/1866/1866-61-c-1869.html.

858 The Linnean Society of London, "Career and Legacy," accessed September 1, 2024, https://www.linnean.org/learning/who-was-linnaeus/career-and-legacy.

859 Carolus Linnaeus, *Systema Naturae*, facsimile of the first edition, trans. M. S. J. Engel-Ledeboer and D. R. Engel (Stockholm: The Hague, 1735), 7.

860 Ibid.

861 Ibid., 18.

862 Ibid., 7.

863 The Linnean Society of London, "Career and Legacy."

864 J. M. Serb and D. J. Eernisse, "Charting Evolution's Trajectory: Using Molluscan Eye Diversity to Understand Parallel and Convergent Evolution," *Evolution: Education and Outreach* 1, no. 4 (2008): 439–447, https://doi.org/10.1007/s12052-008-0084-1.

865 MedlinePlus, "PAX6 Gene," *U.S. National Library of Medicine*, accessed March 15, 2025, https://medlineplus.gov/genetics/gene/pax6/.

866 Nancy A. Moran and Tyler Jarvik, "Lateral Transfer of Genes from Fungi Underlies Carotenoid Production in Aphids," *Science* 328, no. 5978 (2010): 624–627, https://doi.org/10.1126/science.1187113.

867 Yuxin Zhang et al., "Metabolic Enhancement Contributed by Horizontal Gene Transfer Is Essential for Dietary Specialization in Leaf Beetles," *Proceedings of the National Academy of Sciences* 122, no. 1 (2024): e2415717122, https://doi.org/10.1073/pnas.2415717122.

868 OpenStax, "Features Used to Classify Animals," Biology 2e, accessed August 30, 2024, https://openstax.org/books/biology-2e/pages/27-2-features-used-to-classify-animals.

869 Ibid.

870 OpenStax, "Mammals," *Biology 2e*, accessed August 30, 2024, https://openstax.org/books/biology-2e/pages/29-6-mammals.

871 OpenStax, "Features Used to Classify Animals."

872 OpenStax, "Our Ancient Past: The Earliest Hominins," Introduction to Anthropology, accessed August 30, 2024, https://openstax.org/books/introduction-anthropology/pages/4-7-our-ancient-past-the-earliest-hominins.

873 Martindale, M. Q. "Evolution of Development: The Details Are in the Entrails." *Current Biology* 23, no. 1 (January 7, 2013): R25–R28. https://doi.org/10.1016/j.cub.2012.11.023.

874 David Taylor, "Ecosystem Management," Virginia Tech, accessed July 12, 2024, https://dendro.cnre.vt.edu/forsite/2004presentations/taylor/forsite/forsite.html.

875 Qing Yang, Grace L. Salter, and Chloë Schmidt, "Marine Biodiversity and Ecosystem Functioning," *Frontiers in Marine Science*, published June 26, 2023, https://www.frontiersin.org/articles/10.3389/fmars.2023.1260709/full.

876 OpenStax, "Energy Flow through Ecosystems," Concepts of Biology. Accessed September 1, 2023. https://openstax.org/books/concepts-biology/pages/20-1-energy-flow-through-ecosystems.

877 California Wolf Center, "Biodiversity," accessed July 12, 2024, https://www.californiawolfcenter.org/biodiversity.

878 National Geographic Society, "The Role of Keystone Species in an Ecosystem," accessed July 12, 2024, https://education.nationalgeographic.org/resource/role-keystone-species-ecosystem/.

879 Brodie Farquhar, "How the Reintroduction of Wolves Changed Yellowstone," Yellowstone Park, last updated June 22, 2023, https://www.yellowstonepark.com/things-to-do/wildlife/wolf-reintroduction-changes-ecosystem/.

880 Lydia May, "Complex Systems Fail in Complex Ways: The Essentiality of Software Updates," *Sentar*, October 22, 2024, https://www.sentar.com/complex-systems-fail-in-complex-ways-the-essentiality-of-software-updates/.

881 Natural Resources Defense Council (NRDC), "Keystone Species 101," NRDC, September 9, 2019, https://www.nrdc.org/stories/keystone-species-101#what-is.

882 Pollinator Partnership, "Pollinators," accessed July 12, 2024, https://www.pollinator.org/pollinators.

883 Alison N. P. Stevens, "Dynamics of Predation," *Nature Education Knowledge* 3, no. 10 (2010): 46, https://www.nature.com/scitable/knowledge/library/dynamics-of-predation-13229468/.

884 NOAA Ocean Today, "Rainforests of the Sea," accessed July 12, 2024, https://oceantoday.noaa.gov/fullmoon-rainforestsofthesea/welcome.html.

885 Ibid.

886 U.S. Environmental Protection Agency, "Why are Wetlands Important?" accessed July 12, 2024, https://www.epa.gov/wetlands/why-are-wetlands-important.

887 Lake Wilderness Arboretum, "From Forest Floor to Canopy: The Layers of the Forest," accessed July 12, 2024, https://www.lakewildernessarboretum.org/gardens/woodland-garden-2/from-forest-floor-to-canopy-the-layers-of-the-forest/.

888 Qing Yang, Grace L. Salter, and Chloë Schmidt, "Marine Biodiversity and Ecosystem Functioning."

889 Ibid.

890 Ibid.

891 "Biogeochemical Cycles," BYJU's, accessed August 30, 2024, https://byjus.com/biology/biogeochemical-cycles/.

892 Matt Sandy, "The Amazon Rainforest Faces Crossroads as Brazil Shifts Leadership," Foreign Policy, December 9, 2022, https://foreignpolicy.com/2022/12/09/amazon-rainforest-climate-change-deforestation-bolsonaro-lula/.

893 UNESCO World Heritage Centre, "The Old City of Dubrovnik," accessed July 12, 2024, https://whc.unesco.org/en/list/154/.

894 Ibid.

895 National Geographic Society, "Serengeti," accessed July 12, 2024, https://education.nationalgeographic.org/resource/serengeti/.

896 John Archibald Wheeler, "*Information, Physics, Quantum: The Search for Links,*" in *Proceedings of the Third International Symposium on Foundations of Quantum Mechanics in the Light of New Technology*, ed. S. Toyoda (Tokyo: Physical Society of Japan, 1989), 354-368.

897 "What kind of 'superintellect that monkeyed with physics' did Sir Fred Hoyle have in mind given that he was an atheist?" Philosophy Stack Exchange, last edited December 24, 2024, https://philosophy.stackexchange.com/questions/120808/what-kind-of-superintellect-that-monkeyed-with-physics-did-sir-fred-hoyle-have.

898 John Archibald Wheeler, "*Information, Physics, Quantum: The Search for Links,*" 7.

899 Ibid., 18.

900 Ibid., 14.

901 William Rowan Hamilton, "On a General Method in Dynamics."

902 Robert B. Laughlin, *A Different Universe: Reinventing Physics from the Bottom Down* (New York: Basic Books, 2005), preface, xv. Laughlin argues that all physical law emerges from collective origins and not from fundamental laws alone. He claims "control of nature is achieved only when nature allows this through a principle of organization," but leaves unaddressed the question of what—or who—does the organizing.

903 Isaac Newton, *The Mathematical Principles of Natural Philosophy*, trans. Andrew Motte, rev. and corr. with a life of the author by H. W. Chittenden (New York: Daniel Adee, 1846), 74.

904 I. Bernard Cohen, *The First English Version of Newton's Hypotheses non fingo*, Isis 53, no. 3 (1962): 379–388, https://www.journals.uchicago.edu/action/showCitFormats?doi=10.1086%2F349598.

905 Dennis W. Sciama, *On the Origin of Inertia*, *Monthly Notices of the Royal Astronomical Society* 113, no. 1 (1953): 35, https://doi.org/10.1093/mnras/113.1.34.

906 Ibid.

907 Jonathan Fay, *On Sciama (1953)*, January 3, 2024, file://huguenot-fs-01/users/scooper2/Desktop/Jonathan_Fay_On_Sciama_Modified_Inertia.pdf.

908 Jonathan Fay, On the Relativity of Magnitudes, Studies in History and Philosophy of Science Part A 106 (2024): 165–176.

909 D. R. Brill and S. Deser, "Instability of Closed Spaces in General Relativity," *Communications in Mathematical Physics* 32, no. 4 (1973): 298. "Closed flat space is isolated: there are no 'neighbors' of flat space whose volume does not change in time at the initial surface... the initial value problem in a

flat closed manifold is always maximally unstable if the initial and varied 3-geometries are extremal surfaces."

910 Albert Einstein, *The Meaning of Relativity*, 5th ed., 32.

911 Ibid.

912 Albert Einstein, *The Meaning of Relativity*, 5th ed. (Princeton, NJ: Princeton University Press, 1956), 32.

913 Alexandra Witze, "Special Relativity Aces Time Trial," *Nature* (2014), https://doi.org/10.1038/nature.2014.15970. While the article confirms predictions made by general relativity regarding time dilation at different altitudes, its framing is somewhat misleading. The phrase "aces time trial" implies that GR offers a definitive explanation, when it primarily offers a predictive framework. The dilation observed may not be caused by spacetime curvature itself, but rather misinterpreted as such—obscuring deeper structural mechanisms such as coherence alignment or orientation resistance. This reflects a broader tendency in the prevailing paradigm to conflate predictive success with causal insight.

914 Stephan J. G. Gift, "GPS Satellite Clock Corrections without Relativity Theory," *Journal of Applied Mathematics and Physics* 9, no. 10 (2021): 2476–2482, https://doi.org/10.4236/jamp.2021.910158.

915 Albert Einstein, *The Meaning of Relativity: Fifth Edition, Including the Relativistic Theory of the Non-Symmetric Field* (Princeton, NJ: Princeton University Press, 1955), 32.

916 Chengyu Dai, Isaac R. Bruss, and Sharon C. Glotzer, "Phase Separation and State Oscillation of Active Inertial Particles," *Soft Matter* 16, no. 11 (2020): 2712–2720, https://doi.org/10.1039/c9sm01683j, 1-2. "...particle motility is reduced by the interplay between particle inertia and collisions." "...the coupling between AIP velocities and their local environment suffices to generate... oscillations..." These highlight how inertia governs behavior through interaction history, exactly as I propose in my memory model of inertial dynamics (see Appendix).

917 Ibid.

918 Isaac Newton, *The Mathematical Principles of Natural Philosophy*, trans. Andrew Motte.

919 Ibid., 372.

920 Paulinus Chinaenye Eze et al., "Positioning Control of Satellite Antenna for High Speed Response Performance," *IPTEK: The Journal of Engineering* 10, no. 2 (2024): 119–136, https://www.researchgate.net/publication/382937392.

921 Floyd M. Gardner, *Phaselock Techniques*, 3rd ed. (Hoboken, NJ: Wiley, 2005)

922 Richard P. Feynman, "The Development of the Space-Time View of Quantum Electrodynamics," *Nobel Lecture*, December 11, 1965, The Nobel Prize, https://www.nobelprize.org/prizes/physics/1965/feynman/lecture/.

923 Ibid.

924 Ibid.

925 Ibid.

926 Charles Kittel, *Introduction to Solid State Physics*, 8th ed. (Hoboken, NJ: Wiley, 2005), 346–356.

927 Ibid.

928 Ibid.

929 Ibid.

930 Ran Holtzman et al., "The Origin of Hysteresis and Memory of Two-Phase Flow in Disordered Media," *Communications Physics* 3, no. 222 (2020), https://doi.org/10.1038/s42005-020-00492-1. The authors note: "A remarkable property of the model is its ability to capture the memory properties of PS trajectories... The return-point memory (RPM) property is shared by a number of seemingly disparate slowly driven disordered systems... revealing that not only PS trajectories are infinitely multivalued, but also that their actual value depends on the previous sequence of return points... which is stored in the interfacial configuration."

931 Ibid., "Not only PS trajectories are infinitely multivalued, but also that their actual value depends on the previous sequence of return points in the current trajectory, which is stored in the interfacial configuration evolution. ... Quasistatic hysteresis is the macroscale manifestation of micro-scale energy-dissipative events... associated with history dependence, or memory, which reflects the inaccessibility of alternative states (energy minima) due to large energy barriers."

932 Fabio Pulizzi, "Spintronics," *Nature Materials* 11, no. 5 (2012): 367, https://doi.org/10.1038/nmat3327.

933 David Tong, *Gauge Theory* (Cambridge: Department of Applied Mathematics and Theoretical Physics, University of Cambridge, 2005), https://www.damtp.cam.ac.uk/user/tong/gauge.pdf.

934 Ibid., The author states: "The gauge field is a guide, telling the internal, colour degrees of freedom... how to evolve through parallel transport. The gauge field 'connects' these internal degrees of freedom at one point in space to those in another," 206.

935 Ibid.

936 Ibid.

937 Ibid., 176.

938 John D. Jackson, *Classical Electrodynamics*, 3rd ed. (New York: Wiley, 1998), 20-21. Jackson shows how electric fields and surface charges are idealized using directional field constraints and boundary models, reinforcing that electromagnetic behavior is shaped by structural coherence, not classical force.

939 Ibid.

940 David J. Griffiths, *Introduction to Electrodynamics*, 4th ed. (Cambridge: Cambridge University Press, 2017), 367–372. Griffiths explains that electromagnetic waves propagate through vacuum at a fixed speed determined solely by vacuum permittivity and permeability, implying vacuum itself has intrinsic structure.

941 Ibid., 382. Griffiths explains that electromagnetic waves move at a fixed speed through vacuum, a behavior governed by intrinsic properties of space (vacuum permittivity and permeability). This strongly implies that vacuum possesses structural characteristics, even if it's not traditionally described that way.

942 Henri Poincaré, *Science and Hypothesis*, trans. W.J.G. (London: Walter Scott Publishing, 1905), 175, https://www.gutenberg.org/files/37157/37157-pdf.pdf.

943 Eugene Hecht, Optics, 5th ed. (Boston: Pearson, 2017),18.

944 Ibid., 18.

945 Ibid., 330.

946 Ibid., 365.

947 Ibid., 374.

948 Aitchison, I.J.R. and Hey, A.J.G. *Gauge Theories in Particle Physics: A Practical Introduction*, 4th ed., CRC Press, 2012, p. 26.

949 Ibid.

950 Ibid., 27.

951 Ibid., 10.

952 Ibid., 26.

953 CERN. White Rabbit: Sub-Nanosecond Timing Distribution. The White Rabbit Project is an open-source protocol developed by CERN for synchronizing distributed systems with sub-nanosecond accuracy. It achieves this through deterministic timing and phase alignment rather than discrete signal exchange. Official documentation available at: https://white-rabbit.web.cern.ch.

954 Mattia Rizzi et al., *White Rabbit Clock Characteristics* (CERN, Geneva, Switzerland: White Rabbit Project, 2019), https://white-rabbit.web.cern.ch/documents/White_Rabbit_Clock_Characteristics.pdf.

955 Ibid.

956 Ibid.

957 Mark Srednicki, *Quantum Field Theory* (Cambridge: Cambridge University Press, 2007), 494.

958 Ibid., 550.

959 Ibid., 505.

960 Ibid., 503.

961 Ibid., 396.

962 Eugene Hecht, *Optics*, 5th ed. (Boston: Pearson, 2015), 46–48.

963 John D. Jackson, *Classical Electrodynamics*, 3rd ed. (New York: Wiley, 1998), 418–419.

964 Ibid.

965 "Photoelectric Effect – an overview," *ScienceDirect Topics*, accessed May 10, 2025, https://www.sciencedirect.com/topics/earth-and-planetary-sciences/photoelectric-effect.

966 Richard P. Feynman and F. L. Vernon Jr., "The Theory of a General Quantum System Interacting with a Linear Dissipative System," *Annals of Physics* 24 (1963): 170.

967 M. Longair, "'...a paper ...I hold to be great guns': A Commentary on Maxwell (1865) 'A Dynamical Theory of the Electromagnetic Field,'" *Philosophical Transactions of the Royal Society A* 373 (2015): 20140473, https://doi.org/10.1098/rsta.2014.0473.

968 James Clerk Maxwell, "On Physical Lines of Force," *Philosophical Magazine* Series 4, vol. 21 (1861): 162.

969 NASA. "Pointing Control." *NASA Science: Hubble Space Telescope*. Accessed April 6, 2025. https://science.nasa.gov/mission/hubble/observatory/design/pointing-control/.

970 Ibid.

971 Roland Pease, "The Time? About a Quarter Past a Kilogram," *Nature*, August 30, 2013, https://doi.org/10.1038/nature.2013.12191.

972 Ashby, Neil. "Relativity in the Global Positioning System." *Living Reviews in Relativity* 6, no. 1 (2003): 1. https://doi.org/10.12942/lrr-2003-1. The author states: "Although clock velocities are small and gravitational fields are weak near the earth, they give rise to significant relativistic effects... including Doppler frequency shifts and gravitational frequency shifts."

973 Wang, Rui, Xinxin Zhang, Zhi Zhang, and Hesheng Wang. "A Review of Sensor Drift Compensation Algorithms for Electronic Noses." Sensors 19, no. 18 (2019): 3844. https://www.mdpi.com/1424-8220/19/18/3844.

974 Wojciech H. Zurek, "Decoherence and the Transition from Quantum to Classical—Revisited," *arXiv:quant-ph*/0306072v1 (June 2003), 22, https://arxiv.org/abs/quant-ph/0306072.

975 Ibid., 21.

976 Jose Sanjuan, Andriy Sinyukov, Mohammad F. Warrayat, and Francisco Guzman, "Gyro-Free Inertial Navigation Systems Based on Linear Opto-Mechanical Accelerometers," *Sensors* 23, no. 8 (April 19, 2023): 4093, https://doi.org/10.3390/s23084093.

# Illustration Credits

Within chapters: Figures are labeled with the chapter number followed by the figure number within that chapter. For example, "Figure 1.4" refers to the fourth figure in Chapter 1, and "Figure 4.3" refers to the third figure in Chapter 4.

> \* The usage of these photographs, illustrations, or images does not suggest any endorsement by the copyright holders or creators. All images are used under appropriate licenses or permissions.

**Figure 1.1.** 3D model of the bacterial flagellum motor. Image source: Wikimedia Commons, CC International 4.0, https://commons.wikimedia.org/wiki/File:Flagellar_Motor_Assembly.jpg#filelinks.

**Figure 1.2.** Beating pattern of eukaryotic "flagellum" and "cilium." **Image source:** Urutseg, Wikimedia Commons, CC BY 3.0, https://commons.wikimedia.org/wiki/File:Flagellum-beating.svg.

**Figure 1.3.** Rotor of the motor from an A.E.G. Zossen motor car. **Image source:** *Electric Railway Engineering* by H. F. Parshall and H. M. Hobart, 1907. Wikimedia Commons, Public Domain, https://commons.wikimedia.org/wiki/File:340,_341._Rotor_of_Motor_of_A.E.G._Zossen_Motor_Car,_(341_rotor_on_axle).jpg.

**Figure 1.4.** Turbofan engine. **Image source:** Ariadacapo, Wikimedia Commons, CC BY-SA 3.0, https://commons.wikimedia.org/wiki/File:Core_section_of_a_sectioned_Rolls-Royce_Turbom%C3%A9ca_Adour_turbofan.jpg.

**Figure 1.5.** Internal structure of the 1963 Chrysler Turbine Engine. **Image source:** Tm, Wikimedia Commons, CC BY 2.0, https://commons.wikimedia.org/wiki/File:1963_Chrysler_Turbine_Engine_%283106 4178624%29.jpg.

**Figure 1.6.** Bacterial flagellar motor switch complex. **Image source:** Epipelagic, Wikimedia Commons, CC BY-SA 4.0, https://commons.wikimedia.org/wiki/File:Landmarks_in_understanding_the_bacterial _flagellar_motor_switch.png.

**Figure 1.7.** Diagram of a bacterial flagellum. **Image source:** Wikimedia Commons, Public Domain, https://en.wikipedia.org/wiki/Flagellum#/media/File:Flagellum_base_diagram-en.svg

**Figure 3.1** Chinese Kirigami pop-up artwork featuring a detailed pavilion, showcasing precise cuts and folds that reveal an intricate design when unfolded. **Image source:** fdecomite, CC BY 2.0, https://www.flickr.com/photos/fdecomite.

**Figure 4.1.** Ancient metalworking depicted in the Tomb of Rekhmire. **Image source:** Pharos, Wikimedia Commons, CC0 1.0, https://commons.wikimedia.org/wiki/File:Metal_Working,_Tomb_of_Rekhmire_ME T_31.6.22_EGDP017258.jpg.

**Figure 4.2** An ancient Egyptian painting depicting the domestication of cattle, showcasing early agricultural practices. **Image source:** Wikimedia Commons, Public Domain, https://en.wikipedia.org/wiki/Neolithic_Revolution#/media/File:Egyptian_Domestica ted_Animals.jpg.

**Figure 4.3** Cuneiform inscription at Erebuni Fortress, Armenia. **Image source:** File Upload Bot (Magnus Manske), Wikimedia Commons, CC BY-SA 2.0. https://commons.wikimedia.org/wiki/File:Armenia_-_Cuneiform_at_Erebuni_(5034053035).jpg.

**Figure 4.4 (Left).** Brass astrolabe, c. 14th century. **Image source:** MartinPoulter, Wikimedia Commons, CC BY 4.0, https://commons.wikimedia.org/wiki/File:MHS_42649_Astrolabe.jpg.

**Figure 4.5 (Right).** Sextant, a navigational instrument. **Image source:** Fotokannan, Wikimedia Commons, CC BY 3.0, https://commons.wikimedia.org/wiki/File:A_sextant.JPG.

**Figure 5.1.** Engineer fitting prosthetic arm. Image source: ThisIsEngineering from Pexels, CC BY 4.0, https://www.pexels.com/photo/engineer-fitting-prosthetic-arm-3912992/.

**Figure 5.2.** Detailed view of deeper palmar structures of the wrist and hand. **Image source:** Wilfredor, CC BY-SA 3.0, https://commons.wikimedia.org/wiki/File:Wrist_and_hand_deeper_palmar_dissection -en.svg.

**Figure 5.3.** The SynCardia Total Artificial Heart (TAH) is a medical device designed to replace both ventricles and all four heart valves in patients with end-stage biventricular heart failure. **Image source:** Luciasyncardia, Wikimedia Commons, CC BY 4.0, https://commons.wikimedia.org/wiki/File:SynCardia_Total_Artificial_Heart_(TAH).jp g#.

**Figure 5.4.** Sectional anatomy of the human heart, illustrating internal structures including chambers and valves. **Image source:** Wapcaplet http://blausen.com/, Wikimedia Commons, CC BY 3.0, https://en.wikipedia.org/wiki/en:User:Wapcaplet.

**Figure 5.5.** Sectional anatomy of the human eye, illustrating the internal structures such as the cornea, lens, retina, and optic nerve. Image source: Rhcastilhos and Jmarchn, Wikimedia Commons, https://commons.wikimedia.org/wiki/User:Jmarchn.

**Figure 5.6.** Illustration of the human brain from an inferior view, detailing major nerves and structures involved in sensory and motor functions. Image source: Wikimedia Commons, Public Domain, https://commons.wikimedia.org/w/index.php?curid=29135453.

**Figure 11.1.** Intel Core i7-8700 processor with a clock speed of 3.2 GHz. **Image source:** Kzuo, Wikimedia Commons, CC BY-SA 4.0, https://commons.wikimedia.org/wiki/File:2023_Intel_Core_i7_8700_%282%29.jpg.

**Figure 11.2.** Illustration of cosmic expansion, analogous to a cosmic pancake with stars embedded. **Image source:** S.A. Cooper.

**Figure 12.1.** Illustration of the foundational network of dark matter that binds the cosmos together, much like an invisible spider web. **Image source:** S. A. Cooper.

**Figure 12.2.** Wikipedia's framing of Intelligent Design as pseudoscience reflects a broader pattern in which agency-based explanations are both dismissed and misframed—despite relying on the same inferential logic accepted in materialist contexts such as dark matter and cosmic structure. *Wikipedia*. This page was last edited on 25 February 2025, at 04:14 (UTC). https://en.wikipedia.org/wiki/Intelligent_design/.

**Figure 12.3.** Wikipedia's treatment of *irreducible complexity* continues the pattern of framing design-based reasoning as pseudoscience—despite relying on inferential logic that mirrors accepted reasoning in other scientific domains, such as the inference of unseen forces like dark matter. *Wikipedia*. This page was last edited on 9 March 2025, at 19:52 (UTC), https://en.wikipedia.org/wiki/Irreducible_complexity/.

**Figure 12.4.** Wikipedia's framing of *specified complexity* continues the trend of labeling design-based reasoning as pseudoscience, even as it describes logic that closely mirrors accepted inference methods in other fields—such as detecting intention from pattern or ruling out chance based on statistical improbability. *Wikipedia*. This page was last edited on 28 January 2025, at 04:24 (UTC), https://en.wikipedia.org/wiki/Specified_complexity/.

**Figure 13.1.** A partial map of the Internet based on January 15, 2005, data from opte.org. Each line connects two nodes (IP addresses), with line length representing network delay. The structure reveals a web-like topology, resembling natural systems such as neural networks and the cosmic web. **Image source**: The Opte Project, Wikimedia Commons, CC BY 2.5, https://commons.wikimedia.org/wiki/File:Internet_map_1024.jpg.

**Figure 13.2.** Polarization of the Cosmic Microwave Background (CMB) as detected by ESA's Planck satellite over the entire sky. **Image source:** NASA/JPL-Caltech/ESA/Planck Collaboration, NASA ID: PIA18916, https://www.jpl.nasa.gov/images/pia18916-polarization-of-the-cosmic-microwave-background.

**Figure 14.1.** Illustration of a magnetic levitation (maglev) train system with electromagnets that levitate and propel the train. **Image source:** Yasugawa, Wikimedia Commons, CC BY-SA 3.0, https://commons.wikimedia.org/wiki/File:SC_Maglev_%28Electrodynamic_Suspension%29.png.

**Figure 14.2.** The Maglev train at Longyang Road Station in Shanghai, photographed on November 15, 2014. **Image source:** Alex Needham, Wikimedia Commons, CC BY-SA 4.0, https://commons.wikimedia.org/wiki/File:2014.11.15.141113_Maglev_train_Longyang_Road_Station_Shanghai.jpg.

**Figure 15.1.** A geometric pattern depicting the synodic cycle between Venus and Earth as they orbit the Sun, from 2016 to 2023. **Image courtesy** of Guy Ottewell, https://www.universalworkshop.com/venus/.

**Figure 15.2.** Iconic image, referred to as the "Pillars of Creation," captured by the Hubble Space Telescope. **Image source:** NASA, ESA, CSA, STScI; J. DePasquale, A. Koekemoer, A. Pagan (STScI), CC BY 4.0 INT, https://www.esa.int/ESA_Multimedia/Images/2022/10/Webb_s_portrait_of_the_Pillars_of_Creation_NIRCam.

**Figure 15.3.** This image from NASA's James Webb Space Telescope shows neon emissions around the young star SZ Chamaeleontis. **Image source:** NASA/JPL-Caltech, https://www.jpl.nasa.gov/news/webb-follows-neon-signs-toward-new-thinking-on-planet-formation.

**Figure 15.4.** A cross-sectional view of Saturn, illustrating its layered internal structure—including a rocky core, an icy layer, metallic hydrogen, and helium—as well as its atmospheric composition and ring system, highlighting key features such as the Encke gap, Cassini division, and the planet's hexagonal storm at the north pole. **Image Source**: Kelvinsong, Wikimedia Commons, https://commons.wikimedia.org/wiki/User:Kelvinsong.

**Figure 16.1.** Artistic impression of the Moon's formation through the fission theory, suggesting it split from Earth's crust. **Image source:** ESO/José Francisco (josefrancisco.org), CC BY 4.0, https://supernova.eso.org/exhibition/images/0208c_moon_formation_fission/.

**Figure 16.2.** This diagram depicts a smaller Moon on its initial trajectory toward Earth, before being gradually captured and stabilized into orbit around the planet, illustrating one of the hypothesized scenarios for the Moon's origin. **Image source:** S. A. Cooper.

**Figure 16.3.** "This artist's concept shows a celestial body about the size of our moon slamming at great speed into a body the size of Mercury. NASA's Spitzer Space Telescope found evidence that a high-speed collision of this sort occurred a few thousand years ago around a young star, called HD 172555, still in the early stages of planet formation." — NASA. **Image source:** NASA/JPL-Caltech, https://www.nasa.gov/image-article/planetary-smash-up/.

**Figure 17.1.** Layers of the Earth - A diagram illustrating the various geological layers of the Earth, including the crust, mantle, and core. **Image source:** A. Shteiwi, Wikimedia Commons, CC BY-SA 4.0, https://commons.wikimedia.org/wiki/File:Earth_layers_%D8%B7%D8%A8%D9%82%D8%A7%D8%AA_%D8%A7%D9%84%D8%A7%D8%B1%D8%B6.png.

**Figure 17.2.** This image highlights the vivid blue sky typical of Balneário Camboriú, Brazil, offering a backdrop that enhances the city's coastal beauty. **Image source:** Panoramio upload bot, Wikimedia Commons, CC BY-SA 3.0, https://commons.wikimedia.org/wiki/File:Balne%C3%A1rio_Cambori%C3%BA_-_panoramio_%288%29.jpg.

**Figure 17.3.** A vibrant sunset over Dublin's cityscape capturing the fading light of dusk. **Image Source:** Giuseppe Milo, Flickr, Creative Commons 2.0, https://www.flickr.com/photos/giuseppemilo/29028858596/.

**Figure 18.1.** Image of Earth from space, predominantly showing ocean coverage. **Image source:** NASA, https://www.nasa.gov/image-article/water-planet/.

**Figure 18.2.** This image displays Comet 67P/Churyumov-Gerasimenko. **Image source:** ESA/Rosetta/NavCam – NASA, https://sci.esa.int/web/rosetta/-/54523-cometwatch-navcam-images.

**Figure 18.3.** This illustration depicts the early Earth undergoing a bombardment by water-filled comets and asteroids. **Image source:** S. A. Cooper.

**Figure 20.1.** A New Holland E215 Excavator at work. **Image source:** Wikimedia Commons, Public Domain.

**Figure 20.2.** This detailed anatomical illustration, originally created by George Stubbs. **Image source:** Wellcome Collection, CC BY 4.0, https://commons.wikimedia.org/wiki/File:Muscles_and_bones_of_the_shoulder,_arm _and_hand;_three_figur_Wellcome_V0008182EL.jpg.

**Figure 20.3.** An illustration of a HEPA filter. **Image source:** BruceBlaus, CC BY 4.0, https://commons.wikimedia.org/wiki/File:HEPA_Filter.png.

**Figure 20.4.** Diagram of the nasal epithelium showing various specialized cells, including cilia, mucus-producing goblet cells, basal cells, and columnar epithelium. **Image source:** ccconline.org, by Cenveo, CC BY 3.0 US, https://pressbooks.ccconline.org/bio106/chapter/respiratory-levels-of-organization/.

**Figure 20.5.** Head anatomy with olfactory nerve, including labels for the nasal cavity, olfactory nerves, cribriform plate, olfactory bulb, and olfactory. **Image source:** Patrick J. Lynch, medical illustrator, https://commons.wikimedia.org/wiki/File:Head_Olfactory_Nerve_Labeled.png.

**Figure 20.6.** Central nervous system regions that receive information from the olfactory bulb. **Image source**: anatomytool.org, CC BY 4.0, https://anatomytool.org/content/cenveo-drawing-central-nervous-system-regions-receive-information-olfactory-bulb-english.

**Figure 20.7.** A sheriff uses a trained sniffer dog to detect contraband, exemplifying natural technology at work. The dog's advanced olfactory system functions as a vital tool in law enforcement. **Image source**: Bjwhite66212, CC BY-SA 4.0, https://commons.wikimedia.org/wiki/File:PoliceSnifferDog.jpg.

**Figure 20.8.** Anatomy of the ear, showing the outer ear, middle ear, and inner ear structures responsible for hearing and equilibrium. **Image source**: anatomytool.org; "Cenveo - Drawing Anatomy of the Ear - English labels" by Cenveo, CC BY 4.0, https://anatomytool.org/content/cenveo-drawing-anatomy-ear-english-labels.

**Figure 20.9.** Illustration of how the maculae sense linear acceleration, such as gravity acting on a tilting head. **Image source**: "OpenStax AnatPhys fig.14.11 - Maculae and Equilibrium - English labels" by OpenStax, licensed under CC BY, https://openstax.org/details/books/anatomy-and-physiology.

**Figure 20.10.** Illustration of the middle ear showing the reflex action of the stapedius and tensor tympani muscles in response to intense sound. **Image source:** BruceBlaus, CC BY 3.0, https://commons.wikimedia.org/wiki/File:Middle_ear.png.

**Figure 20.11.** Fourth-generation behind-the-ear cochlear implant processor by Advanced Bionics, circa 2021. **Image source:** RespectCE, CC BY-SA 4.0, https://commons.wikimedia.org/wiki/File:Advanced_Bionics_cochlear_implant_proces sor,_brand-named_%22Marvel%22_(fourth-generation_behind-the-ear_processor).jpg.

**Figure 21.1.** Richard Feynman at Paine Mansion Woods, 1984. **Image source:** Tamiko Thiel, Wikimedia Commons, CC BY-SA 3.0, https://commons.wikimedia.org/wiki/File:RichardFeynman-PaineMansionWoods1984_copyrightTamikoThiel_bw.jpg.

**Figure 26.1.** This historic image captures the Wright brothers' first successful flight on December 17, 1903, at Kitty Hawk, North Carolina. **Image source:** The National Park Service (NPS),

'Historic image colorized for the National Park Service,'
https://www.nps.gov/wrbr/learn/december-17-2020.htm.

**Figure 27.1.** A fossil *Drotops armatus* on display in the Sant Hall of Oceans at the Smithsonian Museum of Natural History. *Drotops armatus* is a species of trilobite, an arthropod related to shrimp, lobsters, scorpions, crabs, and crayfish. Trilobites first appeared about 542 million years ago during the early Cambrian period. **Image source:** Tim Evanson, https://www.flickr.com/photos/timevanson/7282110324/in/photostream/.

**Figure 27.2.** A fossil *Cheirurus ingricus* on display in the Sant Hall of Oceans at the Smithsonian Museum of Natural History. **Image source**: Tim Evanson, https://www.flickr.com/photos/timevanson/7282110704/in/photostream/.

**Figure 27.3.** Basic types of Neolithic or Copper Age stone tools—including knives, axes, a hammer-axe, and a hammerstone—from The City of Prague Museum, creatively remixed with modern objects such as an Iphone. **Image source**: Zde, Wikimedia Commons, CC BY-SA 4.0 (Modern phone and watch added for illustrative purposes) https://commons.wikimedia.org/wiki/File:Stone_tools,_Neolithic_or_Copper_Age,_City_of_Prague_Museum,_175541.jpg.

**Figure 28.1.** This is Charles Darwin's famous sketch from 1837, often referred to as his "Tree of Life" diagram, drawn in one of his notebooks as he began developing his theory of evolution by natural selection. **Image source**: Public Domain

**Figure 28.2.** Evolutionary progression from ape to human. This illustration is inspired by the famous "March of Progress" from *Early Man* (1965) by Rudolph Zallinger. **Image source:** S. A. Cooper.

# Index

www.ingramcontent.com/pod-product-compliance
Lightning Source LLC
Chambersburg PA
CBHW061545120626
46550CB00004B/1367